Lecture Notes in Computer Scienc

Commenced Publication in 1973
Founding and Former Series Editors:
Gerhard Goos, Juris Hartmanis, and Jan van Leeuwen

Editorial Board

Jan Vahrenhold (Ed.)

Experimental Algorithms

8th International Symposium, SEA 2009
Dortmund, Germany, June 4-6, 2009
Proceedings

 Springer

Volume Editor

Jan Vahrenhold
Technische Universität Dortmund
Fakultät für Informatik, Informatik XI
Otto-Hahn-Strasse 14, 44227 Dortmund, Germany
E-mail: jan.vahrenhold@cs.tu-dortmund.de

Library of Congress Control Number: Applied for

CR Subject Classification (1998): F.2, E.1, D.2, I.3.5, I.2.8

LNCS Sublibrary: SL 1 – Theoretical Computer Science and General Issues

ISSN   0302-9743
ISBN-10   3-642-02010-0 Springer Berlin Heidelberg New York
ISBN-13   978-3-642-02010-0 Springer Berlin Heidelberg New York

Typesetting: Camera-ready by author, data conversion by Scientific Publishing Services, Chennai, India
Printed on acid-free paper   SPIN: 12682904   06/3180   5 4 3 2 1 0

Preface

This volume contains the papers presented at the 8th International Symposium on Experimental Algorithms (SEA 2009). The symposium was held at the Technische Universität Dortmund, Germany, during June 4–6, 2009.

The main theme of the SEA series is the role of experimentation and of algorithm engineering techniques in the design and evaluation of algorithms and data structures. Contributions are supported by experimental evaluation, methodological issues in the design and interpretation of experiments, the use of (meta-)heuristics, or application-driven case studies that deepen the understanding of a problem's complexity. For each symposium, papers are solicited from all areas of algorithmic engineering research.

Previous meetings, under the name of "Workshop on Experimental Algorithms" (WEA), were held in Riga (Latvia, 2001), Ascona (Switzerland, 2003), Angra dos Reis (Brazil, 2004), Santorini (Greece, 2005), Menorca Island (Spain, 2006), Rome (Italy, 2007), and Provincetown (USA, 2008).

The Program Committee of SEA 2009 received 64 submissions. Each submission was reviewed by at least three Program Committee members and evaluated on its quality, originality, and relevance to the symposium. Overall, the Program Committee wrote 249 reviews with the help of almost 100 trusted external referees. The Committee selected 23 papers, leading to an acceptance rate of 35.9%. The decision process was made electronically using the EasyChair conference management system.

In addition to the accepted contributions, this volume also contains abstracts of the invited talks given by Heinz Bast (Intel), Michael A. Bender (Stony Brook University and Tokutek, Inc.), and Marc Schoenauer (INRIA Saclay – Île-de-France).

We would like to thank all the authors who responded to the call for papers, the invited speakers, the members of the Program Committee, the external referees, and—last but not least—the members of the Organizing Committee.

June 2009   Jan Vahrenhold

Organization

Program Committee

Mark de Berg	Eindhoven University of Technology (The Netherlands)
Gerth S. Brodal	MADALGO, Aarhus (Denmark)
Sándor P. Fekete	Technische Universität Braunschweig (Germany)
Carlos M. Fonseca	University of Algarve (Portugal)
Giuseppe F. Italiano	University of Rome "Tor Vergata" (Italy)
Alejandro López-Ortiz	University of Waterloo (Canada)
Petra Mutzel	Technische Universität Dortmund (Germany)
Panos M. Pardalos	University of Florida (USA)
Mike Preuß	Technische Universität Dortmund (Germany)
Rajeev Raman	University of Leicester (UK)
Mauricio G. C. Resende	AT&T Labs – Research (USA)
Peter Sanders	University of Karlsruhe (Germany)
Matt Stallmann	North Carolina State University (USA)
Laura Toma	Bowdoin College (USA)
Jan Vahrenhold (Chair)	Technische Universität Dortmund (Germany)
Xin Yao	University of Birmingham (UK)

Steering Committee

Edoardo Amaldi	Politecnico di Milano (Italy)
David A. Bader	Georgia Institute of Technology (USA)
Josep Diaz	Universitat Politécnica de Catalunya (Spain)
Giuseppe F. Italiano	University of Rome "Tor Vergata" (Italy)
David Johnson	AT&T Labs – Research (USA)
Klaus Jansen	University of Kiel (Germany)
Kurt Mehlhorn	MPII Saarbrücken (Germany)
Ian Munro	University of Waterloo (Canada)
Sotiris Nikoletseas	University of Patras and CTI (Greece)
José Rolim (Chair)	University of Geneva (Switzerland)
Paul Spirakis	University of Patras and CTI (Greece)

Organizing Committee

Fabian Gieseke	Technische Universität Dortmund (Germany)
Gundel Jankord	Technische Universität Dortmund (Germany)
Norbert Jesse	Technische Universität Dortmund (Germany)
Mike Preuß	Technische Universität Dortmund (Germany)
Jan Vahrenhold	Technische Universität Dortmund (Germany)

Referees

Spyros Angelopoulos
Diego Arroyuelo
Veit Batz
Regina Berretta
Therese Biedl
Vladimir Boginski
Ingo Brinkmeier
Luciana Buriol
Sergiy Butenko
Alberto Caprara
Siu-Wing Cheng
Marco Chiarandini
Markus Chimani
Altannar Chinchuluun
Francisco Claude
Ovidiu Daescu
Reza Dorrigiv
Michael Emmerich
Leah Epstein
Jeff Erickson
Arash Farzan
Paola Festa
Robert Fraser
Ricardo Fukasawa
Robert Geisberger
Fabian Gieseke
Roberto Grossi
Qianping Gu
Carsten Gutwenger
Robert Görke
Peter Hachenberger
Torben Hagerup

Angele Hamel
Jason Hartline
Meng He
Keld Helsgaun
Klaus H. Hinrichs
Cor Hurkens
Jens Jägersküpper
Thomas Sejr Jensen
Xiaoyi Jiang
David Johnson
Tom Kamphans
Maria Kandyba
Jyrki Katajainen
Karsten Klein
Alexander Kröller
Ravi Kumar
Piyush Kumar
Luigi Laura
Peter Lewis
Antonio Loureiro
Dennis Luxen
Manuel López-Ibáñez
Maarten Löffler
Marco Lübbecke
Tobias Marschall
Marcel Martin
Claudio Meneses
Matthias
    Müller-Hannemann
Frank Neumann
Andreas Nüchter
Rasmus Pagh

Luís Paquete
Wolfgang Paul
Artur Alves Pessoa
Marc Pfetsch
Sven Rahmann
Jens Rasmussen
Srinivasa Rao
Marcus Ritt
Alejandro Salinger
Nils Schweer
Ricardo Silva
Johannes Singler
Christian Sohler
Roberto Solis-Oba
Rob van Stee
Dawn Strickland
Thomas Stützle
Tiow Seng Tan
Frederik Transier
Sebastián Urrutia
Mikael
    Vejdomo Johansson
Dalessandro Vianna
Berthold Vöcking
Renato Werneck
Carsten Witt
Alexander Wolff
Hoi-Ming Wong
Bernd Zey
Uwe Zimmermann

Table of Contents

Parallelism in Current and Future Processors – Challenges and Support for Designing Optimal Algorithms

(Invited Talk)

Heinz Bast

Intel GmbH, Dornacher Strasse 1, 85622 Feldkirchen/München, Germany

Abstract. Both explicit usable and implicit transparent parallelism is nothing really new in processor technology but has been restricted to high-end computer systems accessible to only a few developers. In recent years, however, parallelism on all levels has found its way into even the cheapest desktop and notebook system and thus every algorithm being developed today should reflect this change to optimally exploit theses additional resources.

This session will outline the parallelism offered by current Intel processors and some new parallel enhancements of future architectures including the many-core approach implemented by Larrabee and the coming improvements for data-parallel (SIMD, SSE) execution. Some of these enhancements will introduce new challenges to the algorithm designer and developer. This includes the massive number of available hardware-threads, the increased size of vector operations, and non-uniform memory access. Intel actively looks for new parallel programming models to tackle these challenges including CT, Software Transactional Memory and Concurrent Collections for C++. While these models might make it into future program development environments, there are multiple developer tools for parallel program development as mature products available today – including compilers, libraries, thread checker/debugger, performance analysis tools using hardware performance counters etc. The talk will outline how some of these tools as offered by Intel and how they can facilitate the complete development cycle for parallel program development.

J. Vahrenhold (Ed.): SEA 2009, LNCS 5526, p. 1, 2009.

From Streaming B-Trees to Tokutek:
How a Theoretician Learned to be
VP of Engineering
(Invited Talk)

Michael A. Bender

Department of Computer Science, Stony Brook University,
Stony Brook, NY 11794-4400, USA
Tokutek^{TM} Inc., 146 West 29th St., New York, NY 10001, USA

Abstract. I present the cache-oblivious streaming B-tree, a high performance alternative to the traditional B-tree. Modern databases and file systems are based on B-trees. As a result, they can exhibit performance cliffs and unpredictable run times. Replacing B-trees with cache-oblivious streaming B-trees can remove some of these performance deficiencies.

I explain some of the technical issues that we needed to address to turn the streaming B-tree prototype into an industrial-strength product. Then I explain how legacy and established practice influenced our engineering decisions.

J. Vahrenhold (Ed.): SEA 2009, LNCS 5526, p. 2, 2009.

Experimental Comparisons of Derivative Free Optimization Algorithms
(Invited Talk)

A. Auger[1,2], N. Hansen[1,2], J.M. Perez Zerpa[1], R. Ros[1], and M. Schoenauer[1,2]

[1] TAO Projetct-Team, INRIA Saclay – Île-de-France
LRI, Bat 490 Univ. Paris-Sud 91405 Orsay Cedex France
[2] Microsoft Research-INRIA Joint Centre
28 rue Jean Rostand, 91893 Orsay Cedex, France

Abstract. In this paper, the performances of the quasi-Newton BFGS algorithm, the NEWUOA derivative free optimizer, the Covariance Matrix Adaptation Evolution Strategy (CMA-ES), the Differential Evolution (DE) algorithm and Particle Swarm Optimizers (PSO) are compared experimentally on benchmark functions reflecting important challenges encountered in real-world optimization problems. Dependence of the performances in the conditioning of the problem and rotational invariance of the algorithms are in particular investigated.

1 Introduction

Continuous Optimization Problems (COPs) aim at finding the global optimum (or optima) of a real-valued function (aka *objective* function) defined over a (subset of) a real vector space. COPs commonly appear in everyday's life of many scientists, engineers and researchers from various disciplines, from physics to mechanical, electrical and chemical engineering to biology. Problems such as model calibration, process control, design of parameterized parts are routinely modeled as COPs. Furthermore, in many cases, very little is known about the objective function. In the worst case, it is only possible to retrieve objective function values for given inputs, and in particular the user has no information about derivatives, or even about some weaker characteristics of the objective function (e.g. monotonicity, roughness, ...). This is the case, for instance, when the objective function is the output of huge computer programs ensuing from several years of development, or when experimental processes need to be run in order to compute objective function values. Such problems amount to what is called *Black-Box Optimization* (BBO).

Because BBO is a frequent situation, many optimization methods (aka *search algorithms*) have been proposed to tackle BBO problems, that can be grossly classified in two classes: (i) deterministic methods include classical derivative-based algorithms, in which the derivative is numerically computed by finite differences, and more recent Derivative Free Optimization (DFO) algorithms [1], like pattern search [2] and trust region methods [3]; (ii) stochastic methods rely on random variables sampling to better explore the search space, and include recently introduced bio-inspired algorithms (see Section 3).

J. Vahrenhold (Ed.): SEA 2009, LNCS 5526, pp. 3–15, 2009.

However, the practitioner facing a BBO problem has to choose among those methods, and there exists no theoretical solid ground where he can stand to perform this choice, first because he does not know much about his objective function, but also because all theoretical results either make simplifying hypotheses that are not valid for real-world problems, or give results that do not yield any practical outcome. Moreover, most of BBO methods require some parameter tuning, and here again very little help is available for the practitioner, who is left with a blind and time-consuming test-and-trial approach.

In such context, this paper proposes an experimental perspective on BBO algorithms comparisons. Rigorous procedures to compare the results of different BBO algorithms have been proposed [4], taking into account the stochastic nature of many of them, and giving fair chances to each one of them. However, a critical issue in such experiments is that of the benchmark suite. And because no set of real-world problems can be guaranteed to cover all possible cases of difficult COPs, the approach that has been chosen here is to build artificial test functions with some precise characteristics that are known to be possible sources of difficulty for optimization (e.g. ill-conditioning, non-separability, non-convexity, ruggedness, ...). Such experimental results could then be cautiously generalized, leaving only a few good-performing algorithms in each specific context.

Of course, in real-life BBO situations, it is assumed that nothing is known about the objective function. However, the user sometimes has some partial information (e.g. because his problem is known to be similar to other better-known problems) that might lead him to decide for a BBO method that is (experimentally) known to perform well, 'in vitro', in his precise situation. But on the other hand, assuming absolutely nothing is known in advance about the objective function, running the champion algorithms as identified in perfectly controlled environment might give him some information about his function (e.g. if numerical gradient-based algorithms perform 100 times better than all other methods, his problem is probably very similar to a quadratic problem). This paper is a first step in aiming such 'in vitro' results.

Next, in Section 2, some characteristics of the objective function are surveyed that are known to make the corresponding BBO problem hard. Section 3 introduces the algorithms that will be compared here. Section 4 then introduces the test bench that illustrates the different difficulties highlighted in Section 2, as well as the experimental conditions of the comparisons. The results are presented and discussed in Section 5, and the paper ends with some conclusions in Section 6.

2   What Makes a Search Problem Difficult?

In this section, we discuss problem characteristics that are especially challenging for search algorithms.

2.1   Ill-Conditioning

The conditioning of a problem can be defined as the range (over a level set) of the maximum improvement of objective function value in a ball of small radius centered on a given level set. In the case of convex quadratic functions ($f(x) = \frac{1}{2}x^T H x$ where H is a symmetric definite matrix), the conditioning can be exactly defined as the condition number of the Hessian matrix H, i.e., the ratio between the largest and smallest eigenvalue. Since level sets associated to a convex quadratic function are ellipsoids, the condition number corresponds to the squared ratio between the largest and shortest axis lengths of the ellipsoid.

Problems are typically considered as ill-conditioned if the conditioning is larger than 10^5. In practice we have seen problems with conditioning as large as 10^{10}. In this paper we will quantitatively assess the performance dependency on the conditioning of the objective function.

2.2   Non-separability

An objective function $f(x_1, \ldots, x_n)$ is separable if the optimal value for any variable x_i can be obtained by optimizing $f(\widetilde{x}_1, \ldots, \widetilde{x}_{i-1}, x_i, \widetilde{x}_{i+1}, \ldots, \widetilde{x}_n)$ for any fixed choice of the variables $\widetilde{x}_1, \ldots, \widetilde{x}_{i-1}, \widetilde{x}_{i+1}, \ldots, \widetilde{x}_n$. Consequently optimizing an n-dimensional separable objective function reduces to optimizing n one-dimensional functions.

Functions that are additively decomposable, i.e., that can be written as $f(x) = \sum_{i=1}^{n} f_i(x_i)$ are separable. One way to render a separable test function non-separable is to rotate first the vector x, which can be achieved by multiplying x by an orthogonal matrix B: if $x \mapsto f(x)$ is separable, the function $x \mapsto f(Bx)$ might be non-separable for all non-identity orthogonal matrices B. In this paper we will investigate separable and non-separable problems.

2.3   Non-convexity

Some BBO methods implicitly assume or exploit convexity of the objective function. Composing a convex function $f \in \mathbb{R}$ to the left with a monotonous transformation $g : \mathbb{R} \to \mathbb{R}$ can result in a non-convex function, for instance the one-dimensional convex function $f(x) = x^2$ composed with the monotonous function $g(.) = |.|^{1/4}$ becomes the non-convex function $\sqrt{|.|}$. In this paper we will assess performance dependency on convexity.

3   Algorithms Tested

This section introduces the different algorithms that will be compared in this paper. They have been chosen because they are considered to be the champions in their category, both in the deterministic optimization world (BFGS and NEWUOA) and in the stochastic bio-inspired world (CMA-ES, DE and PSO). They will also be a priori discussed here with respect to the difficulties of continuous optimization problems highlighted in previous Section 2.

3.1   The Algorithms

BFGS. is a well-known quasi-Newton (i.e. gradient-based) method: from the current point, it computes a 'descent direction' using an approximation of the inverse of the

Hessian matrix of the objective function applied to its gradient, and performs a line-search (1D optimization) along this direction. It then updates the approximate inverse Hessian. BFGS method is a local method: it has a proven convergence to a stationary point... provided the starting point is close enough from the solution, and the objective function is regular. The Matlab$^{\circledR}$ version of BFGS (Matlab function fminunc) will be used here, because it is blindly used by many scientists facing optimization problems. Default parameters were used except for stopping criteria: the algorithms stops if the function value improvement in one iteration is less than 10^{-25}.

In BBO context, the gradients have to be computed numerically (an option in Matlab BFGS), which might be a source of possible numerical problems.

NEWUOA. (NEW Unconstrained Optimization Algorithm) has been proposed by Powell [3]: it is a DFO algorithm using the trust region paradigm. The trust region is a ball, centered on the current best point. NEWUOA computes a quadratic interpolation of the objective function within the current trust region, based on known values of the objective, and then performs a truncated conjugate gradient minimization of the interpolated model in the trust region. It then updates either the current best point or the radius of the trust region, based on the a posteriori interpolation error, and some thresholds on the trust region size. Here, the implementation by Matthieu Guibert posted at http://www.inrialpes.fr/bipop/people/guilbert/newuoa/newuoa.html has been used.

An important parameter of NEWUOA is the quadratic model to use for the interpolation, or, equivalently, the number of points that are necessary to compute the interpolation. As recommended by Powell [3], $2n + 1$ points have been used here (n is the dimension of the search space). Other critical parameters are the initial and final radii of the trust region: the initial radius governs the granularity of the objective function that the algorithm will 'see' and the final radius tunes the amount of local search that will performed. Here the initial and final values 100 and 10^{-15} were used, after some preliminary experiments.

CMA-ES. is an Evolution Strategy (ES) [5,6] algorithm: from a set of 'parents' (potential solutions), 'offspring' are created by sampling Gaussian distributions, and the best of the offspring (according to the objective function values) become the next parents. The art of Evolution Strategies lies in the way the parameters of the Gaussian distributions are updated: the Covariance Matrix Adaptation [7] uses the path that has been followed by evolution so far to (i) adapt the step-size, a scaling parameter that tunes the granularity of the search, comparing the actual path length to that of a random walk; and (ii) modify the covariance matrix of the multivariate Gaussian distribution by modifying its eigenvectors in order to increase the likelihood of recent beneficial moves. A single Gaussian distribution is maintained, its mean being a linear combination of the parents. Besides the population size, CMA-ES is parameter-free. The population size has been set to its default value $4 + \lfloor 3\log(n) \rfloor$, but it needs to be increased in order to tackle highly rugged search landscapes. The initial step-size has been set to a third of the parameters' range. The version used in this paper (Scilab 0.92) implements weighted recombination and rank-μ update [8] (version 0.99 is available at http://www.lri.fr/~hansen/cmaes_inmatlab.html)

PSO. (Particle Swarm Optimization) [9] is a bio-inspired algorithm that recently raised a lot of interest, thanks to several published good results, and the simplicity of its implementation. The biological paradigm is that of a swarm of particles that 'fly' over the objective landscape, exchanging information about the best locations (i.e. potential solutions) they have seen. More precisely, each particle updates its velocity, stochastically twisting it toward the direction of the best positions so far visited by (i) itself and (ii) the whole swarm; it then updates its position according to its velocity and computes the new value of the objective function.

A Scilab transcription of the Standard PSO 2006, that is still available on the main page of *PSO Central* http://www.particleswarm.info/, was used here, with default settings.

Differential Evolution. (DE [10]) borrows from Evolutionary Algorithms (EAs) a population of potential solutions that evolves subject to objective-function based selection. However, the main operator used to generate new solutions, that somehow replaces mutation, is specific to DE (and the source for its name): the difference between two points in the population is added to a third one. Uniform crossover is used with some probability. The implementation posted by the original authors at http://www.icsi. berkeley.edu/~storn/code.html was used here. However, the authors themselves confess, in their guidance to DE parameter tuning, that the results might be very dependent on the parameters. They propose in the code 6 possible settings, and extensive experiments (3×288 trials) on a moderately ill-conditioned problem lead us to consider the "*DE/local-to-best/1/bin*" strategy, where a single difference vector, computed between a random point and the best point in the population, is used to generate the new points. In those preliminary experiments, the use of crossover seemed to have little beneficial impact on the results, so no crossover was used, thus making DE rotationally invariant (see below). Those preliminary experiments also resulted in values of the other parameters of DE: the population size was set to the recommended value of $10n$, a weighting factor to $F = 0.8$.

3.2 Invariances

Some a priori comparisons can be made about those algorithms, related to the notion of *invariance*. Indeed, invariances add to the robustness of an algorithm: functions belonging to the same equivalence class with respect to some invariance property will look exactly the same for an algorithm that is invariant under the transformation defining this equivalence class.

Two sets of invariance properties are distinguished, whether they regard transformations of the objective function value or transformations of the search space. First, all comparison-based algorithms are invariant under monotonous transformations of the objective function, as comparisons are unaltered if the objective function f is replaced with some $g \circ f$ for some monotonous function g. All bio-inspired algorithms used in this paper are comparison-based, while the BFGS and NEWUAO are not (see Section 2.3).

Regarding transformations of the search space, all algorithms are trivially invariant under translation of the coordinate system. But let us consider some orthogonal rotations: BFGS is coordinate-dependent due to the computation of numerical gradients.

NEWUOA is invariant under rotation when considering the complete quadratic model, i.e. built with $\frac{1}{2}(n+1)(n+2)$ points. This variant is however often more costly compared to the $2n+1$ one – but the latter is not invariant under rotation. The rotational invariance of CMA-ES is built-in, while that of DE depends whether or not crossover is used – as crossover relies on the coordinate system. This was one reason for omitting crossover here. Finally, PSO is (usually) not invariant under rotations, as all computations are done coordinate by coordinate [11, 12].

4   Test Functions and Experimental Setup

4.1   Test Functions

The benchmark functions tested are given in Table 1. The functions are tested in their original axis-parallel version (i.e. B is the identity and $y = x$), and in rotated versions, where $y = Bx$. The orthogonal matrix B is chosen such that each column is uniformly distributed on the unit hypersphere surface [7], fixed for each run.

Table 1. Test functions with coordinate-wise initialization intervals and target function value, where $y := Bx$ implements an angle-preserving, linear transformation, *i.e.* B is orthogonal

Function	α	Initialization	f_{target}
$f_{elli}(x) = \sum_{i=1}^{n} \alpha^{\frac{i-1}{n-1}} y_i^2$	$[1, 10^{10}]$	$[-20, 80]^n$	10^{-9}
$f_{Rosen}(x) = \sum_{i=1}^{n-1} \left(\alpha(y_i^2 - y_{i+1})^2 + (y_i - 1)^2\right)$	$[1, 10^{8}]$	$[-20, 80]^n$	10^{-9}
$f_{elli}^{1/4}(x) = \left(\sum_{i=1}^{n} \alpha^{\frac{i-1}{n-1}} y_i^2\right)^{1/4}$	$[1, 10^{10}]$	$[-20, 80]^n$	10^{-9}

The ellipsoid function f_{elli} is a convex-quadratic function where the parameter α is the condition number of the Hessian matrix that is varied between 1 and 10^{10} in our experiments. If $\alpha = 1$ the ellipsoid is the isotropic separable sphere function. The function $f_{elli}^{1/4}$ has the same contour lines (level sets) as f_{elli}, however it is neither quadratic nor convex. For $\alpha \neq 1$, the functions f_{elli} and $f_{elli}^{1/4}$ are separable if and only if $B = I$.

The Rosenbrock function f_{Rosen} is non-separable, has its global minimum at $x = [1, 1, \ldots, 1]$ and, for large enough α and n, has one local minimum close to $x = [-1, 1, \ldots, 1]$, see also [13]. The contour lines of the Rosenbrock function show a bent ridge that guides to the global optimum (the Rosenbrock is sometimes called banana function) and the parameter α controls the width of the ridge. In the classical Rosenbrock function α equals 100. For smaller α the ridge becomes wider and the function becomes less difficult to solve. We vary α between one and 10^8.

4.2   Experimental Setup

For each algorithm tested we conduct 21 independent trials of up to 10^7 function evaluations. If, for BFGS, no success was encountered, the number of trials was extended to 1001.

We quantify the performance of the algorithms using the success performance $SP1$ used in [14], analyzed in [15], and also denoted as Q-measure in [16]. The $SP1$ equals the average number of function evaluations for successful runs divided by the ratio of successful experiments, where an experiment is successful if the f_{target} is reached before 10^7 function evaluations are exceeded. The $SP1$ is an estimator of the expected number of function evaluations to reach f_{target} if the algorithm is restarted until a success (supposing infinite time horizon) and assuming that the expected number of function evaluations for unsuccessful runs equals the expected number of evaluations for successful runs.

5 Results

Results are shown for dimension 20. Results for 10 and 40D reveal similar tendencies and are displayed in Appendix A.

Ellipsoid functions: dependencies. Figure 1 shows $SP1$ (search costs, expected running time in number of function evaluations) versus condition number on all ellipsoidal functions. A remarkable dependency on the condition number can be observed in most

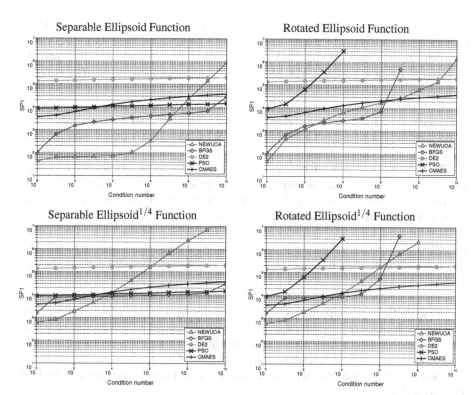

Fig. 1. All ellipsoidal functions in 20D. Shown is $SP1$ (the expected running time or number of function evaluations to reach the target function value) versus condition number.

cases. The two exceptions are PSO on the separable functions and DE. In the other cases the performance declines by at least a factor of ten for very ill-conditioned problems as for CMA-ES. The overall strongest performance decline is shown by PSO on the rotated functions. NEWUOA shows in general a comparatively strong decline, while BFGS is only infeasible for high condition numbers in the rotated case, reporting some numerical problems. The decline of CMA-ES is moderate.

For CMA-ES and DE the results are (virtually) independent of the given ellipsoidal functions, where CMA-ES is consistently between five and forty times faster than DE. For PSO the results are identical on Ellipsoid and Ellipsoid$^{1/4}$, while the performance decline under rotation (left versus right figures) is very pronounced. PSO performs well only on separable or very well-conditioned functions. A similar strong decline under rotation can be observed for NEWUOA on the Ellipsoid function for moderate condition numbers. BFGS, on the other hand, shows a strong rotational dependency on both functions only for large condition numbers $\geq 10^6$.

Switching from Ellipsoid (above) to Ellipsoid$^{1/4}$ (below) only effects BFGS and NEWUOA. BFGS becomes roughly five to ten times slower. A similar effect can be seen for NEWUOA on the rotated function. On the separable Ellipsoid function the effect is more pronounced, because NEWUOA performs exceptionally well on the separable Ellipsoid function.

Ellipsoid functions: comparison. On the separable Ellipsoid function up to a condition number of 10^6 NEWUOA clearly outperforms all other algorithms. Also BFGS performs still better than PSO and CMA-ES while DE performs worst. On the separable Ellipsoid$^{1/4}$ function BFGS, CMA-ES and PSO perform similar. NEWUOA is faster for low condition numbers and slower for large ones. For condition number larger than 10^6, NEWUOA becomes even worse than DE.

On the rotated functions, the performance of PSO declines fast with increasing condition number. For numbers larger than 10^3, PSO is remarkably outperformed by all other algorithms. On the rotated Ellipsoid function for moderate condition numbers BFGS and NEWUOA perform best and outperform CMA-ES by a factor of five, somewhat more for low condition numbers, and less for larger condition numbers, while PSO and DE are much worse. For large condition numbers CMA-ES becomes superior and DE is within a factor of ten of the best performance.

On the rotated Ellipsoid$^{1/4}$ BFGS and CMA-ES perform similar up to condition of 10^6. NEWUOA performs somewhat better for lower condition numbers up to 10^4. For larger condition numbers BFGS and NEWUOA decline and CMA-ES performs best.

Rosenbrock function. On the Rosenbrock function NEWUOA is the best algorithm (Figure 2). NEWUOA outperforms CMA-ES roughly by a factor of five, vanishing for very large values for the conditioning parameter α. For small α, BFGS is in-between, and for $\alpha > 10^4$ BFGS fails. DE is again roughly ten times slower than CMA-ES. Only PSO shows a strong dependency on the rotation of the function and it reveals the strongest performance decline with increasing α, while it never competes with the best three algorithms.

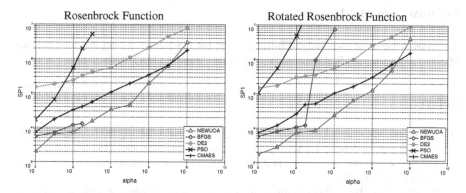

Fig. 2. Rosenbrock function. Shown is $SP1$ (the expected running time or number of function evaluations to reach the target function value) versus conditioning parameter α.

Scaling behaviors. The scaling of the performance with search space dimension is similar for all functions (see Appendix A for the data). CMA-ES, NEWUOA and PSO show the best scaling behavior. They slow down by a factor between five and ten in 40D compared to 10D. For BFGS the factor is slightly above ten, while for DE the factor is thirty or larger, presumably because the default population size increase linearly with the dimension.

6 Summary

In this paper we have conducted a comparison of BFGS, NEWUOA, and three stochastic bio-inspired optimization methods in a black-box optimization scenario. The empirical study was conducted on smooth functions with varying condition number. Aside from gradients being not provided, we consider these functions as the favorite playgrounds of BFGS and NEWUOA. We find that NEWUOA performs exceptional on separable quadratic functions, it performs in all cases very well with moderate condition numbers, but shows a comparatively steep performance decline with increasing ill-conditioning. BFGS performs well overall, but shows a strong decline on very ill-conditioned non-separable functions. For DE, the parameters are difficult to tune and yet it performs overall poorly with the single best parameter setting on our small function set. With the chosen parameters, DE shows the strongest robustness to ill-conditioning though. PSO performs similar to CMA-ES on the separable problems, with an even weaker dependency on the conditioning. On non-separable problems PSO exhibits a strong performance decline with increasing conditioning and performs very poorly even on moderately ill-conditioned functions. Finally, CMA-ES generally outperforms DE and PSO, while up to a moderate function conditioning BFGS and NEWUOA are significantly faster in most cases. Due to their invariance properties, the performance results of CMA-ES and DE are the most stable ones and most likely to generalize to other functions.

Acknowledgements

We would like to acknowledge Philippe Toint for his kind suggestions, and Nikolas Mauny for writing the Scilab transcription of the Standard PSO 2006 code.

References

1. Scheinberg, K., Conn, A.R., Toint, P.L.: Recent progress in unconstrained nonlinear optimization without derivatives. Mathematical Programming 79(3), 397–415 (1997)
2. Torczon, V.: On the convergence of pattern search algorithms. SIAM Journal on optimization 7(1), 1–25 (1997)
3. Powell, M.J.D.: The NEWUOA software for unconstrained optimization without derivatives. In: Large Scale Nonlinear Optimization, pp. 255–297 (2006)
4. Suganthan, P.N., Hansen, N., Liang, J.J., Deb, K., Chen, Y.P., Auger, A., Tiwari, S.: Problem definitions and evaluation criteria for the CEC 2005 special session on real-parameter optimization. Technical report, Nanyang Technological University, Singapore and KanGAL Report Number 2005005 (Kanpur Genetic Algorithms Laboratory, IIT Kanpur) (May 2005)
5. Rechenberg, I.: Evolutionsstrategie: Optimierung Technischer Systeme nach Prinzipien der Biologischen Evolution. Frommann-Holzboog (1973)
6. Schwefel, H.-P.: Evolution and Optimum Seeking. Sixth-Generation Computer Technology Series. John Wiley & Sons, Chichester (1995)
7. Hansen, N., Ostermeier, A.: Completely derandomized self-adaptation in evolution strategies. Evolutionary Computation 9(2), 159–195 (2001)
8. Hansen, N.: The CMA evolution strategy: a comparing review. In: Lozano, J.A., Larranaga, P., Inza, I., Bengoetxea, E. (eds.) Towards a new evolutionary computation. Advances on estimation of distribution algorithms, pp. 75–102. Springer, Heidelberg (2006)
9. Kennedy, J., Eberhart, R.: Particle swarm optimization. In: Proceedings of IEEE International Conference on Neural Networks, 1995, vol. 4, pp. 1942–1948 (1995)
10. Storn, R., Price, K.: Differential evolution – a simple and efficient heuristic for global optimization over continuous spaces. J. of Global Optimization 11(4), 341–359 (1997)
11. Hansen, N., Ros, R., Mauny, N., Schoenauer, M., Auger, A.: PSO facing non-separable and ill-conditioned problems. Research Report RR-6447, INRIA (2008)
12. Wilke, D.N., Kok, S., Groenwold, A.A.: Comparison of linear and classical velocity update rules in particle swarm optimization: Notes on scale and frame invariance. Int. J. Numer. Meth. Engng. 70, 985–1008 (2007)
13. Shang, Y.-W., Qiu, Y.-H.: A note on the extended rosenbrock function. Evol. Comput. 14(1), 119–126 (2006)
14. Hansen, N., Kern, S.: Evaluating the CMA evolution strategy on multimodal test functions. In: Yao, X., Burke, E.K., Lozano, J.A., Smith, J., Merelo-Guervós, J.J., Bullinaria, J.A., Rowe, J.E., Tiňo, P., Kabán, A., Schwefel, H.-P. (eds.) PPSN 2004. LNCS, vol. 3242, pp. 282–291. Springer, Heidelberg (2004)
15. Auger, A., Hansen, N.: Performance evaluation of an advanced local search evolutionary algorithm. In: Proceedings of the IEEE Congress on Evolutionary Computation (2005)
16. Feoktistov, V.: Differential Evolution: In Search of Solutions. In: Optimization and Its Applications. Springer, New York (2006)

A All Results

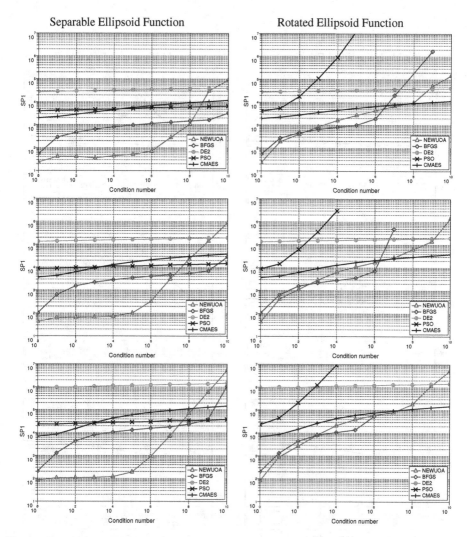

Fig. 3. Ellipsoid function. Shown is *SP*1 (the expected running time or number of function evaluations to reach the target function value) versus condition number.

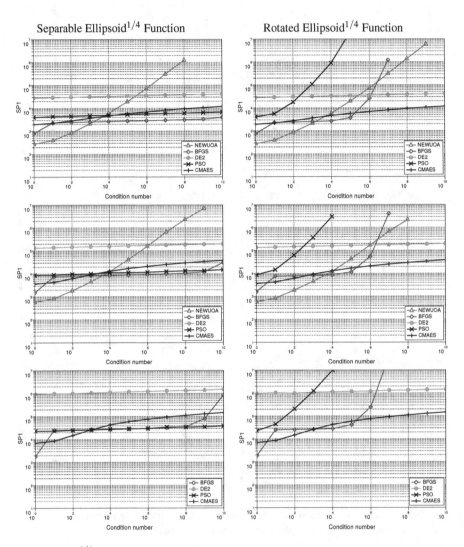

Fig. 4. Ellipsoid$^{1/4}$ function. Shown is $SP1$ (the expected running time or number of function evaluations to reach the target function value) versus condition number.

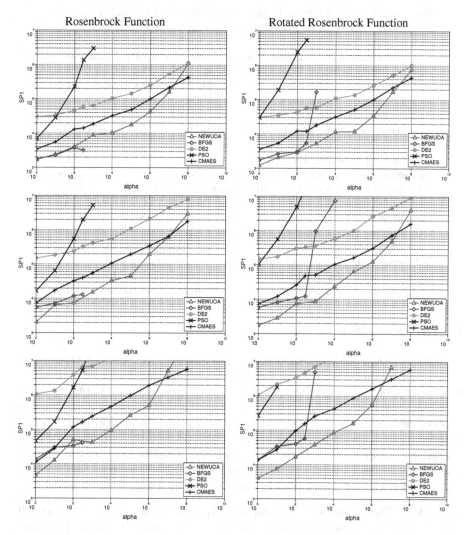

Fig. 5. Rosenbrock function. Shown is $SP1$ (the expected running time or number of function evaluations to reach the target function value) versus conditioning parameter α.

On Computational Models for Flash Memory Devices[*]

Deepak Ajwani[1], Andreas Beckmann[2], Riko Jacob[3], Ulrich Meyer[2],
and Gabriel Moruz[2]

[1] Department of Computer Science, Aarhus University, Denmark
[2] Institut für Informatik, Goethe-Universität Frankfurt am Main, Germany
[3] Computer Science Department, TU München, Germany

Abstract. Flash memory-based solid-state disks are fast becoming the
dominant form of end-user storage devices, partly even replacing the
traditional hard-disks. Existing two-level memory hierarchy models fail
to realize the full potential of flash-based storage devices. We propose
two new computation models, the general flash model and the unit-cost
model, for memory hierarchies involving these devices. Our models are
simple enough for meaningful algorithm design and analysis. In particular, we show that a broad range of existing external-memory algorithms
and data structures based on the merging paradigm can be adapted
efficiently into the unit-cost model. Our experiments show that the theoretical analysis of algorithms on our models corresponds to the empirical
behavior of algorithms when using solid-state disks as external memory.

1 Introduction

In many practical applications, one needs to compute on data that exceeds the
capacity of the main memory of the available computing-device. This happens
in a variety of settings, ranging from small devices, such as PDAs, to high-
performance servers and large clusters. In such cases, the cost of data transfers
between disk and the main memory often proves to be a critical bottleneck in
practice, since a single disk transfer may be as time-costly as millions of CPU
operations. To capture the effect that memory transfers have on the running
time of algorithms, several computational models have been proposed over the
past decades. One of the most successful of these models is the *I/O-model*.

I/O-model. The I/O-model, as defined in [1], is a two-level memory model. It
consists of a CPU, a fast internal memory of size M and a slow external-memory
of infinite size. The CPU can access only data stored in the internal memory, and
data transfers between the two memories are performed in chunks of B consecutive data items. The I/O-complexity of an algorithm is given by the number of
memory transfers, or I/Os, performed. Many problems have been studied in this
model and efficient algorithms have been proposed. For comprehensive overviews
we refer the interested reader to [2,3].

[*] Partially supported by the DFG grant ME 3250/1-1, and by MADALGO – Center for
Massive Data Algorithmics, a Center of the Danish National Research Foundation.

Flash memories. In the recent years, a new trend has emerged in the storage device technology – that of solid-state disks based on flash memory. Flash memories are non-volatile, reprogrammable memories. Flash memory devices are lighter, more shock resistant and consume less power. Moreover, since random accesses are faster on solid-state disks compared to traditional mechanical hard-disks, flash memory is fast becoming the dominant form of end-user storage in mobile computing. Many recent notebook and netbook models have already replaced traditional mechanical hard-disks by flash memory disks. Market research company In-Stat predicted in July 2006 that 50% of all mobile computers would use flash (instead of hard-disks) by 2013.

Flash memory devices typically consist of an array of memory cells that are grouped into *pages* of consecutive cells, where a fixed amount of consecutive pages form a *block*. Reading a bit is performed by reading the whole page containing the given bit. When writing, we distinguish between changing bits from 1 to 0 and from 0 to 1. To change a bit from 0 to 1, the device first "erases" the entire block containing the given bit, i. e. all the bits in the block are set to 1. However, changing a bit from 1 to 0 is done by writing only the page containing it, and each page can be programmed only a small number of times before it must be erased again. Reading and writing pages is relatively fast, whereas erasing a block is significantly slower. Each block can sustain only a limited number of erasures. To prevent blocks from wearing prematurely, flash devices usually have a built-in micro-controller that dynamically maps the logical block addresses to physical addresses to even out the erase operations sustained by the blocks.

Related work. Recently, there has been an increased interest in using flash memories to improve the performance of computer systems. This includes the experimental use of flash memories in database systems [4,5,6], using flash memories as caches in hard-disks (e. g. Seagate's Momentus 5400 PSD hybrid drives), Windows Vista's ReadyBoost, i. e. using USB flash memories as a cache, or integrating flash memories into motherboards or I/O-buses, e. g. Intel's Turbo Memory technology [7].

Most previous algorithmic work on flash memories deals with wear leveling, i. e. block-mapping and flash-targeted file systems (see [8] for a comprehensive survey). There exists very little work on algorithms designed to exploit the characteristics of flash memories. Wu et al. [9,10] proposed flash-aware implementations of B-trees and R-trees without file system support by explicitly handling block-mapping. More recently, efficient dictionaries on flash disks have been engineered [11]. Other works include the use of flash memories for model checking [12] or route planning on mobile devices [13,14].

Our contributions. Owing to the lack of good computation models to help exploiting the particular characteristics of flash devices, there is no firm theoretical foundation for comparing algorithms. In this paper, we propose two computational models for flash devices that exploit their constructive characteristics – the general flash model and the unit-cost flash model. These models can be used as a basis for a theoretical comparison between different algorithms on flash memory devices. While the general flash model is very generic and is especially suitable for studying lower bounds, the unit-cost flash model is appealing for the design and analysis of algorithms. In particular, we show that a large number of external-memory algorithms can be easily adapted to give efficient algorithms in the unit-cost flash model. Interestingly, we observe that external-memory algorithms based on the merging paradigm are easy to adapt in the unit-cost flash model, while this is not true for algorithms based on the distribution paradigm. We conduct experiments on several algorithms exhibiting various I/O-access patterns, i.e. random and sequential reads, as well as random and sequential writes. Our experiments confirm that the analysis of algorithms on our models (particularly, the unit-cost flash model) predicts the observed running-times much better than the I/O model. Our experiments also show that the adaptations of these algorithms improve their running-times on solid-state disks.

2   Models for Flash Memory

In this section we propose and discuss models for flash memories. We first discuss the practical behavior of flash memories. We then propose two models of computation, a *general flash model* and a *unit-cost flash model*. They are both based on the I/O-model, but use a different block size for reading than for writing.

Flash memory behavior. Due to constructive characteristics, in practice flash memories have a significantly different behavior compared to hard disks [15,16,17]. In Figure 1 we give empirical results showing the dependence of throughput on the block size when performing random reads and writes, as well as sequential reads and writes. We used two different disks: a 64 GB Hama SSD drive and a Seagate Barracuda 7200 rpm 500 GB hard-drive. The main difference concerns the relative performance of random reads and random writes. For hard-disks random reads and random writes provide similar throughput, whereas for the SSD drive random reads provide significantly more throughput than random writes, especially for small block sizes. Furthermore, the throughput of random

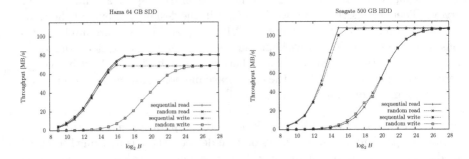

Fig. 1. Performance summary of solid-state disks (left) vs. hard disks (right). The x-axis shows the block size (in bytes), in logarithmic scale.

accesses converges to the throughput of the corresponding sequential accesses at different block sizes, implying different block sizes for reading and writing. Also, the throughput provided by sequential reads is nearly the same as the throughput provided by sequential writes for most flash devices [15].

The key characteristic of the flash devices that we model is the different block sizes for reading and writing. For the general flash model we also consider different throughput for reading and writing. To keep our computation models simple enough for algorithm design, we abstract away the other flash-memory characteristics, such as effects of misalignment, limited endurance etc.

General flash model. The general model for flash memory devices is similar to the I/O model, with the exception that read and write block sizes are different and that they incur different costs. The general flash model assumes a two-level memory hierarchy, with a fast internal memory of size M and a slow external flash memory of infinite size. The input and output data reside on the external flash memory, and computation can only be done on data residing in the internal memory. Read and write I/Os from and to the flash memory occur in blocks of consecutive data of sizes B_r and B_w respectively. The complexity of algorithms is $x + c \cdot y$, where x and y are the number of read and write I/Os respectively, and c is a penalty factor for writing. Similarly to the I/O-model, the parameters M, B_r, B_w, and c are known to the algorithms. Typically, we assume $B_r \leq B_w < M \ll N$ and $c \geq 1$. We note that the I/O-model is a particular case of this general model, when $B_r = B_w = B$ and $c = 1$.

Unit-cost flash model. The fact that in the general flash model c may take arbitrary values implies arbitrary relative costs between read and write I/Os. This complicates the reuse of existing external-memory algorithms and algorithmic techniques. In [15] it was shown that for most flash devices the throughput provided by reads and writes is nearly the same, assuming proper block sizes, i.e. B_r and B_w are set so that the maximum throughput is achieved on random I/Os. This means that, in spite of different read and write block sizes, the access time per element is nearly the same. The unit-cost flash model is the general flash model augmented with the assumption of an equal access time per element for reading and writing. This simplifies the model considerably, since it becomes significantly easier to adapt external-memory results. For the sake of clarity, the cost of an algorithm performing x read I/Os and y write I/Os is given by $xB_r + yB_w$, where B_r and B_w denote the read and write block sizes respectively. Essentially, the cost of an algorithm in this model is given by the total amount of items transferred between the flash-disk and the internal memory.

For both models, we note that "items transfered" refers to all the B_r (B_w) elements moved during a read (write) I/O and not just the useful elements transfered. Also, our models can be adapted to obtain hardware-oblivious models.

Relating unit-cost models to external-memory models. We turn to exploring the relation between the unit-cost models and the external-memory models.

Lemma 1. *Any algorithm designed in the unit-cost flash model which transfers* $f(N, M, B_r, B_w)$ *items can be simulated by an external-memory algorithm with* $B = B_r$ *which performs* $f(N, M, B_r, B_w)/B_r$ *I/Os.*

Consider some algorithm A in the unit-cost flash model, which transfers $f(N, M, B_r, B_w)$ items. Denote by $f_r(N, M, B_r, B_w)$ the total cost for read I/Os and let $f_w(N, M, B_r, B_w)$ be the total cost for write I/Os. The algorithm is executed as an external-memory algorithm with a block size $B = B_r$ as follows. Read operations are done in blocks of size B_r and therefore the reads incur $f_r(N, M, B_r, B_w)/B_r$ I/Os, whereas writes are done in blocks of size B_w which implies that each write incurs B_w/B_r I/Os. We obtain that all the writes take $(f_w(N, M, B_r, B_w)/B_w) \cdot (B_w/B_r) = f_w(N, M, B_r, B_w)/B_r$ I/Os.

The simulation in Lemma 1 provides an efficient mechanism for obtaining lower bounds in the unit-cost flash model, as stated in Lemma 2.

Lemma 2. *A problem that requires* $\Omega(L(N, M, B))$ *I/Os in the I/O-model requires* $\Omega(B_r \cdot L(N, M, B_r))$ *items transferred in the unit-cost flash model.*

3   Algorithms for the Unit-Cost Flash model

Typical external-memory algorithms manipulate buffers using various operations, such as merging and distributing. Given that in the unit-cost flash model the block sizes for reads and writes are different, algorithms can merge $\mathcal{O}(M/B_r)$-ways and distribute $\mathcal{O}(M/B_w)$-ways. Since $M/B_r > M/B_w$, merging is preferred to distributing because more buffers can be manipulated simultaneously. A surprisingly large body of merging-based external-memory algorithms (and data structures) can be easily adapted to get efficient and sometimes even optimal algorithms (and data structures) in the unit-cost flash model, sometimes by simply setting the block size B to B_r. In this section we show a few typical examples of how simple changes lead to efficient algorithms in the unit-cost flash model.

3.1   Sorting

Sorting N records in the I/O-model requires $\Omega(N/B \log_{M/B} N/B)$ I/Os [1]. Using Lemma 2, we obtain that sorting N elements needs $\Omega(N \log_{M/B_r} N/B_r)$ items to be transfered in the unit-cost flash model.

To sort in the unit-cost flash model, we use multi-way mergesort, which is optimal in the I/O-model, and we show that it achieves optimality also in the unit-cost flash model. The algorithm splits the input into $\Theta(M/B)$ subsequences, recursively sorts them, and in the end merges the (sorted) subsequences. The I/O-complexity is $\Theta(N/B \log_{M/B} N/B)$ I/Os. For the unit-cost flash model, different costs are achieved depending on the number of subsequences the input is split into. Splitting the input in $\Theta(M/B_w)$ subsequences yields an algorithm that transfers $\mathcal{O}(N \log_{M/B_w} N/B_w)$ items, whereas splitting $\Theta(M/B_r)$-ways yields the optimal $\Theta(N \log_{M/B_r} N/B_r)$ cost.

Lemma 3. *Sorting N elements can be done by transferring* $\Theta(N \log_{M/B_r} N/B_r)$ *items in the unit-cost flash model.*

3.2 Data Structures

In this section we give brief descriptions of efficient implementations for search trees and priority queues in the unit-cost flash model.

Search trees. For searching, we show how to adapt the B-trees used in the I/O-model to obtain an efficient implementation in the unit-cost flash model. We employ a two-level structure. The primary data structure is a B-tree with a fan-out of $\Theta(B_w)$; each node of the primary structure is stored also as a B-tree, but with nodes having a fan-out of $\Theta(B_r)$. Searches and updates transfer $\mathcal{O}(B_r \log_{B_r} N)$ items.

Priority queues. Several optimal external-memory priority queues have been proposed [18,19,20,21]. Each of them takes amortized $\mathcal{O}(1/B \log_{M/B} N/B)$ I/Os per operation. However, only the cache-oblivious priority queue in [20] translates directly into an optimal priority queue in unit-cost flash model, taking amortized $\mathcal{O}(\log_{M/B_r} N/B_r)$ items transfered per operation. This is because it only merges buffers, whereas the other priority queues also employ distribution and achieve only amortized $\mathcal{O}(\log_{M/B_w} N/B_w)$ transfered items. We note that priority queues are the core of time forward processing, a technique widely employed to achieve efficient external memory graph algorithms.

3.3 BFS

For BFS on undirected graphs $G(V,E)$ in the unit-cost flash model, we focus on the randomized external-memory algorithm by Mehlhorn and Meyer [22]. For ease of exposition, we restrict ourselves to sparse graphs, i.e. $|E| = \mathcal{O}(|V|)$. The algorithm starts with a preprocessing phase, in which the input graph is rearranged on disk. This is done by building $|V|/\mu$ disjoint clusters of small diameter ($\mathcal{O}(\mu \cdot \log |V|)$ with high probability (whp.)) that are laid contiguously on disk. In the BFS phase, the algorithm exploits the fact that in an undirected graph, the edges from a node in BFS level t lead to nodes in BFS levels $t-1$, t or $t+1$ only. Thus, in order to compute the nodes in BFS level $t+1$, the algorithm collects all neighbors of nodes in level t, removes duplicates and removes the nodes visited in levels $t-1$ and t. For collecting the neighbors of nodes efficiently, the algorithm spends one random read I/O (and possibly, some further sequential read accesses depending on the cluster size) for loading a whole cluster as soon as a first node of it is visited and then keeps the cluster data in some efficiently accessible data structure (hot pool) until all nodes in the cluster are visited. The preprocessing and BFS phases together require $\mathcal{O}(\text{scan}(|V|) \cdot \mu \cdot \log |V| + \text{sort}(|V|))$ I/Os (reading and writing) whp. plus another $\mathcal{O}(|V|/\mu)$ read-I/Os. In the I/O-model, choosing $\mu = \Theta\left(\sqrt{B/\log |V|}\right)$ implies a total cost of $\mathcal{O}(|V| \cdot \sqrt{\log |V|/B} + \text{sort}(|V|))$ I/Os whp. In the unit-cost flash model this means a total cost of $\mathcal{O}(|V| \cdot \mu \cdot \log |V| + |V| \cdot \log_{M/B_r} \frac{|V|}{B_r} + |V| \cdot B_r/\mu)$, which is minimized by choosing $\mu = \Theta\left(\sqrt{\frac{B_r}{\log |V|}}\right)$.

Lemma 4. *Computing undirected BFS on sparse graphs ($|E| = \mathcal{O}(|V|)$) in the unit-cost flash model requires $\mathcal{O}(|V| \cdot \sqrt{B_r \cdot \log |V|} + |V| \cdot \log_{M/B_r}(|V|/B_r))$ item transfers.*

4   Experimental Results

The main goal of our experimental study is to verify the suitability of the proposed unit-cost flash model for predicting the running-time of algorithms using SSD as an external-memory. We want to check how well the behavior of the algorithms on SSDs correspond to their theoretical analysis on the unit-cost flash model. In particular, we look at the improvements from the adaptation process as predicted theoretically on the unit-cost flash model and ascertain if these gains are actually observed in practice. We consider three algorithms which present various I/O-patterns and have very different complexities in the I/O model. First, we consider sorting, which takes $\text{sort}(N) = \mathcal{O}(N/B \log_{M/B} N/B)$ I/Os and performs mainly sequential I/Os. We then move to BFS, which requires $\mathcal{O}(|V| \cdot \sqrt{\log |V|/B} + \text{sort}(|V|))$ I/Os whp. for sparse graphs and causes both sequential and random reads, but no random writes. Finally, the classical DFS implementation performs $\mathcal{O}(|V|)$ I/Os on sparse graphs and does a large number of random reads and writes. We observe the performance of these algorithms when using a SSD as external-memory.

Experimental setup. For algorithms and data structures designed in the I/O-model we use implementations already existent in the STXXL library [23] wherever possible. We show results where the size of blocks in which data is transferred between the internal memory and the flash device is set to both the read and write block sizes of the device. According to our flash models, algorithms read blocks of size B_r and write blocks of size B_w. To comply with this requirement, we implement a translation layer similar to Easy Computing Company's MFT (Managed Flash Technology) [24]. The translation layer prevents random writes of blocks of size B_r by buffering B_r-sized blocks into blocks of size B_w that provide optimal throughput when written to the disk. When using the translation layer, an algorithm reads and writes pages of size B_r. Oblivious to the algorithm, the translation layer logically groups B_w/B_r pages into a block of size B_w, which is written to the flash disk. To do so, $O(1)$ B_w-sized buffers are reserved in the memory, so that when one such buffer gets full it is immediately written to the flash disk. To keep track of the data used, this layer maintains a mapping of the logical addresses of the pages viewed by the algorithm to their actual address on the flash disk. Since this mapping occupies little space and is used only to manage temporary data, the translation layer is stored in main memory throughout the execution of the algorithm. Additionally, the translation layer is responsible for keeping track of the free pages and blocks.

Due to its simplicity and generality, we view the translation layer as a generic easy-to-implement adaptation of I/O algorithms to algorithms in the unit-cost flash model. However, we note that there exist cases where the translation layer

Table 1. The read volume (RDV), write volume (WRV), and the running time (RT) for sorting N random integers (taking the specified volume) when using the translation layer (TL), setting the block size to B_r and to B_w respectively. RDV and WRV are measured in GB, and RT is measured in seconds.

input		TL			B_r			B_w		
$\log_2 N$	volume [GB]	RDV [GB]	WRV [GB]	RT [s]	RDV [GB]	WRV [GB]	RT [s]	RDV [GB]	WRV [GB]	RT [s]
25	0.12	0.20	0.25	9.10	0.25	0.25	9.35	0.25	0.25	9.13
26	0.25	0.49	0.50	16.73	0.50	0.50	16.72	0.50	0.50	17.10
27	0.50	0.99	1.00	32.25	1.00	1.00	31.29	1.00	1.00	33.58
28	1.00	1.99	2.00	62.35	2.00	2.00	60.96	3.00	3.00	93.46
29	2.00	3.99	4.00	120.82	4.00	4.00	118.84	6.00	6.00	192.98
30	4.00	8.00	8.00	240.24	8.00	8.00	238.74	12.00	12.00	387.16
31	8.00	16.00	16.00	478.46	16.00	16.00	475.11	32.00	32.00	1002.95
32	16.00	32.00	32.00	946.88	32.00	32.00	950.04	64.00	64.00	2029.41

can not be employed, e.g. extremely large inputs when the translation layer may no longer fit into the main memory.

Our experiments were conducted on a standard Linux machine, with an Intel Core 2 Quad 2.4 GHz CPU, 8 GB RAM out of which algorithms are restricted to use only 512 MB, and a 64 GB HAMA flash disk. The smallest block sizes where the disk reaches optimal performance for random reads and random writes are 128 KB and 16 MB respectively, see e. g. Figure 1, and consequently we set B_r and B_w to these values. The code was compiled using GCC version 4.3.

Sorting. For sorting we consider the STXXL implementation, which is based on (cache-aware) multi-way mergesort. The results in Table 1 show that when the block size is set to B_w, the running time is larger than when the block size equals B_r, and the volume of data read and written by the algorithm is larger as well. This behavior is easily explained theoretically by the larger number of recursion levels in the former case, noticeable by the relative ratio between the read/write volumes and the input volume. Also, when using the translation layer we obtain very similar results to when setting the block size to B_r. This behavior is also in line with the theoretical analysis in unit-cost flash model, since the algorithm essentially writes data sequentially, and in this case writing blocks of size B_r yields the same throughput as when writing blocks of size B_w (when using the translation layer). Such a behavior would be inexplicable in the I/O-model, which assumes reads and writes in equally sized blocks for reading and writing. We note that, due to the limited size of the flash disk, we could not sort larger sequences.

BFS. We perform experiments on square grid graphs as they have proven to be a difficult graph class [25] for the external-memory BFS algorithm. As shown in Table 2, using the translation layer yields only a small benefit compared to the read block size. This is explained by the fact that the algorithm performs no random writes, while random and sequential reads are not affected by the layer.

For preprocessing, using a smaller block size, and consequently a smaller μ, results in smaller running time, since the computed clusters tend to contain fewer

Table 2. Read/write volumes (in GB) and running times (in seconds) for external-memory BFS with randomized preprocessing on square grid graphs, separated into preprocessing phase (pp) and BFS phase, using block sizes B_r, B_w and the translation layer (TL)

$\log_2 |V|$	TL			B_r			B_w		
	pp	bfs	Σ	pp	bfs	Σ	pp	bfs	Σ
READ VOLUME [GB]									
20	0.194	0.000	0.194	0.670	1.924	2.594	0.406	0.094	0.500
22	2.423	5.968	8.391	2.709	8.080	10.789	1.500	0.188	1.688
24	26.943	60.406	87.350	27.187	61.660	88.848	91.922	457.750	549.672
26	108.953	316.341	425.294	109.726	320.881	430.607	364.578	2621.047	2985.625
WRITE VOLUME [GB]									
20	0.594	0.000	0.594	0.560	0.009	0.569	0.250	0.172	0.422
22	2.281	0.094	2.375	2.271	0.104	2.375	1.016	0.234	1.250
24	9.344	1.656	11.000	9.251	1.654	10.905	22.734	0.812	23.547
26	36.750	5.531	42.281	36.783	5.531	42.313	89.938	1.203	91.141
RUNNING TIME [s]									
20	21.5	744.5	766.0	31.5	768.4	799.9	40.5	381.4	421.9
22	95.0	1668.4	1763.4	100.0	1697.0	1797.0	76.2	1126.0	1202.2
24	609.8	4581.2	5191.0	632.9	4570.4	5203.3	1738.2	9184.6	10922.8
26	2426.8	15755.4	18182.2	2524.2	15778.9	18303.1	6824.8	43329.1	50153.9

nodes and have a smaller diameter. Comparing the preprocessing times for B_r and B_w on the square grid graph in Table 2 confirms this, as preprocessing using B_w takes up to three times as long as when B_r is used.

For the BFS phase, choosing a larger block size reduces the number of random I/Os needed to load clusters, but at the same time potentially increases the size of the hot pool because clusters with bigger diameter tend to stay longer in the pool. This affects the performance adversely if the hot pool no longer fits in internal memory as can be seen in Table 2 for $|V| \geq 2^{24}$. At that point the algorithm using B_w is outperformed by the one using B_r.

DFS. For DFS, we use a straightforward non-recursive implementation of the text-book RAM algorithm. The algorithm explores the graph by visiting for each node the first not yet visited neighbor, and to do so we use two data structures: a vector to mark the nodes visited and a stack to store the nodes for which not all the neighbors have been visited. The key particularity of this algorithm is that it performs extensive random reads to access many adjacency lists, as well as extensive random writes to mark the nodes. For a graph $G = (V, E)$ the unit-cost of the algorithm is given by $\mathcal{O}(|E| \cdot B_r + |V| \cdot B_w)$, since there are $|E|$ read accesses to the adjacency lists and $|V|$ write accesses to mark the vertices visited. The costs for accessing the stack are much smaller since both reads and writes can be buffered. We note that when transferring data in chunks of size B_r the cost of the algorithm remains $\mathcal{O}(|E| \cdot B_r + |V| \cdot B_w)$, but when the block size is set to B_w the cost increases to $\mathcal{O}(|E| \cdot B_w + |V| \cdot B_w)$.

We conduct experiments which show the running time of DFS when transferring chunks of B_r and B_w consecutive data between the memory and the flash disk, as well as on using the translation layer. Due to extensive running times, we restrict to square grid graphs. We noted that for all input sizes using the translation layer yields better running times than when doing I/Os in blocks of size B_r, which is due to writing many blocks of size B_r at random locations. When the graph fits into the main memory the algorithm is extremely fast. For $|V| \leq 2^{20}$, the running times were below two seconds. However, when the graph no longer fits into the main memory, the running times and the I/O-traffic increase significantly.

For $|V| = 2^{22}$, the running times were of $4\,180$, $4\,318$, and $610\,000$ seconds for the translation layer, B_r, and B_w block sizes respectively. The huge running time for the B_w block size is explained by the huge volume of read data, of about 46 TB, compared to 336 GB read when using B_r-sized blocks and 311 GB when using the translation layer. The volume ratio between B_w and B_r approximately matches $\frac{B_w}{B_r} = 128$. However, the volume of data written was significantly low (less that 300 MB in each experiment). This is due to vector marking the visited nodes completely residing in memory.

Therefore we used another approach and stored the visited information with each node, effectively scattering the bits over a larger range of external memory. Internal memory was further restricted to cache at most half of an external memory data structure. Comparable experiments with block size B_w are not possible in these settings because the internal memory cannot store a required minimal amount of blocks. For $|V| = 2^{21}$ the DFS using the translation layer took $6\,064$ seconds reading 250 GB and writing 146 GB of data. Using block size B_r instead, the running time increased to $11\,352$ seconds and read volume of 421 GB, while write volume was 145 GB. The translation layer could serve a fraction of the read requests directly from its write buffers explaining the increase in read volume. While the written volume and write throughput rate were nearly unchanged (145 GB, 77-80 MB/s), the read throughput dropped from 69 MB/s to 46 MB/s. The subobptimal block size used for writing obviously triggers reorganization operations in the flash device that block subsequent operations (reads in our case). This accounts for the major part of the additional running time showing a clear benefit for the translation layer bundling these small random write requests.

5   Conclusions and Future Research

We proposed two models that capture the particularities of the flash memory storage devices, the general flash model and the unit-cost flash model. We show that existing external-memory algorithms and data structures, based on the merging paradigm, can be easily translated into efficient algorithms in the unit-cost flash model. Relevant examples include sorting, search trees, priority queues, and undirected BFS. We conduct experiments that the unit-cost flash model predicts correctly the running times of several algorithms that present various I/O-patterns.

For the general flash model, an interesting future direction concerns obtaining lower bounds for fundamental problems, such as sorting or graph traversals, even for extreme cases when we set the penalty factor c to a very large value that allows the algorithm to write only the output. Future investigations in this model include engineering fast algorithms for basic problems, such as sorting.

For the unit-cost flash model, possible topics for future research include identifying problems for which the best external memory upper bounds cannot be matched in the unit-cost flash model.

Promising directions also include introducing relevant computational models that capture other characteristics of the flash devices and yet allow meaningful algorithm design.

References

1. Aggarwal, A., Vitter, J.S.: The Input/Output complexity of sorting and related problems. Communications of the ACM 31(9), 1116–1127 (1988)
2. Meyer, U., Sanders, P., Sibeyn, J.F. (eds.): Algorithms for Memory Hierarchies, Advanced Lectures [Dagstuhl Research Seminar], March 10-14, 2002. Springer, Heidelberg (2003)
3. Vitter, J.S.: Algorithms and Data Structures for External Memory. Now Publishers (2008)
4. Lee, S.W., Moon, B.: Design of flash-based DBMS: an in-page logging approach. In: SIGMOD Conference, pp. 55–66 (2007)
5. Lee, S.W., Moon, B., Park, C., Kim, J.M., Kim, S.W.: A case for flash memory ssd in enterprise database applications. In: Proc. ACM SIGMOD international conference on Management of data, pp. 1075–1086 (2008)
6. Myers, D.: On the use of NAND flash memory in high-performance relational databases. Master's thesis, Massachussets Institute of Technology (2008)
7. Matthews, J., Trika, S., Hensgen, D., Coulson, R., Grimsrud, K.: Intel® turbo memory: Nonvolatile disk caches in the storage hierarchy of mainstream computer systems. ACM Transactions on Storage 4(2), 1–24 (2008)
8. Gal, E., Toledo, S.: Algorithms and data structures for flash memories. ACM Computing Surveys 37(2), 138–163 (2005)
9. Wu, C.H., Chang, L.P., Kuo, T.W.: An efficient R-tree implementation over flash-memory storage systems. In: Proc. 11th ACM International Symposium on Advances in Geographic Information Systems, pp. 17–24 (2003)
10. Wu, C.H., Kuo, T.W., Chang, L.P.: An efficient B-tree layer implementation for flash-memory storage systems. ACM Transactions on Embedded Computing Systems 6(3) (2007)
11. Li, Y., He, B., Luo, Q., Yi, K.: Tree indexing on flash disks. In: Proc. 25th International Conference on Data Engineering (2009) (to appear)
12. Barnat, J., Brim, L., Edelkamp, S., Sulewski, D., Šimeček, P.: Can flash memory help in model checking? In: Proc. 13th International Workshop on Formal Methods for Industrial Critical Systems, pp. 159–174 (2008)
13. Goldberg, A.V., Werneck, R.: Computing point-to-point shortest paths from external memory. In: Proc. 7th Workshop on Algorithm Engineering and Experiments, pp. 26–40 (2005)
14. Sanders, P., Schultes, D., Vetter, C.: Mobile route planning. In: Halperin, D., Mehlhorn, K. (eds.) ESA 2008. LNCS, vol. 5193, pp. 732–743. Springer, Heidelberg (2008)

15. Ajwani, D., Malinger, I., Meyer, U., Toledo, S.: Characterizing the performance of flash memory storage devices and its impact on algorithm design. In: McGeoch, C.C. (ed.) WEA 2008. LNCS, vol. 5038, pp. 208–219. Springer, Heidelberg (2008)

16. Birrell, A., Isard, M., Thacker, C., Wobber, T.: A design for high-performance flash disks. ACM SIGOPS Operating Systems Review 41(2), 88–93 (2007)

17. Bouganim, L., Jónsson, B.P., Bonnet, P.: uFLIP: Understanding Flash IO Patterns. In: Proc. 4th biennial conference on innovative data systems, CIDR (2009)

18. Arge, L.: The buffer tree: A technique for designing batched external data structures. Algorithmica 37(1), 1–24 (2003)

19. Arge, L., Bender, M.A., Demaine, E.D., Holland-Minkley, B., Munro, J.I.: An optimal cache-oblivious priority queue and its application to graph algorithms. SIAM J. Comput. 36(6), 1672–1695 (2007)

20. Brodal, G.S., Fagerberg, R.: Funnel heap - a cache oblivious priority queue. In: Bose, P., Morin, P. (eds.) ISAAC 2002. LNCS, vol. 2518, pp. 219–228. Springer, Heidelberg (2002)

21. Brodal, G.S., Katajainen, J.: Worst-case efficient external-memory priority queues. In: Arnborg, S. (ed.) SWAT 1998. LNCS, vol. 1432, pp. 107–118. Springer, Heidelberg (1998)

22. Mehlhorn, K., Meyer, U.: External-memory breadth-first search with sublinear I/O. In: Proc. 10th Annual European Symposium on Algorithms, pp. 723–735 (2002)

23. Dementiev, R., Kettner, L., Sanders, P.: STXXL: standard template library for XXL data sets. Software: Practice and Experience 38(6), 589–637 (2008)

24. Easy Computing Company: Managed flash technology, http://www.easyco.com/mft/

25. Ajwani, D., Meyer, U., Osipov, V.: Improved external memory BFS implementation. In: Proc. 9th Workshop on Algorithm Engineering and Experiments, pp. 3–12 (2007)

Competitive Buffer Management with Stochastic Packet Arrivals

Kamal Al-Bawani and Alexander Souza

Institute for Computer Science, Albert-Ludwigs-Universität Freiburg, Germany
{albawani,souza}@informatik.uni-freiburg.de

Abstract. We study a variant of Naor's [23] online packet buffering model: We are given a (non-preemptive) FIFO buffer (e.g., in a network switch or a router) and packets that request transmission arrive over time. Any packet has an intrinsic value R and we have to decide whether to accept or reject it. In each time-step, the first packet in the buffer (if any) is transmitted and our benefit of it is equal to its intrinsic value minus the time it spent in the buffer. The objective is to maximize the total benefit. From a worst-case perspective, Fiat et al. [14] gave a threshold algorithm with a competitive ratio equal to the golden ratio $\phi \approx 1.618$. Due to the insensitivity of the algorithms towards the input, it was conjectured that this competitive ratio is too pessimistic for packet sequences occurring in practice.

In this paper, we treat this conjecture from an analytical and experimental point of view. In the analytical part, we assume Poisson arrivals and compute a threshold for this algorithm depending on the arrival rate λ and the value R of the packets. This also yields bounds on the (expected) competitive ratio of the algorithm. We discover the phenomenon that the ratio converges to one if R grows or λ moves away from one. Thus (for fixed R) we have that the largest competitive ratios occur for $\lambda = 1$. In that case, the bound is essentially $R/(R-\sqrt{R})$ and gives values smaller than ϕ for $R \geq 8$.

In a second, experimental, part of our study, we compared the competitive ratios achieved by the two threshold algorithms on actual network traffic with our theoretical prediction (which assumes Poisson arrivals). It turns out that the prediction and the measured ratios for our threshold are consistent, where the prediction even tends to be pessimistic. Furthermore, the measured ratios with our threshold where substantially smaller than ϕ and even almost everywhere below the ratios achieved with the threshold of [14].

1 Introduction

In the problem of online packet buffering for network switches, streams of data packets merge at some connection point (e.g., a router or a switch), and request forward transmission through an outgoing link. Due to limited link capacity, before transmission, these packets are stored in a buffer at the connection point. Thus, in general, packets are delayed or even lost in case of an overflowing

J. Vahrenhold (Ed.): SEA 2009, LNCS 5526, pp. 28–39, 2009.
© Springer-Verlag Berlin Heidelberg 2009

buffer. In the past, several approaches for modeling such a scenario theoretically have been proposed, e.g. the multiqueue unit-value model [5,3,2], the two-value model [1,21,4,17], and the bounded delay model [10,19,12]. Another one is Naor's model [23] for packet buffering.

Naor's Model in Competitive Analysis. The system under consideration is a single FIFO (First-In First-Out) queue of unbounded capacity and a stream of packets that request transmission arrives over time. An incoming packet is either admitted to the queue or discarded, where preemption is not allowed, i.e., all enqueued packets are eventually transmitted in the order of their arrivals. Time is discretized into *slots* of unit length and transmission times are deterministic: when the queue is not empty at the end of any time slot, exactly one packet is transmitted.

Each packet arrives with an intrinsic value. Here, we consider only the simpler case of *homogeneous* packets, i.e., all packets are assigned the same value $R > 1$. Packets are latency-sensitive, i.e., a packet loses one unit of its initial value for every unit-time delay. Since one transmission occurs in each time slot, the delay of a packet is equal to the number of packets that are in the queue when it is admitted to the queue. The *benefit* of transmitting a packet is equal to its value minus its value loss. The total benefit of the sequence is the sum of individual benefits of the transmitted packets and the objective is to maximize the total benefit.

The model is attractive for several reasons. First, it uses a non-preemptive FIFO queue, which is easier to maintain than queues of arbitrary regime. Second, it does not need explicit deadlines. Instead, a packet with large value is attractive for early transmission and can survive for a long time in the network, thus increasing the likeliness of reaching its destination.

A *threshold algorithm* ALG for this model determines whether to accept or reject an incoming packet based on the current size of the buffer, denoted by B, and a selected integer *threshold* $n \geq 1$. A packet is accepted only if $B < n$. The ultimate goal of any threshold policy analysis is to find a value for n that maximizes the total benefit of the sequence. Intuitively, n must not exceed the packet value R, otherwise packets will be transmitted without any benefit.

The problem of packet buffering is online by nature. A sequence of incoming packets arrives over time and decisions upon these arrivals are made without knowing future arrivals. In competitive analysis, inputs are generated by a malicious *adversary* and the performance of an online algorithm ALG is measured by comparing its benefit to the benefit of an offline optimal algorithm OPT, i.e., one that knows the entire input sequence beforehand. Specifically, an (online) algorithm ALG is called *c-competitive* if for any input sequence the inequality OPT/ALG $\leq c$ holds, where ALG and OPT denote the respective total benefits.

Recently, Fiat et al. [14] gave a competitive analysis of threshold algorithms in this model. In case of homogeneous packets, i.e., the value of each incoming packet is equal, they proved a lower bound equal to the golden ratio $\phi \approx 1.618$ for any deterministic or randomized algorithm, and they gave an optimal threshold algorithm that matches this lower bound. For heterogeneous packets, i.e., arbitrary values, they give a simple threshold algorithm with a competitive ratio equal to 5.25, and show a lower bound of 3 for any deterministic algorithm and a lower bound of 4.1 for deterministic and memoryless algorithms.

Our Contribution. Due to the insensitivity of the algorithms towards the input, it was conjectured [22] that these competitive ratios are too pessimistic for packet sequences actually occurring. In this paper, we aim at bridging the gap between theory and practice with a stochastic analysis. There, input sequences are generated by a probability distribution D and we measure the average-case performance of the online algorithm by comparing its expected benefit to the expected offline optimal benefit, where the expectations are taken with respect to the distribution D. We are interested in the long-term behaviour of the system, i.e., in the limit of t, where t is the number of time slots considered. Analogously, an algorithm is *c-competitive on average* if $\lim_{t \to \infty} \mathbb{E}\left[\text{OPT}\right]/\mathbb{E}\left[\text{ALG}\right] \leq c$.

Our results fall into two categories: analytical and experimental. In the (first) analytical part of the paper we assume that the packets arrive according to a Poisson process at a certain (known) rate $\lambda > 0$ and have all the same intrinsic real value $R > 1$.

(1) We give closed-form expressions for bounds on the expected competitive ratio of the threshold algorithm of Fiat et al. [14]. We have to distinguish the cases $\lambda = 1$ and $\lambda \neq 1$. For the case $\lambda = 1$ and $R \geq 3$ we obtain the bound

$$c \leq \frac{R}{R - \sqrt{R - \frac{3}{4}} - \frac{1}{2}}.$$

Thus the ratio decreases as R increases. In particular, the bound improves upon the one of Fiat et al. [14] for $R \geq 8$. For $1 < R \leq 3$ the bound is $c \leq 2R/(R-1)$. The same phenomenon is true for the case $\lambda \neq 1$. (We also have a closed form for $\lambda \neq 1$, but due to space limitations, we have to defer the treatment to the full version of the paper.)

(2) For any fixed value of R, our bound on the competitive ratio achieves the maximum for $\lambda = 1$, i.e., when the arrival rate of the packets equals their send rate. There is an intuitive reason for this: First suppose that λ is very small, then both, the online algorithm and the offline optimum, will accept (essentially) every incoming packet, since these are rare. On the other extreme, if λ is large, both algorithms will reject all but one incoming packet. This is because there will be an incoming packet in the next slot (with some large probability). In the remaining case $\lambda = 1$ both effects can occur: phases with relatively few, and phases with a lot of arrivals. Thus, the threshold of the online algorithm has to be chosen so that it stores several packets in the buffer when many of them arrive for compensating fewer arrivals later on. It turns out that the right amount is roughly $\sqrt{R}$. This also explains the above bound since the optimal benefit can be up to R per packet and the average benefit of the online algorithm is about $R - \sqrt{R}$ per packet.

In the experimental part of this study, we found that the theoretical predictions (assuming Poisson arrivals) are consistent with the competitive ratios measured with real network traffic. (We used the same data sets as Albers and Jacobs [2] for their experimental study for the multiqueue unit-value model.) We defined three classes of experiments. In the first class, we evaluate the competitive ratio experimentally for several values of R and variable λ. Also experimentally, the largest competitive ratios are attained for values of λ between one and two. In the second class, we fixed λ and varied R. As predicted, the competitive ratio decreases as R grows. In the third class, we relate the observed competitive ratio with the threshold. It turns out that our choice is close to the optimal choice. So, in total, it seems that the assumption of Poisson arrivals is suitable for obtaining predictions coming close to practical experience in this model.

Discussion and Conclusion. In this paper we treated the homogeneous case of the problem, only. It would be interesting to extend the analysis and the experiments to the inhomogeneous version also.

Packet arrivals are often assumed to be Poisson in probabilistic models of networks. This assumption has recently lost some of its attractiveness after discovering the self-similar characteristics of the network traffic [18,24]. That is, packets arrive in bursts rather than in smooth Poisson-like flows as it is commonly assumed. For this reason, in a seminal work, Kesselman et al. [17] avoided stochastic analysis for the problem of packet buffering, and since then, the competitive analysis has been a measure of choice when approaching this problem. However, we have returned to stochastic analysis (with Poisson arrivals): In a recent experimental study for the multiqueue unit-value model, Albers and Jacobs [2] showed that experimentally observed competitive ratios are much smaller than their counterparts in theory. Also, strategies with smaller theoretical ratios do not necessarily perform better in practice. In the light of this result, we gave an average-case competitive analysis and experimental study for Naor's model.

Our assumption of Poisson arrivals is debatable and its removal is also subject to future work. However, the discovery of the self-similarity of network traffic does not completely rule out Poisson modeling, see [20,25]: It was observed, e.g., in [20], that traffic burstiness occurs over long timescales, but arrivals tend to be Poisson over short timescales (within bursts). To make the present arrival model more realistic, i.e., so that the bursty arrival-characteristic of the network traffic is captured, one may consider the so-called Interrupted Poisson Process (IPP) [15]. In such a model, two periods of time are defined with randomly varying and exponentially distributed lengths. Poisson arrivals occur only during one period known as the active or busy period. The second period, known as the inactive period, witnesses no arrivals.

2 Stochastic Analysis

In this section, we derive the average benefits of ALG and OPT per time slot. It turns out that ALG is equivalent to a $M/D/1/K$ queue (see below). The main part of the work is to find a "good" threshold n_0 (which depends on λ and R). Further, OPT can be estimated directly.

Table 1. n_0 and c for selected values of λ and R

	$\lambda = 0.05$		$\lambda = 0.1$		$\lambda = 0.5$		$\lambda = 1.0$		$\lambda = 2.0$		$\lambda = 3.0$		$\lambda = 5.0$
R	n_0	c	n_0	c	n_0	c	n_0	c	n_0	c	n_0	c	n_0	c
2	1	2.099	1	2.2	1	2.999	1	4.0	1	2.999	1	2.666	1	2.399
5	4	1.249	4	1.25	3	1.3	2	2.05	1	1.874	1	1.666	1	1.5
10	9	1.111	9	1.111	5	1.13	3	1.548	2	1.326	1	1.481	1	1.333
20	19	1.052	18	1.052	10	1.061	4	1.323	2	1.21	2	1.165	2	1.137
30	28	1.034	27	1.034	15	1.04	5	1.245	3	1.126	2	1.129	2	1.1
50	47	1.02	45	1.02	25	1.023	7	1.176	3	1.088	3	1.071	2	1.073
80	76	1.012	72	1.012	40	1.014	9	1.133	4	1.056	3	1.05	2	1.057
100	95	1.01	90	1.01	50	1.011	10	1.116	4	1.047	3	1.043	3	1.035
110	104	1.009	99	1.009	55	1.01	10	1.11	4	1.044	3	1.04	3	1.032
115	109	1.008	103	1.008	58	1.01	11	1.107	4	1.042	3	1.039	3	1.031

For sake of exposition and space limitations, we restrict our attention to the case $\lambda = 1$ in the analysis below. The case $\lambda \neq 1$ is similar but with more technical complication and thus deferred to the full version of the paper. But notice that we do have closed-form approximations for both cases. An evaluation of the respective competitive ratios on average is given in Table 1.

Theorem 1. *For the arrival rate $\lambda = 1$ we have*

$$\lim_{t \to \infty} \frac{\mathbb{E}\left[\text{OPT}\right]}{\mathbb{E}\left[\text{ALG}\right]} \leq \begin{cases} \frac{2R}{R-1} & \text{for } 1 < R < 3, \\ \frac{R}{R - \sqrt{R - \frac{3}{4}} - \frac{1}{2}} & \text{for } R \geq 3, \end{cases}$$

where our threshold n_0 for ALG *is given through*

$$n_0 = \begin{cases} 1 & \text{for } 1 < R < 3, \\ \sqrt{R - \frac{3}{4}} + \frac{1}{2} & \text{for } R \geq 3. \end{cases}$$

Average Benefit of OPT. Here we are interested in the average benefit of OPT per time slot:

Lemma 1. *For $\lambda \geq 1$, the average benefit of* OPT *per time slot is at most R.*

Obviously, the average benefit of OPT in each time slot can never be more than R since this is the maximum benefit per packet. However, it turns out that this simple upper bound is sufficient for the case $\lambda \geq 1$. Why? On average, OPT will accept one packet per time slot, which is then transmitted without delay. We have to be more careful for the case $\lambda < 1$, as treated in the full version.

Average Benefit of ALG. Although ALG has a buffer with unbounded capacity, the threshold n acts as a regulator on its size. In each time slot, ALG accepts packets as long its buffer size has yet not reached n. Thus, our model coincides with the well-known $M/D/1/K$ queue model in queueing theory. In that queue, we have a single server, that can store up to K packets, i.e., a queue with bounded length, and each packet takes deterministic service time T. (We have $K = n$ and we will substitute accordingly.) The arrivals of the packets are Poisson with rate λ. The *service utilization factor* ρ in this model is defined as $\rho = \lambda T$. In our case we have $T = 1$ and $K = n$. Our $M/D/1/n$ queue is ergodic, i.e., a unique stationary distribution (also called steady state) $p = (p_1, \ldots, p_n)$ exists and is guaranteed to be assumed for any λ and independent of the initial distribution [23]. From now on we argue about this queue in its stationary distribution.

Lemma 2. *For $\lambda = 1$, if the threshold of* ALG *is n, in the steady state, the average benefit per time slot of* ALG *is: $R/2 - 1/2$ for $n = 1$ and $R - (R + n^2 - 1)/(2n - 1)$ for $n > 1$.*

Let A denote the expected number of accepted packets and L the expected number of transmitted packets in a time slot when the queue is in the stationary distribution. In the steady state A and L are equal. Notice that (in the steady state) exactly one packet is sent in each time slot with probability $1 - p_0$, where p_0 is the probability that the queue is empty. The average number of transmitted packets in a time slot is thus $L = 1 - p_0$ and hence also $A = 1 - p_0$ holds.

Let q be the average queue size in the steady state. Thus, the average delay experienced by enqueued packets in time slot is equal to q. Since the value loss of a packet is equal to its delay, the average benefit of ALG in a time slot i (in the steady state) is

$$RA - q = R(1 - p_0) - q \tag{1}$$

Exact solutions for p_0 and q are derived by Brun and Garcia [8]:

$$p_0 = \frac{1}{1 + \rho b_{n-1}}, \tag{2}$$

$$q = n - \frac{\sum_{k=0}^{n-1} b_k}{1 + \rho b_{n-1}}, \tag{3}$$

where n is the queue capacity, and b_k is defined by

$$b_0 = 1, \quad \text{and} \quad b_k = \sum_{i=0}^{k} \frac{(-1)^i}{i!} (k - i)^i e^{(k-i)\rho} \rho^i, \quad \forall k \geq 1.$$

We will approximate p_0 and q by approximating the b_k for $\rho = 1$ in (2) and (3). To this end, we use an alternative form of b_k given also in [8]. It holds that

$$b_k = \sum_{i=0}^{k} a_i, \tag{4}$$

where $a_0 = 1$, $a_1 = e^\rho - 1$, and $a_i = e^\rho a_{i-1} - e^\rho \sum_{j=1}^{i-1} \alpha_j a_{i-j} - \alpha_{i-1} a_0$, for $i \geq 2$, with $\alpha_j = \rho^j e^{-\rho}/j!$.

Proposition 1. *For $\rho = 1$, $a_0 = 1$, $a_1 = 1.72$, $a_2 = 1.95$, and $a_i \to 2$ as $i \to \infty$.*

The proof is omitted due to space limitations but is derived by expanding the series of the a_i. This yields

$$b_k \approx \begin{cases} 1, & k = 0, \\ 2k, & k > 0. \end{cases} \tag{5}$$

Lemma 2 is implied from plugging the following approximations for (2) and (3) in (1). These approximations are strong enough for our purposes.

$$p_0 \approx \begin{cases} \frac{1}{2}, & n = 1, \\ \frac{1}{2n-1}, & n > 1, \end{cases} \quad \text{and} \quad q \approx \begin{cases} \frac{1}{2}, & n = 1, \\ \frac{n^2-1}{2n-1}, & n > 1. \end{cases} \tag{6}$$

Threshold Optimization. Equation (1) gives the average benefit of ALG implicitly as a function of its threshold n, R, and λ. The next step is to find an optimal value n_0 of n that maximizes ALG for fixed R and λ. We do so by differentiation.

The average benefit of ALG is a unimodal function of n in the interval $[1, R]$. Thus we seek the roots of the first derivative. Lemma 2 tells us that, for $\lambda = 1$, we either want to choose $n = 1$, which yields average benefit of $R/2 - 1/2$ or to set $n > 1$ such that $R - (R + n^2 - 1)/(2n - 1)$ is maximized. Thus we have the first candidate $n_0^{(1)} = 1$. The two other candidates

$$n_0^{(2)} = \frac{1 + \sqrt{1 + 4(R - 1)}}{2} \quad \text{and} \quad n_0^{(3)} = \frac{1 - \sqrt{1 + 4(R - 1)}}{2}$$

are the roots of the first derivative of $R - (R + n^2 - 1)/(2n - 1)$. The solution $n_0^{(3)}$ is excluded because it is non-positive for any $R \geq 1$. Furthermore, one can easily verify that we want to choose $n_0 = n_0^{(1)}$ for $1 \leq R < 3$ and $n_0 = n_0^{(2)}$ for $R \geq 3$. This implies the statement on n_0 of Theorem 1. See also Table 1.

Competitive Ratio. In the case of $\lambda = 1$, if we use the threshold n_0 (see above) for ALG, by Lemma 2, we have that the average benefit per time slot is $R/2 - 1/2$ for $1 \leq R < 3$ and $R - \sqrt{R - 3/4} - 1/2$ for $R \geq 3$. As argued in Lemma 1, the benefit of OPT per time slot is bounded by R. These bounds imply Theorem 1.

3   Experiments

The main results of the experiments with our threshold strategy on real-life traffic are twofold. First, we observe that it outperforms the R/ϕ^2-threshold algorithm given in [14] experimentally. Second, we find that the theoretical average-case bounds (with Poisson assumptions) are consistent with the competitive ratios measured on real-life data.

3.1   Settings

We carry out our experiments on the same data sets as Albers and Jacobs [2] taken from the Internet Traffic Archive [16], see Table 2. The four DEC-PKT-n traces cover one-hour wide-area TCP traffic between Digital Equipment Corporation (DEC) and the outside world. The two LBL-PKT-n and the LBL-TCP-3 data sets capture wide-area TCP traffic between the Lawrence Berkeley Laboratory (LBL) and the rest of the world.

Table 2. TCP traces used in the experiments

Name	Date	♯ Packets	First Time	Last Time	Place
DEC-PKT-1	08.03.1995 22.00-23.00	2,153,462	0.416754	3600.360144	DEC
DEC-PKT-2	09.03.1995 02.00-03.00	2,661,931	0.658800	3600.460672	DEC
DEC-PKT-3	09.03.1995 10.00-11.00	2,873,589	0.341603	3600.136640	DEC
DEC-PKT-4	08.03.1995 14.00-15.00	3,862,336	0.406992	3600.193249	DEC
LBL-PKT-4	21.01.1994 14:00-15:00	862,946	0.002268	3599.995458	LBL
LBL-PKT-5	28.01.1994 14:00-15:00	677,846	0.013738	3599.998492	LBL
LBL-TCP-3	20.01.1994 14:00-16:00	1,789,995	0.008185	7199.999857	LBL

Parameters. Each experiment is defined by a set of parameters: (1) trace data set D; (2) threshold value n; (3) arrival rate λ; and (4) packet's value R. Similarly to [2], we define the average arrival rate of a trace data set D by $\lambda_D =$ (number of packets in D)/(length of time horizon of D). Here, λ_D is defined over the time unit of the data set. To allow different values of λ during the experiments, we simply alter the time unit as required. Thus, the length of a time slot in an experiment is given by λ/λ_D. On each trace data set D, we carry out three classes of experiments.

Class 1: Variable λ. In the first class, we fix the packet value R and vary the arrival rate λ. At each λ, we run two online algorithms on D, our n_0-threshold algorithm (denoted by ALG$_0$) and the R/ϕ^2-threshold algorithm of [14] (denoted by ALG$_\phi$ in experiments). We then compare the competitive ratios of these algorithms and the theoretical competitive ratio that we derived in section 2 (denoted by ALG$_{avg}$ in experiments). We add ALG$_{avg}$ for comparing the average-case analysis with the observed competitive ratios.

Class 2: Variable R. The second class of experiments is similar to the first one except that we fix λ and vary R. We also depict ALG$_0$, ALG$_{avg}$, and ALG$_\phi$.

Class 3: Variable n. In the third class, we fix both λ and R and run $\lfloor R \rfloor$ m-threshold algorithms on D, where $1 \leq m \leq \lfloor R \rfloor$. This is the set of all threshold algorithms for the packet value R. In this experiment, one can compare the competitive ratios between the optimal threshold on the given data set D, our threshold n_0 and the worst-case optimal threshold R/ϕ^2.

3.2 Results

All our results are available online [13]. Generally speaking, all seven data sets give rather similar results. The observed competitive ratio of ALG_0 is below ALG_ϕ almost everywhere. We further observe that, especially for values of $R \geq 20$, the predicted competitive ratio ALG_{avg} and the observed ALG_0 are consistent. For small values of R, our analysis is not accurate enough to give meaningful predictions.

Class 1: Variable λ. Figure 1 exhibits the results of two experiments with $R = 50$ on the data sets DEC-PKT-1 and LBL-TCP-3. However, we have found consistent results for the other traces and values of R between two and more than 100 as well. Especially the data set LBL-TCP-3 is interesting because it was used by Paxson and Floyd [24], where they found that packet arrivals of several wide-area protocols such as FTP and TELNET are not well modeled by Poisson processes.

First, the algorithm ALG_0 yields relatively small competitive ratios experimentally (1.08 at the highest). Second, ALG_0 behaves roughly as predicted in the average-case analysis. It achieves the highest competitive ratios for λ between one and two and this ratio decreases as λ tends towards 0 or ∞. As discussed in Section 2, this behaviour of ALG_0 is due to the response of its threshold value on changes in the packet arrival rate. When arrival is intensive, the threshold is decreased until it becomes 1, and when arrival is light, the threshold is increased until it reaches R. The algorithm reflected by ALG_ϕ does not adapt and hence yields increasing competitive ratios as λ increases. Furthermore, our theoretical bounds ALG_{avg} are close to those measured experimentally, but having the peak in the region around $\lambda = 1$.

Class 2: Variable R. In Figure 2 we have depicted the results for $\lambda = 1$ and $\lambda = 3$ and variable R. The competitive ratio of ALG_0 and the prediction ALG_{avg} decrease and converge to each other as R increases. (As mentioned already, our analysis lacks accuracy for small values of R – hence the large values of ALG_{avg} in that region.)

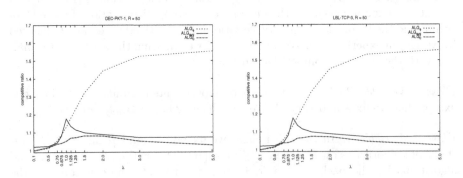

Fig. 1. Class 1: Competitive ratios for $R = 50$ on DEC-PKT-1 and LBL-TCP-3

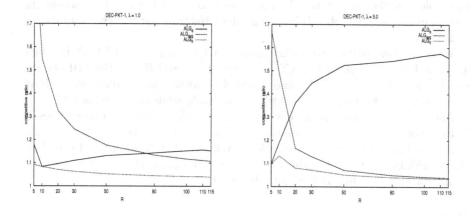

Fig. 2. Class 2: Competitive ratio for $\lambda = 1$, and $\lambda = 3$ on DEC-PKT-1

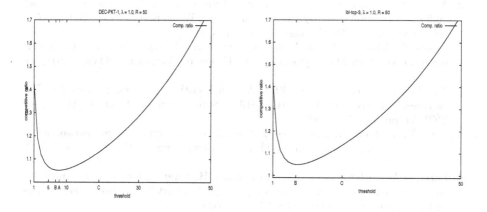

Fig. 3. Class 3: Dependence of the competitive ratio on the threshold for $\lambda = 1$ and $R = 50$ on the traces DEC-PKT-1 and LBL-TCP-3

By contrast, the competitive ratio of ALG_ϕ tends to increase as R increases; and the faster, the larger λ. For example, at $\lambda = 1$, it increases relatively slowly, while at $\lambda = 3$ it increases faster. The threshold of this algorithm is independent of λ but proportional to R. Thus, ALG_ϕ accepts more packets as R increases. This is a virtue when the intensity of arrival is low because the algorithm would save these packets for future periods with no arrivals. However, this rigid strategy fails at high arrival intensity since, in this case, the optimal behavior is to accept few packets so that these packets are sent with almost full benefit.

Class 3: Variable n. The following notation is used in Figure 3: A is the optimal threshold on the given trace data set, $B = n_0$ and $C = R/\phi^2$. We try all possible thresholds on the given trace data set and take A as the one that scores the minimum competitive ratio. B has been calculated from our analysis summarized in Table 1.

We depict the competitive ratios again for DEC-PKT-1 and LBL-TCP-3 with $\lambda = 1$ and $R = 50$. For DEC-PKT-1 we have $A = 8$ and $B = 7$. For LBL-TCP-3 we even have that $A = B = 7$ and hence A is missing in the diagram. In both cases ALG$_0$ yields a competitive ratio around 1.05, while ALG$_\phi$ yields 1.15.

Due to space limitations we could not include the results for other values of λ. However, our observations are that, as λ decreases, the value of C converges to A (and hence the competitive ratio of ALG$_\phi$ to that of an optimal threshold algorithm). In contrast, the value for B is somewhat larger than C. If λ increases, C diverts from A, while B is slightly below A.

References

1. Aiello, W., Mansour, Y., Rajagopolan, S., Rosen, A.: Competitive queue policies for differentiated services. In: Proc. INFOCOM, pp. 431–440 (2000)
2. Albers, S., Jacobs, T.: An experimental study of new and known online packet buffering algorithms. In: Arge, L., Hoffmann, M., Welzl, E. (eds.) ESA 2007. LNCS, vol. 4698, pp. 754–765. Springer, Heidelberg (2007)
3. Albers, S., Schmidt, M.: On the performance of greedy algorithms in packet buffering. In: Proc. 36th ACM Symposium on Theory of Computing (STOC 2004), pp. 35–44 (2004)
4. Andelman, N., Mansour, Y., Zhu, A.: Competitive queueing policies for QoS switches. In: Proc. of 14th ACM-SIAM Symposium on Discrete Algorithms (SODA), pp. 761–770 (2003)
5. Azar, Y., Richter, Y.: Management of multi-queue switches in QoS networks. In: Proc. 35th ACM Symposium on Theory of Computing (STOC 2003), pp. 82–89 (2003)
6. Barry, D.A., Parlange, J.-Y., Li, L., Prommer, H., Cunningham, C.J., Stagnitti, F.: Analytical approximations for real values of the Lambert W-function. Mathematics and Computers in Simulation 53, 95–103 (2000)
7. Borodin, A., El-Yaniv, R.: Online computation and competitive analysis. Cambridge University Press, Cambridge
8. Brun, O., Garcia, J.: Analytical solution of finite capacity $M/D/1$ queues. J. Appl. Prob. 37, 1092–1098 (2000)
9. Cao, J., Ramanan, K.: A Poisson limit for buffer overflow probabilities. In: IEEE Infocom (2002)
10. Chin, F.Y.L., Fung, S.P.Y.: Online scheduling with partial job values: Does time-sharing or randomization help? Algorithmica 37(3), 149–164 (2003)
11. Corless, R.M., Gonnet, G.H., Hare, D.E.G., Jeffrey, D.J., Knuth, D.E.: On the Lambert W function. Adv. Computational Maths. 5, 329–359 (1996)
12. Englert, M., Westermann, M.: Considering suppressed packets improves buffer management in qos switches. In: Proc. of 18th ACM-SIAM Symposium on Discrete Algorithms (SODA) (2007)
13. Experimental results, http://www.informatik.uni-freiburg.de/~souza
14. Fiat, A., Mansour, Y., Nadav, U.: Competitive Queue Management for Latency Sensitive Packets. In: Proc. of 19th ACM-SIAM Symposium on Discrete Algorithms (SODA), pp. 228–237 (2008)

15. Gelenbe, E., Pujolle, G.: Introduction to queueing networks. John Wiley, cop., Chichester (1998)
16. Internet Traffic Archive, http://ita.ee.lbl.gov/
17. Kesselman, A., Lotker, Z., Mansour, Y., Patt-Shamir, B., Schieber, B., Sviridenko, M.: Buffer overflow management in QoS switches. SIAM J. on Computing 33(3), 563–583 (2004)
18. Leland, W., Taqqu, M., Willinger, W., Wilson, D.: On the Self-Similar Nature of Ethernet Traffic (Extended Version). IEEE/ACM Transactions on Networking 2(1), 1–15 (1994)
19. Li, F., Sethuraman, J., Stein, C.: An optimal online algorithm for packet scheduling with agreeable deadlines. In: Proc. of 16th ACM-SIAM Symposium on Discrete Algorithms (SODA) (2005)
20. Li, Y., Qiu, H., Gu, X., Lan, J., Yang, J.: Design and Buffer Sizing of TCAM-Based Pipelined Forwarding Engines. In: Proc. of the 21st International Conference on Advanced Networking and Applications, pp. 769–776 (2007)
21. Lotker, Z., Patt-Shamir, B.: Nearly optimal FIFO buffer management for DiffServ. In: Proc. 21st ACM Symposium on Principles of Distributed Computing, pp. 134–143 (2002)
22. Nadav, U.: Personnal communication (2007)
23. Naor, P.: The regulation of queue size by levying tolls. Econometrica 37(1), 15–24 (1969)
24. Paxson, V., Floyd, S.: Wide-Area Traffic: The Failure of Poisson Modeling. IEEE/ACM Transactions on Networking 3(3), 226–244 (1995)
25. Raina, G., Wischik, D.: Buffer sizes for large multiplexers: TCP queueing theory and instability analysis. In: EuroNGI (2005)

Fast and Accurate Bounds on Linear Programs*

Ernst Althaus and Daniel Dumitriu**

Johannes Gutenberg University
Institut für Informatik
55099 Mainz, Germany
{althaus,dumitriu}@informatik.uni-mainz.de

Abstract. We present an algorithm that certifies the feasibility of a linear program while using rational arithmetic as little as possible. Our approach relies on computing a feasible solution of the linear program that is as far as possible from satisfying an inequality at equality. To realize such an approach, we have to detect the set of inequalities that can only be satisfied at equality.

Compared to previous approaches for this problem our algorithm has a much higher rate of success.

1   Introduction

As solvers for linear programs (LPs) became more and more efficient in recent years, LPs are solved in various areas of applications. In particular, they are used in software and hardware verification [1,2,3], where the safety of a system is certified by solving LPs. Often, a problem is not handled by solving a single LP, but a search tree is computed, where in each node of it an LP is solved and the (in)feasibility of the LP, or bounds computed on the value of the LP are used to guide or stop the search (compare, e.g., branch-and-cut algorithms).

State-of-the-art LP solvers rely on floating point arithmetic and hence can have wrong results. Our goal is to certify the (in)feasibility of an LP or compute safe bounds on its value efficiently. Notice that we are considering LPs which are solved during parsing a search tree and hence they are not numerically too complicated, and we want to find a certified answer with a bound close to the correct answer in most of the cases almost without overhead. Only in the few remaining examples a slow exact solver has to be used (or the algorithm has to deal without a certified solution in some other way).

In the following we always assume that we are considering a minimization problem. Previous work in this area goes into two different directions:

Dhiflaoui et al. [4] propose to take the basis computed by the LP solver and certify its feasibility and/or optimality with rational arithmetic. This approach has later been implemented more efficiently by Koch [5] and Applegate et al. [6].

---

* This work is partially supported by the German Research Foundation (DFG), as part of its priority program "SPP 1307: Algorithm Engineering", grant AL 1139/1-1.

** Part of this work has been done while the authors were at the Max-Planck-Institut für Informatik, Saarbrücken.

It has two major drawbacks. First, if the basis is infeasible, nothing can be said on the feasibility of the LP and one has to switch to an exact solver. Second, it takes much more time to solve the basis system with rational arithmetic than solving the LP (for large and complicated LPs). On the positive side, it computes very accurate bounds, e.g., if the floating-point solver happens to find the correct basis (as it is the case for most problems) it finds and certifies the optimal solution.

Another approach is to compute bounds purely with floating point arithmetic. This approach was introduced by Neumeier and Shcherbina [7] to compute lower bounds for LPs where all variables are lower and upper bounded and later improved by Keil and Jansson [8]. Still for most of the LPs from Netlib [9] no bound could be computed (for only 40% of them an upper bound could be computed).

We want to extend the floating point approach to be able to handle LPs with unbounded variables that are numerically not too complicated, thereby using rational arithmetic whenever necessary without solving large systems of linear equations.

With our approach we are able to compute bounds for 90% of the Netlib instances and the overhead in running time is negligible (see Section 5).

2   Certifying Feasibility of an LP

In this section, we describe our approach to certify the feasibility of an LP. More precisely, we describe an algorithm that either states that the given LP is feasible or it answers that it cannot certify feasibility. In the second case one should try to certify infeasibility with a similar approach (see Section 3). If this fails too, the (in)feasibility of the LP cannot be certified with our approach and one has to use an exact solver.

2.1   The General Idea

Let us first fix some notation. We are considering a linear system of the form $\{x \in \mathbb{R}^n \mid A'x \le b', A''x = b''\}$. Let $A = \binom{A'}{A''}$ and $b = \binom{b'}{b''}$. For a subset of the (in)equalities B let A_B be the rows of A corresponding to the (in)equalities in B and b_B the corresponding rows of b.

We use $\approx, \preceq, \not\approx$ to denote equations that are evaluated with floating point arithmetic, i.e., we make our decisions depending on the outcome of some expression evaluated with floating point arithmetic and typical ϵ-comparisons but we have no guarantee that the computed values are correct.

Assume we want to certify the feasibility of the linear system $\{A'x \le b', A''x = b''\}$. We use a state-of-the-art LP solver to compute a feasible solution. Typically, we get basis B, i.e., a subset of the (in)equalities of A' containing all equations such that $x^* = A_B^{-1}b_B$ has a unique solution that satisfies all inequalities. As we use floating-point arithmetic, we get $x^\approx$ with $x^\approx \approx A_B^{-1}b_B$. Then we use a solver for systems of linear equations that gives safe error bounds (see Section 2.2) to obtain a vector of intervals $x^{\square} = (x^\ell, x^u) \in (\mathbb{R} \times \mathbb{R})^n$ with $A_B^{-1}b_B \in x^{\square}$.

If all vectors in $x^{[]}$ satisfy all remaining inequalities, we are done. This can be checked easily using interval arithmetic. As the vector of intervals contains typically rather small intervals, we can certify the feasibility, if the true vector x^* satisfies the inequalities strictly. Unfortunately, many LPs are degenerated, i.e., some further inequalities are satisfied with equality.

Therefore, we use the LP solver to find an optimal basis B for the LP

$$\begin{aligned} \max \quad & \delta \\ \text{s.t. } A'x + \mathbb{1}\delta \leq \ & b' \\ A''x = \ & b'' \end{aligned}$$

i.e., we try to find a feasible point of the linear system that is as far from tight at the inequalities as possible.

This modified approach typically works if $\delta \succ 0$, but clearly fails, if the optimal value of the LP is 0. Notice that the dual of the LP is

$$\begin{aligned} \min \quad & p^T b' + q^T b'' \\ \text{s.t. } p^T A' &+ q^T A'' = 0 \\ p^T \mathbb{1} &= 1 \\ p &\geq 0 \end{aligned}$$

and hence we get a dual solution of the linear program above that gives us a vector $(p^\approx, q^\approx)$ with $p^{\approx T} A' + q^{\approx T} A'' \approx 0$, $p^{\approx T} b' + q^{\approx T} b'' \approx 0$, $p^\approx \not\succeq 0$. If there were no floating point error, we would know that all inequalities with non-zero multiplier have to be satisfied with equality in any solution of the LP. This can be seen as follows: assume (p^*, q^*) is a dual feasible solution and x^* is a feasible solution that satisfies at least one inequality strictly, i.e., $a_i^T x^* < b_i$ for some i with $p_i^* \neq 0$. Then we have $0 = \sum_i p_i^* A_i x^* + \sum_j q_j^* A_j x^* < \sum_i p_i^* b_i + \sum_j q_j^* b_j = 0$, a contradiction.

We simply assume the correctness and transform the inequalities with non-zero dual value $p^\approx$ to equalities and iterate. Notice that a wrong result of the floating point computations can lead to transforming too many inequalities to equalities, i.e., we reduce the feasible region of the LP. Hence these fixings can lead to an infeasible LP although the original LP is feasible, but not the other way around.

Notice that at least one of the new equalities is redundant and we want to remove it. In order to make this removal safe (i.e., that we do not make an infeasible linear system feasible), we have to check that these equalities are indeed redundant with rational arithmetic. The same holds for redundant equalities that appear in the initial LP formulation.

As typically $(p^\approx, q^\approx)$ contains only a very small number of non-zero entries (see Section 5), the rational arithmetic is not too costly in this step.

Furthermore, if the LP solver returns a basis that contains an equality, this could be caused by a linear dependency among the linear equations (either in the original linear program or due to the transformation) and we have to certify whether this is the case before we remove these equalities (see Section 2.3 for details). For the Netlib instances this is the case for 24 of the 94 LPs.

2.2 Computing Safe Intervals for the Solution of a System of Linear Equations

In this section we discuss how we compute safe intervals for the solution of a system of linear equations for the case when we are given a non-singular square matrix A_B. In the next section we discuss the case when our system contains redundant equalities.

In a state-of-the-art LP solver, a highly tuned (in efficiency and numerical stability) algorithm for computing the LU decomposition is implemented. As the LP solver SoPlex allows us access to its LU decomposition, we want to use it in our computations.

We do so by the following approach given in [10], where different estimates for $||L^{-1}||$ of an approximate LU decomposition are used to obtain rigorous error bounds. We use the fact that $||L^{-1}|| \leq ||M(L)^{-1}||$,

$$M(L)_{ij} = \begin{cases} |L_{ij}| & \text{for } i = j \\ -|L_{ij}| & \text{for } i \neq j \end{cases}$$

and that $||M(L)^{-1}||$ can be computed easily. For further details, we refer to the original paper.

2.3 Overdetermined Linear Systems

In order to use the method outlined in Section 2.2, we have to detect and remove all redundant equalities, i.e., a subset of equalities that are linearly dependent on the remaining equalities such that the remaining equalities are linearly independent.

Before removing an equality from our system, we have to make sure that it is indeed redundant. For this check, floating point arithmetic is not sufficient, since a wrong answer would make us remove an equality that is not redundant, and hence possibly enlarge the feasible region. Therefore we have to test redundancy with rational arithmetic.

To show that $Ax = b$ contains a redundant equality one typically transforms $(A\ b)$ into row echelon form. This can be a quite expensive operation, especially when performed with rational arithmetic (since it is as expensive as solving a system of linear equations, i.e., as certifying the feasibility of a basis with rational arithmetic). Therefore we have to use as few equations as possible for which we compute the row echelon form.

Consider an LP of the form

$$\begin{aligned} \max \quad & \delta \\ \text{s.t.} \quad & A'x + \mathbb{1}\delta \leq b \\ & A''x = b'' \end{aligned}$$

and assume that its objective function value is zero. Solving this LP, we get a dual solution (p^*, q^*) with objective function value 0. The (in)equalities with non-zero dual value are transformed into equalities and the resulting set of equalities contains at least one redundant equality (compare to Section 2.1). Furthermore this set is expected to be very small and hence we can afford to compute the row echelon form with rational arithmetic.

Now assume that the objective function value of the above LP is different from zero, hence we are in the last iteration of our algorithm. If the LP solver does not indicate that an equality is redundant (by having the equality in the basis), we are fine. Otherwise, we have to transform the set of all equations into row echelon form. If there are many equalities in the last LP, this can take a while (see Section 5). We remove all equalities that are proved to be redundant and use the method outlined in Section 2.2 to prove that the solution of the remaining basis satisfies all basic inequalities. Notice that we have to recompute the LU decomposition.

To improve the running time we can use a heuristic to select a subset of the equations so that this subset has as many redundant equalities as the set of all equalities. If the method outlined in Section 2.2 still certifies a vector of intervals such that the solution of the remaining equations lies within this interval, we know that we detected all redundant equalities.

Taking all equalities that are marked basic by the LP solver seems to be a good choice (see Section 5) because it often succeeds and the running time for computing the row echelon form is very small for this set.

2.4   Decreasing the Number of Iterations

The first experiments with our approach have shown that δ remains zero for many iterations (see Section 5), resulting in a long running time of our approach. Yet we realized that in many iterations, a single inequality implies all bounds of the occurring variables. In this case, we simplify the LP, i.e., we set all occurring variables to the respective bounds and remove the inequality.

More exactly, let $\ell_i \leq x_i \leq u_i$ be the bound on x_i (with $\ell_i, u_i \neq \pm\infty$). Then a constraint $a_1 x_1 + a_2 x_2 + \cdots + a_n x_n \sim b$ can be simplified:

- when $\sim$ is $\leq$, if $\min(a_1\ell_1, a_1u_1) + \cdots + \min(a_n\ell_n, a_nu_n) = b$; we have

$$\min(a_i\ell_i, a_iu_i) = \begin{cases} \ell_i, & \text{if } a_i > 0 \\ u_i, & \text{if } a_i < 0 \end{cases}$$

- when $\sim$ is $\geq$, if $\max(a_1\ell_1, a_1u_1) + \cdots + \max(a_n\ell_n, a_nu_n) = b$; we have

$$\max(a_i\ell_i, a_iu_i) = \begin{cases} u_i, & \text{if } a_i > 0 \\ \ell_i, & \text{if } a_i < 0 \end{cases}$$

- when $\sim$ is $=$, if either of $\leq$ or $\geq$ can be simplified as above.

These verifications are first performed using floating-point arithmetic; in case the answer is positive, a double-check using rational arithmetic takes place, and only if it also ends up positive, the constraint is deemed as simplifiable.

For any constraint that could be simplified after the verifications were carried as before, the variables contained are set to their respective bound used in the simplification check. The constraint is thus redundant and finally removed.

3 Certifying Infeasibility and Computing Bounds

A linear system $\{Ax \leq b\}$ is infeasible if and only if the system $\{y^T A \leq 0, y \geq 0, y^T b = -1\}$ is feasible (by Farkas' Lemma). Hence, we can use the approach described above to certify infeasibility of an LP.

To prove an upper bound for the LP we do the following: At the end of our algorithm for certifying the feasibility, we have transformed some inequalities to equalities. Let

$$\max \quad \delta$$
$$\text{s.t. } A'x + \mathbb{1}\delta \leq b'$$
$$A''x = b''$$

be the final LP and $x^{[]}$ be the certified interval for the solution. We solve the LP

$$\min \ c^T x$$
$$\text{s.t. } A'x \leq b'$$
$$A''x = b''$$

and compute a certified solution $\bar{x}^{[]}$. We consider the segments from $\bar{x}^{[]}$ to $x^{[]}$, i.e., $(1-\lambda)\bar{x}^{[]} + \lambda x^{[]}$ for $0 \leq \lambda \leq 1$, and compute the smallest λ^* so that for each possible resulting point $x^* \in (1-\lambda^*)\bar{x}^{[]} + \lambda^* x^{[]}$ on one of the segments satisfies $A'x' \leq b$. Finally we compute the largest objective function value of a point in $(1 - \lambda^*)\bar{x}^{[]} + \lambda^* x^{[]}$.

We have not implemented these algorithms so far and hence cannot report on experimental results.

4 Implementation

We implemented our algorithm in C++ using SoPlex version 1.4.0 [11,12]. For compilation we used g++ 4.1.2 with the optimization flag -O3 turned on. We used GMP 4.2.1 [13] as multi-precision library, and boost 1.33.1 [14] for the interval arithmetic and for basic linear algebra.

Our application proceeds in two phases: a preprocessing step and an iterated routine, for which we provide more details in the following subsections.

4.1 Preprocessing Phase

The LP problem is first transformed into a δ-modified problem:

1. The objective function is changed to $\max \delta$.
2. A new bound, $0 \leq \delta \leq 1$, is introduced.
3. The inequality constraints are transformed as follows: $C \leq rhs$ becomes $C + \delta \leq rhs$, and $C \geq rhs$ becomes $C - \delta \geq rhs$.
4. Bounds referring to a variable that only appeared in the old objective function are removed, to prevent the solver from raising errors related to variables not declared previously.

5. The remaining bounds are transformed to so-called "monomial constraints", i.e., $x_i \leq b_k$ becomes $x_i + \delta \leq b_k$, and $x_i \geq b_k$ becomes $x_i - \delta \geq b_k$; the bound for δ and the fixed bounds are not affected.

6. For any variable x_i that does not appear in bounds, a new bound $x_i - \delta \geq 0$ is introduced.

4.2   Iteration Phase

We proceed then in successive iterations. The program stops either when a problem could not be solved to optimality, or when we obtain a non-zero solution for δ, or when the preset maximum number of iterations is reached.

An iteration of the program consists of the following steps:

File input. The problem is read from file (for the first iteration, the δ-modified version of the original problem is used). The coefficients and bounds in the file are processed and stored both as regular `doubles` and as GMP `mpq_class` rationals.

Transforming constraints to bounds. Monomial constraints as defined above are transformed to bounds. A constraint of form $ax_i \pm \delta \sim b_k$ or $ax_i \sim b_k$ (with $a \neq 0$) is replaced with the bound $x_i \sim \frac{b_k}{a}$; depending on the sign of a, the inequality sense could be reversed.

Problem simplification. We check whether the problem can be simplified by setting the variables to their respective bounds as described in Section 2.4.

Transforming bounds to constraints. Bounds are again transformed to monomial constraints as in Section 4.1.

Solving with SoPlex. We make use of the SoPlex API to load the problem transformed through the previous operations and solve it.

Tightening constraints. If the objective value (i.e., δ) is zero, the constraints corresponding to dual variables with non-zero value are made tight and δ is removed from them; one such constraint is removed, since it is redundant (see Section 2.3).

File output. The problem is written to a file on disk, ready to be used for the next iteration.

Note that reading and writing the file to disk has only historical reasons and will be changed soon.

5   Experiments

The experiments were conducted on a computer with a dual-core AMD Opteron 2220 SE processor at 2.8 GHz and 16 GB of RAM, running Debian Etch with Linux kernel version 2.6.24. As input we used the 94 problems from the Netlib LP collection [9].

In Table 1 we summarize our main results. For each of the Netlib instances (for which we give the name and some statistics on its size), we report on the number of iterations of our approach, on the running time to solve the LPs and the running time to compute the reduced row echelon form of all equations and of the equations that are marked basic in the last LP solved with SoPlex. Furthermore we report whether the heuristic finds all redundant equalities.

In the same table, we report on the number of iterations our approach needs without the simplification of the LPs and whether the algorithm of Keil and Jansson [8] is able to prove the feasibility of the LP.

By computing the row echelon form of all equalities we can prove the feasibility of all but 3 LPs. If we compute the row echelon form of the equalities that are marked basic in the last LP, we can prove the feasibility of all but 9 LPs. In this case, the time spent using rational arithmetic is very small compared to the running time to solve the LPs in all but 3 LPs. Our simplification yields a big reduction in the number of calls to the LP solver.

Table 1. Experimental results on the Netlib problems

It — iterations in our approach
It_0 — iterations without simplification of LP
Last objv — value of objective function (i.e. δ) in the last iteration
t_{solve} — total time to solve LPs in all iterations (in seconds)
tr_1 — time to compute reduced row echelon form of SoPlex basic equalities
tr_2 — time to compute reduced row echelon form of all equalities
Red — does heuristic find all redundant equalities?
KJ — is feasibility proven by Keil and Jansson in [8]? (*miss* = problem missing)

Name	Rows	Cols	It	It_0	Last objv	t_{solve}	tr_1	tr_2	Red	KJ
25FV47	822	1571	1	1	−2.0000e−01	1.51	0.01	269.35	yes	no
80BAU3B	2263	9799	4	2	−2.7881e−04	24.89	—	—	—	yes
ADLITTLE	57	97	1	2	−1.8717e−01	0.01	—	—	—	yes
AFIRO	28	32	1	1	−1.0000e+00	0.01	—	—	—	yes
AGG	489	163	1	71	−4.1273e−02	0.01	—	—	—	no
AGG2	517	302	1	3	−1.0000e+00	0.01	—	—	—	no
AGG3	517	302	1	3	−1.0000e+00	0.01	—	—	—	no
BANDM	306	472	1	22	−2.3164e−02	0.05	—	—	—	no
BEACONFD	174	262	1	79	−7.5000e−02	0.01	—	—	—	no
BLEND	75	83	1	1	−7.3015e−02	0.01	—	—	—	yes
BNL1	644	1175	1	81	−8.5027e−04	0.50	0.01	49.29	yes	no
BNL2	2325	3489	1	112	−4.1131e−03	2.81	—	—	—	no
BOEING1	351	384	2	2	−3.3333e−01	0.15	—	—	—	miss
BOEING2	167	143	5	2	−1.3938e−02	0.03	—	—	—	miss
BORE3D	234	315	2	140	−3.1051e−02	0.02	0.01	8.52	no	no
BRANDY	221	249	1	44	−1.9080e−02	0.03	0.01	5.94	yes	no
CAPRI	272	353	1	1	−1.6844e−01	0.03	—	—	—	yes
CYCLE	1904	2857	1	333	−4.4062e−03	2.93	0.01	3126.66	yes	no
CZPROB	930	3523	1	348	−4.4515e−04	1.86	—	—	—	no
D2Q06C	2172	5167	3	2	−1.6667e−02	8.19	—	—	—	no
D6CUBE	416	6184	1	1	−1.7064e−02	20.13	0.01	635.58	yes	no
DEGEN2	445	534	12	7	infeasible	0.01	—	—	—	no
DEGEN3	1504	1818	8	8	infeasible	19.41	—	—	—	miss
DFL001	6072	12230	6	13	−4.3478e−02	777.57	—	—	—	miss

Table 1. (*continued*)

Name	Rows	Cols	It	It$_0$	Last objv	t_{solve}	tr_1	tr_2	Red	KJ
E226	224	282	1	31	−2.9723e−03	0.04	—	—	—	yes
ETAMACRO	401	688	1	48	−5.6667e−03	0.07	—	—	—	no
FFFFF800	525	854	1	20	−1.6501e−02	0.23	—	—	—	no
FINNIS	498	614	1	29	−1.2500e−05	0.03	—	—	—	yes
FIT1D	25	1026	1	1	−3.6269e−02	0.89	—	—	—	yes
FIT1P	628	1677	1	1	−1.0000e+00	0.22	—	—	—	yes
FIT2D	26	10500	1	1	−1.0000e+00	6.74	—	—	—	yes
FIT2P	3001	13525	1	26	−1.0000e+00	1.71	—	—	—	yes
FORPLAN	162	421	2		−1.4952e−01	0.06	—	—	—	miss
GANGES	1310	1681	1	101	−1.0000e+00	0.15	—	—	—	no
GFRD-PNC	617	1092	1	27	−2.2331e−01	0.19	0.01	176.18	yes	yes
GREENBEA	2393	5405	1	2	−2.8563e−02	12.69	—	—	—	no
GREENBEB	2393	5405	1	1180	−2.8563e−02	13.18	0.01	27746.72	yes	no
GROW15	301	645	1	1	−1.0000e+00	0.06	—	—	—	yes
GROW22	441	946	1	1	−1.0000e+00	0.13	—	—	—	yes
GROW7	141	301	1	1	−1.0000e+00	0.02	—	—	—	yes
ISRAEL	175	142	1	1	−1.0000e+00	0.01	—	—	—	yes
KB2	44	41	1	1	−1.0000e+00	0.01	—	—	—	yes
LOTFI	154	308	1	1	−1.0000e+00	0.02	—	—	—	yes
MAROS	847	1443	5	525	−6.3361e−02	2.68	0.15	267.63	yes	no
MAROS-R7	3137	9408	1		−1.0000e+00	8.90	—	—	—	yes
MODSZK1	688	1620	1		−1.0000e+00	0.20	0.01	419.23	no	no
NESM	663	2923	1	1	−1.8000e−02	1.20	—	—	—	miss
PEROLD	626	1376	1	40	−5.0000e−06	0.98	—	—	—	no
PILOT	1442	3652	1	79	−3.4335e−03	21.03	—	—	—	miss
PILOT4	411	1000	1	7	−1.0800e−01	0.70	—	—	—	no
PILOT87	2031	4883	1		−5.0000e−03	26.80	—	—	—	no
PILOT.JA	941	1988	1	82	−2.5000e−06	1.52	—	—	—	no
PILOTNOV	976	2172	2	58	−2.5000e−06	3.52	—	—	—	no
PILOT.WE	723	2789	1	32	−2.5000e−06	1.76	—	—	—	no
RECIPE	92	180	1	18	−1.0000e+00	0.01	—	—	—	no
SC105	106	103	1	2	−1.0000e+00	0.01	—	—	—	yes
SC205	206	203	1	3	−1.0000e+00	0.01	—	—	—	yes
SC50A	51	48	1	2	−1.0000e+00	0.01	—	—	—	yes
SC50B	51	48	1	3	−1.0000e+00	0.01	—	—	—	yes
SCAGR25	472	500	1	1	−1.0000e+00	0.06	—	—	—	yes
SCAGR7	130	140	1	1	−1.0000e+00	0.01	—	—	—	yes
SCFXM1	331	457	1	15	−5.1809e−02	0.04	—	—	—	no
SCFXM2	661	914	1	29	−5.1809e−02	0.15	—	—	—	no
SCFXM3	991	1371	1	43	−5.1809e−02	0.29	—	—	—	no
SCORPION	389	358	7	67	−1.3904e−04	0.14	0.09	14.78	no	no
SCRS8	491	1169	1	42	−1.2346e−02	0.15	—	—	—	no
SCSD1	78	760	1	1	−1.0000e+00	0.04	—	—	—	yes
SCSD6	148	1350	1	1	−1.0000e+00	0.13	0.01	25.13	yes	yes
SCSD8	398	2750	1	1	−1.0000e+00	0.34	—	—	—	no
SCTAP1	301	480	1	1	−3.3333e−01	0.01	—	—	—	no

Table 1. (*continued*)

Name	Rows	Cols	It	It$_0$	Last objv	t_{solve}	tr_1	tr_2	Red	KJ
SCTAP2	1091	1880	1	1	−3.3333e−01	0.15	—	—	—	yes
SCTAP3	1481	2480	1	1	−3.3333e−01	0.22	—	—	—	yes
SEBA	516	1028	1	133	−1.9000e−01	0.13	—	—	—	miss
SHARE1B	118	225	1	1	−1.0000e+00	0.02	—	—	—	yes
SHARE2B	97	79	1	1	−1.2239e−01	0.01	—	—	—	yes
SHELL	537	1775	1	1	−1.0000e+00	0.12	0.22	575.62	no	no
SHIP04L	403	2118	1	202	−4.1177e−03	0.32	0.01	292.17	yes	no
SHIP04S	403	1458	1	90	−5.9474e−03	0.17	0.01	116.05	yes	no
SHIP08L	779	4283	3	1126	−2.6206e−03	1.84	0.01	4789.71	yes	no
SHIP08S	779	2387	3	552	−8.4205e−03	0.53	0.01	1193.42	yes	no
SHIP12L	1152	5427	1	1067	−1.4757e−03	1.34	0.01	8761.91	yes	no
SHIP12S	1152	2763	1	361	−1.5544e−03	0.23	0.01	1750.12	yes	no
SIERRA	1228	2036	1	1	−1.6667e−01	0.39	0.63	288.24	no	no
STAIR	357	467	1	1	−9.9600e−02	0.15	—	—	—	yes
STANDATA	360	1075	1	55	−2.0000e−01	0.27	—	—	—	no
STANDGUB	362	1184	1		−2.0000e−01	0.28	—	—	—	no
STANDMPS	468	1075	1	55	−1.9928e−02	0.39	—	—	—	no
STOCFOR1	118	111	1	1	−1.6719e−02	0.01	—	—	—	yes
STOCFOR2	2158	2031	1	1	−9.2721e−03	0.44	—	—	—	yes
TRUSS	1001	8806	1	1	−1.0000e+00	7.20	0.01	4693.82	yes	yes
TUFF	334	587	1	30	−2.3999e−04	0.13	0.01	32.48	yes	no
VTP.BASE	199	203	1	172	−1.0000e+00	0.01	—	—	—	no
WOOD1P	245	2594	1	793	−3.4126e−06	2.83	—	—	—	no
WOODW	1099	8405	1	?	−6.2228e−06	13.18	—	—	—	no

6 Conclusion

We presented an algorithm that certifies the feasibility of a linear program while using rational arithmetic as little as possible. For most of the LPs in Netlib, our algorithm certifies the feasibility by solving with floating-point arithmetic only a few LPs of the same size as the given LPs; the time spent in the rational arithmetic is typically very small compared to the solving time of the LPs.

Our algorithm can be extended to compute safe bounds on the objective function value of an LP; we want to implement this approach and conduct experiments. Furthermore, we want to integrate our algorithm in the branch-and-cut framework SCIP [15] and the DPLL framework iSAT [16].

For LPs that are numerically more complicated, it could be beneficial to improve our approach to find a safe vector of intervals for the solution of a system of linear equations. There are several methods proposed in the literature and we have to check which one gives the best compromise between accuracy and running time.

References

1. Devulder, S., Lambert, J.L.: A comparative study between linear programming validation (LPV) and other verification methods. In: Automated Software Engineering (ASE), pp. 299–302 (1999)
2. Brinkmann, R., Drechsler, R.: RTL-Datapath verification using integer linear programming. In: Design Automation Conference (ASP-DAC), pp. 741–746. IEEE, Los Alamitos (2002)
3. Dellacherie, S., Devulder, S., Lambert, J.L.: Software verification based on linear programming. In: Woodcock, J.C.P., Davies, J., Wing, J.M. (eds.) FM 1999. LNCS, vol. 1709, pp. 1147–1165. Springer, Heidelberg (1999)
4. Dhiflaoui, M., Funke, S., Kwappik, C., Mehlhorn, K., Seel, M., Schömer, E., Schulte, R., Weber, D.: Certifying and repairing solutions to large LPs how good are LP-solvers? In: Symposium of Discrete Algorithms (SODA), pp. 255–256 (2003)
5. Koch, T.: The final NETLIB-LP results. Oper. Res. Lett. 32(2), 138–142 (2004)
6. Applegate, D., Cook, W., Dash, S., Espinoza, D.: Exact solutions to linear programming problems. Oper. Res. Lett. 35(6), 693–699 (2007)
7. Neumaier, A., Shcherbina, O.: Safe bounds in linear and mixed-integer linear programming. Math. Program. 99(2), 283–296 (2004)
8. Keil, C., Jansson, C.: Computational experience with rigorous error bounds for the netlib linear programming library. Reliable Computing 12(4), 303–321 (2006)
9. Netlib: A Linear Programming Library, http://www.netlib.org/lp/
10. Higham, N.J.: A survey of condition number estimation for triangular matrices. SIAM Review 29(4), 575–596 (1987)
11. SoPlex: The Sequential object-oriented simplex, http://soplex.zib.de/
12. Wunderling, R.: Paralleler und objektorientierter Simplex-Algorithmus. Ph.D thesis, Technische Universität Berlin (1996),
    http://www.zib.de/Publications/abstracts/TR-96-09/
13. GMP: The GNU Multiple Precision Arithmetic Library, http://gmplib.org/
14. Boost: C++ Libraries, http://www.boost.org/
15. Achterberg, T.: SCIP – a framework to integrate constraint and mixed integer programming. Technical Report 04-19, Zuse Institute Berlin (2004),
    http://www.zib.de/Publications/abstracts/ZR-04-19
16. Fränzle, M., Herde, C., Ratschan, S., Schubert, T., Teige, T.: Efficient solving of large non-linear arithmetic constraint systems with complex boolean structure. JSAT Special Issue on Constraint Programming and SAT 1, 209–236 (2007)

Batch Dynamic Single-Source Shortest-Path Algorithms: An Experimental Study*

Reinhard Bauer and Dorothea Wagner

Karlsruhe Institute of Technology (KIT), Germany
{rbauer,wagner}@ira.uka.de

Abstract. A dynamic shortest-path algorithm is called a batch algorithm if it is able to handle graph changes that consist of multiple edge updates at a time. In this paper we focus on fully-dynamic batch algorithms for single-source shortest paths in directed graphs with positive edge weights. We give an extensive experimental study of the existing algorithms for the single-edge and the batch case, including a broad set of test instances. We further present tuned variants of the already existing SWSF-FP-algorithm being up to 15 times faster than SWSF-FP. A surprising outcome of the paper is the astonishing level of data dependency of the algorithms. More detailed descriptions and further experimental results of this work can be found in [1].

1 Introduction

The *single-source shortest-path problem* is a fundamental graph problem with many real-world applications, such as routing in road networks, routing/data harvesting in sensor networks and internet routing using link state protocols (for example OSPF and IS-IS). In these applications shortest-path trees are stored and have to be updated whenever the underlying graph undergoes changes [2,3,4,5].

Algorithms that update the trees without a full recomputation from scratch are called *dynamic single-source shortest-path algorithms*. Such algorithms slightly differ in the type of their output. Some store only the distances from the source, while others additionally store a shortest-path tree or the shortest-path subgraph. Some of the algorithms known in the literature are only able to cope with the update of one edge at a time, while others can perform *batch updates*, i.e. update the shortest-path information after multiple edges have simultaneously changed their weight.

We consider edge insertions and deletions as special cases of weight changes: Deletions correspond to weight increments to infinity, while insertions are weight decrements from infinity. An algorithm is called *fully dynamic* if both weight increases and decreases are supported, and *semi-dynamic* if only weight decreases or only increases are supported.

---

* Partially supported by the Future and Emerging Technologies Unit of EC (IST priority – 6th FP), under contract no. FP6-021235-2 (project ARRIVAL) and the DFG (project WAG54/16-1).

J. Vahrenhold (Ed.): SEA 2009, LNCS 5526, pp. 51–62, 2009.

In this paper we focus on fully-dynamic batch updates for directed graphs with positive edge weights. In order to compare the different approaches, the only requirement that we make regarding the tested algorithms is that they update the distance vector. We furthermore demand that the algorithms be able to cope with edge insertions and deletions. For our experimental study, we apply integer edge weights.

Related work. Ramalingam and Reps [6] introduce the batch algorithm SWSF-FP, Narvaez et al. [2] propose the NARVÁEZ-framework containing six single-edge update algorithms and a modification to the framework leading to the according batch algorithms. Pure single-edge update algorithms are RR [7] (due to Ramalingam and Reps) and FMN [8] (by Frigioni et al). Buriol et al [9] present a heuristic technique to speed up RR-like approaches. The technique is similar to techniques used in the NARVÁEZ-framework but does not support edge insertions or deletions. Furthermore, in [9] the RR algorithm is adapted to maintain a special (shortest-path) tree proposed in [10].

There is no algorithm known in the literature for which the worst case is asymptotically better than recomputing the new solution from scratch. In the original works the algorithms described in Section 3 are theoretically analyzed with respect to different measures. These measures mostly depend on the size of the subgraph for which the shortest-path subgraph changes.

There is some work on the variant of the problem where edge weights may also be negative. In [7] the algorithm RR is adapted to cope with the existence of negative cycles, in [11] the same is done for the algorithm FMN. In [12] Demetrescu gives some algorithms for that problem. These algorithms use the reweighting technique, which incorporates a complete Dijkstra run on the graph (with changed edge weights). Hence, this approach is impractical for the problem with non-negative edges.

A well-studied related problem is the *fully dynamic all-pairs shortest-path problem*, in which the distances between all pairs of nodes have to be maintained while the graph undergoes changes. See [13] for a survey on the problem.

There is only few experimental work on this topic, all concentrating on single-edge updates. In [14] the algorithms FMN, RR and a full recomputation from scratch are compared on two instance classes: Erdös-Rényi graphs, where updates are chosen uniformly at random and a graph representing the internet on the AS-level, where updates simulate the failure and recovery of the links. In [2] the algorithms of the NARVÁEZ-framework are evaluated on graphs originating from a generator. This generator randomly places nodes on a grid and connects them by edges with probability that exponentially decreases with the distance of the nodes. The generator does not seem to be available any more. In [15] the algorithms SWSF-FP, RR, FMN, NARVÁEZ and full recomputation from scratch using DIJKSTRA, BELLMAN FORD, D'ESOPO PAPE are evaluated with single-edge updates on Erdös-Rényi-like graphs. In [4] one algorithm of the NARVÁEZ-framework is evaluated on random single-edge updates on a graph representing the road-network of Western Europe. In [9], the algorithm RR as well as seven variants thereof are evaluated on a real world AT&T IP network, synthetic internet-related graphs and a large set of other synthetic instances, namely those of [16] with non-negative edge lengths.

Overview. This paper is organized as follows. Section 2 states basic definitions and formally introduces the problem. Section 3 reviews the existing algorithms. Section 4 presents our tuned variants of the SWSF-FP-algorithm, while an extensive experimental study of these algorithms on synthetic and real-world data is given in Section 5. The paper ends with a conclusion in Section 6.

2   Problem Statement

Let $G = (V, E)$ be a directed graph with n nodes and m edges and a non-negative length function len : $V \times V \to \mathbb{R}^+ \cup \{\infty\}$. Let $s \in V$ be an arbitrary but fixed *source*. With $d(v)$ we denote the length of a shortest s-v-path in G for any $v \in V$.

A *batch update* is a set of *edge modifications* on G which can be edge insertions, edge deletions, edge weight increases and edge weight decreases (that keep the length function non-negative). We want to maintain a distance vector $D[]$ containing $d(v)$ for each node v in a dynamic environment where G is undergoing batch updates. After each batch update, $D[]$ (and possible required auxiliary data needed by the recomputation algorithm) has to be updated accordingly.

Throughout the text, we will cope with the recomputation of $D[]$ and the auxiliary data when *one* concrete batch update is given (because of the recomputation of the auxiliary data the algorithms are able to handle following updates). We write len_{old} for length function and d_{old} for distance before the update. Accordingly we write len for length function and d for distance after the update. For notational convenience, we consider inserted or deleted edges to be existing in the original and the updated graph and set the edge length to infinity, if necessary.

Some of the following algorithms are designed to handle only one edge modification at a time. Obviously, repeated application of these algorithms also solves the batch case. We call such algorithms *iterative algorithms* while the others are called *batch algorithms*. Iterative algorithms can be split into two parts: the *incremental* part handles edge insertions and weight decreases while the *decremental* part handles edge deletions and weight increases. This terminology can be unintuitive on a first glance but originates from the point of view that the graph increases when edges are inserted.

3   Overview of Algorithms

In this section, we give an overview on the algorithms evaluated in our experimental study. Each algorithm includes a main phase in which a min-based priority queue Q is used to recompute the distances in a Dijkstra-like fashion but on a smaller subgraph.

RR. Ramalingam and Reps [7] describe the iterative algorithm RR that handles only edge insertions and deletions. It can be directly transferred to an algorithm that works with weight increases and decreases. We will use this variant.

FMN. The FMN-algorithm of Frigioni et al. [8] is an iterative algorithm similar to the algorithm RR that uses more complex auxiliary data to obtain better theoretical worst case bounds. The approach relies on the existence of a *k-bounded accounting function* on G, which is a mapping $K : E \to V$ such that for each edge (u, v) the node $K(u, v)$ is either u or v and such that for each node n, no more than k edges are n-valued. We use the constructive 2-approximation algorithm described in [11] for finding a k-bounded accounting function on G.

Narváez. Narvaez et al. [2] propose a batch algorithm incorporating two degrees of freedom. One degree of freedom is the choice of Q which does not necessarily need to be a priority queue but only has to maintain the operations INSERT and EXTRACT. Narvaez et al propose a FIFO queue (Bellman-Ford like approach), a heap (implemented as binary heap or linked list) and a D'Esopo-Pape like approach. The other degree of freedom consists of two different variants for the main phase of the algorithm which we will describe below. We will refer to the diffent variants as NAR{1st, 2nd}{HEAP, BF, PAP}. The main idea of the NARVÁEZ-framework is to early-propagate distance changes through the tentative shortest-path tree.

4   Tuning SWSF-FP

In this section we will review the algorithm SWSF-FP which is due to Ramalingam and Reps [6] and give some tuned variants of it.

The input of the algorithms is the outdated distance vector $D[]$, the graph G, the original length function len_{old}, the batch update $U = (u_1, \ldots u_k)$ and some auxiliary data which will be described for each algorithm separately. The output is the updated distance vector $D[]$ and the updated auxiliary data.

Notation. Given the outdated distance vector $D[]$, we say we *relax* an edge (u, v) when we check if $D[v] > D[u] + \text{len}(u, v)$. We say we *relax and update* an edge (u, v) when we set $D[v] := \min\{D[v], D[u] + \text{len}(u, v)\}$. An edge (u, v) is said to be *consistent* if $D[v] = \text{len}(u, v) + D[u]$ and *underconsistent* if $D[v] > \text{len}(u, v) + D[u]$. The *consistent value* $\text{con}(v)$ of a node v is

$$\text{con}(v) := \begin{cases} \min_{(u,v) \in E} \{D[u] + \text{len}(u, v)\} \,, v \neq s \\ 0 \qquad\qquad\qquad\qquad\qquad , v = s \end{cases}$$

A node is said to be *consistent* if $D[v] = \text{con}(v)$ and to be *over-consistent* if $D[v] > \text{con}(v)$. As convention, we use $\min \emptyset := \infty$.

SWSF-FP. For each node v, a label $d[v]$ is given. Initially, $d[]$ equals $D[]$ (in order to save time for the copy process we implemented $d[]$ as auxiliary data). We say we *adjust* an inconsistent node v when we set $d[v] := \text{con}(v)$ and insert v with priority $\min(D[v], d[v])$ in Q. In case v is already in Q we only update its priority. We adjust a consistent node v when we remove it from Q. If v is not in Q we do nothing.

Initially, we adjust each node which is target of an edge in U. *Main Phase.* While Q is not empty, we perform as follows: We extract and delete the minimum node w from Q. If $d[w] < D[w]$ we set $D[w] := d[w]$ and adjust each outgoing neighbor of w. If $d[w] > D[w]$ we set $D[w] := \infty$ and adjust w and each of its outgoing neighbors.

Tuned SWSF. This algorithms basically works like the SWSF-FP-algorithm, but with less computational effort. When performing SWSF-FP we have to relax all incoming edges of a node n in order to compute con(n). TUNED SWSF relaxes fewer of such incoming edges: When we adjust an outgoing neighbor v of a node w with $d[w] < D[w]$, we compute con(v) by $\min\{d[w] + \text{len}(v,w), d[v]\}$. The same strategy works in the initialization phase when we compute con(n) for a node n that is the target node of an edge with decreased edge weight. When we adjust an outgoing neighbor v of a node w with $d[w] > D[w]$, we set $D_{old} := D[w]$ and $D[w] := \infty$. We can skip v when $D_{old} + \text{len}(w,v) \neq d[v]$. The same strategy holds in the initialization phase for target nodes of edges with increased weight.

Tuned SWSF RR. This variant enhances the algorithm TUNED SWSF with a technique adapted from the RR-algorithm. For each node v, a label indegree(v) is given indicating the number of edges (u,v) with $D[u] + \text{len}(u,v) = d[v]$. Further, for each edge (u,v) a boolean label $DAG(u,v)$ is given indicating if $D[u] + \text{len}(u,v) = d[v]$. The labels indegree and DAG are directly updated whenever len, $D[]$ or $d[]$ change. The algorithm performs like TUNED SWSF with the following difference: After a node v with $d[v] > D[v]$ is extracted from Q only those edges (v,w) have to processed for which indegree[w] = 0.

Tuned SWSF NAR. This variant enhances the algorithm TUNED SWSF with a technique adapted from the NARVÁEZ-algorithm. For each node v that is not the source, a label $P(v)$ is given pointing at another node, such that $D[P(v)] + \text{len}(P(v), v) = d(v)$. At the beginning a shortest-path tree T on the original graph is given implicitly by this label. The main phase of the algorithm works like the main phase of TUNED SWSF. The initialization phase works as follows: First, we update the edge weights. We denote by A the set of all nodes that lie behind a target node of an updated edge. Then, we update the distances $D[]$ of nodes in T according to the new edge weights (but to the original shortest-path tree T). This can be implemented such that for each node $v \in A$, the distance $D[w]$ is updated at most once. Then, we set $d[v] = \text{con}(v)$ for each node v which either is contained in A or has a neighbor in A. Finally we insert each node with $d[v] \neq D[v]$ with priority $\min\{d[v], D[v]\}$ in Q.

5 Experiments

In this section, we present an experimental evaluation of the algorithms described above. Our implementation is written in C++ (using the STL at some points). Our tests were executed on one core of an AMD Opteron 2218, running SUSE Linux 10.3. The machine is clocked at 2.6 GHz, has 32 GB of RAM and 2 x 1 MB of L2 cache. The program was compiled with GCC 4.2, using optimization level 3.

For each experiment, 1000 update instances were generated. To properly measure the speed-ups, a full Dijkstra run is performed directly after each update and the speed-up compared to that run (i.e. the time needed by Dijkstra's algorithm divided by the time needed by the update algorithm) is computed. Finally we compute the mean value of these speed-ups. Thus, measurement disturbances due to background processes etc are avoided as much as possible. For Tables 1-4 we showed in bold letters all algorithms whose performance was at least 85% of the best observed performance.

In our experiments we evaluated all previously described algorithms. We did not include the heuristic of Buriol et al [9] because it does not support edge insertions or deletions. Further, we did not include the D'Esopo-Pape variants of the NARVÁEZ-framework because pretests had revealed some instances with extremely bad performance with this approach. To gain further insights in the performance of the batch-algorithms (NARVÁEZ and TUNED SWSF), we executed these two times: one time with processing the edges in batch, as stated originally and one time with iteratively processing the edges one after another. We refer to these approaches as ITNAR and ITTUNED SWSF. Note that we refer to the NARVÁEZ-framework as a batch algorithm while it actually does not perform updates completely in a batch: its initialization phase handles edge updates iteratively but the following main phase handles all updates in a batch.

5.1 Graph Instances

UNIT DISK. Given n and m, a unit disk graph is generated by randomly assigning each of the n nodes to a point in the unit square of the Euclidean plain. Two nodes are connected by an edge in case their Euclidean distance is below a given radius. This radius is adjusted such that the resulting graph has approximately m edges. As edge weights we use the Euclidean distance to the power of 0 (hop length), 1 (Euclidean distance) and 2 (energy). All tested graphs consist of 15 000 nodes.

RAILWAY. The graph RAIL represents the condensed railway network of Europe, based on timetable information, provided by the company HaCon [17] for scientific use. Nodes represent stations while edges represent direct connections between the stations. The edge weight corresponds to the average travel time between two stations. The graph has 29 578 nodes and 159 914 edges.

AS-GRAPH. The graph AS-HOP represents the internet as of 2008/3/26 on the AS-level, i.e. each node corresponds to an autonomous system and edges represent connections between autonomous systems. This graph is taken from the Routeviews project page [18]. It has 27 909 nodes and 114 474 edges. The edge weight is 1 for each edge. The same graph with edge weights chosen uniformly at random from the interval [1, 1000] is called AS-RAN.

CAIDA. This dataset represents the internet on the router level, i.e. nodes are routers and edges represent connections between routers. The network is taken from the CAIDA webpage [19] and has 190 914 nodes and 1 215 220 edges. The edge weight is 1 for each edge.

ROAD. We evaluate three road networks provided by the PTV AG [20]. DEU represents Germany with 4 378 447 nodes and 10 968 884 edges, NLD the Netherlands with 946 632 nodes and 2 358 226 edges and LUX represents Luxembourg with 30 647 nodes and 75 576 edges. The edge weights are the corresponding travel times with speed profile 'slow car'.

GRID. These are fully synthetic graphs based on two-dimensional square grids. The nodes of the graph correspond to the crossings in the grid. There is an edge between two nodes if these are neighbors on the grid. Edge weights are randomly chosen integer values between 1 and 1000. GRID 100 is a 100x100 grid graph while GRID 300 is a 300x300 grid graph.

5.2   Assessing the Performance of the Algorithms

Let $U = \{u_1, \ldots, u_k\}$ be a set of updated edges. By $\Delta(G, U)$ we denote the number of vertices in V for which the distance from the source changes due to the update. The *expected speed-up* of an update is the number of vertices in the graph divided by $\Delta(G, U)$. This value is roughly the speed-up we expect from a good update algorithm. Of course, speed-ups can even be higher for special instances. It experimentally turned out that when the topology of the original shortest-path tree does not change, the propagation of the updated edge's weights through the tree can gain a large speed-up.

When we want to measure the difficulty of an update for an iterative algorithm we consider $U = (u_1, \ldots, u_k)$ to be ordered. We perform the updates u_i iteratively in the given ordering (always additional to the former updates) obtaining a sequence of graphs $G = G_0, G_1, \ldots, G_k$. We write $\delta(G, (u_1, \ldots, u_k)) := \sum_{i=0}^{k-1} \Delta(G_i, \{u_{i+1}\})$. We have the following hypothesis: the smaller the difference between $\Delta(G, U)$ and $\delta(G, U)$ is, the less do the contained single-edge updates interfere and it is reasonable to use an iterative algorithm for the update. If the difference is great, an iterative algorithm would change the distance of many nodes multiple times. Hence, it is more appropriate to use a batch algorithm. The experimental evaluation will support our hypothesis.

5.3   Space-Saving Implementation of RR

The algorithm RR needs to maintain the shortest-path subgraph. This subgraph is implicitly given by each edge (u, v) with $d[u] + len(u, v) = d(v)$. We implemented the algorithm doubly. One time with explicitly storing the subgraph (RR DAG) and one time with reconstructing it when needed (RR). It turned out that there are only small differences between both implementations, with no variant being clearly superior. We therefore only report the results for the space-saving implementation RR.

5.4   Single-Edge Update Experiments

We start our experimental study by single edge updates. Because of space restrictions and a different focus of our paper we do not carry out a separate analysis

Table 1. Speed-ups of experiments with single-edge updates

	LUX	NLD	DEU	RAIL	CAIDA	AS-HOP	AS-RAN	GR100	GR300	UNIT H	UNIT E
FMN	42	1504	29087	151	22702	1624	2182	25	142	327	36
SWSF-FP	112	3759	65404	366	12429	416	691	59	351	1613	31
tun SWSF-FP	152	**5140**	84873	**562**	16406	893	**3442**	105	598	**2436**	**186**
tun SWSF-NAR	147	3354	70245	215	9306	614	695	94	523	748	129
tun SWSF-RR	118	3798	66068	412	26093	2148	**3766**	74	430	2096	102
RR	155	**4666**	74857	**510**	**34586**	**2599**	**4057**	103	568	**2519**	137
Nar-1st BF	**284**	**5335**	**100944**	357	6578	417	305	**138**	**784**	1176	20
$\Delta(G,U)$	130.42	140.52	70.72	30.68	0.21	0.41	0.74	59	113	0.01	93
expected speed-up	236	6762	62549	986	inf	inf	inf	169	804	inf	163

for the decremental and the incremental case. An update consists of choosing an edge uniformly at random and multiplying its weight by a random value in $(0, 2)$. The results can be seen in Table 1.

We observe that the algorithms of the NARVÁEZ-framework have only tiny differences in performance with Nar-1st BF being slightly (but not significantly) faster most times. There is no such uniform behavior for the SWSF-FP-like algorithms. TUNED SWSF is always faster (between 1.3 and 6 times) than SWSF-FP. The algorithm TUNED SWSF RR is always at least as fast as SWSF-FP and up to 5.5 times faster. The algorithm TUNED SWSF NAR seems to be very volatile being between half as fast and 4 times faster than SWSF-FP.

Comparing the different classes of algorithms, we find the algorithms to perform quite differently, but within the same order of magnitude. The algorithm FMN is most times much slower than the other ones. This is due to the overhead caused by maintaining and reading the priority queues used by this algorithm. The technique used in this algorithm can pay off in case nodes with high degree exists (for which many edge-relaxations can be saved). This is not the case for the test instances used. Exceptions are the INTERNET instances CAIDA, AS-HOP and AS-RAN. Here, the gap to the other algorithms is much smaller, (which meets the theoretical considerations). Hence, it is to be expected that there are dense graph classes for which FMN is the superior algorithm. On the ROAD and GRID instances, the NARVÁEZ-framework is superior. This is because the structure of the shortest-path tree stored by the algorithm hardly changes on these experiments. Therefore, the early-propagation of the weight change works well. On the INTERNET instances, RR is the fastest algorithm. Looking at the small value of $\Delta(G, U)$, we can see that updates hardly have any impact on these instances, which favors the RR-algorithm with its small computational overhead and the early detection of edge weight increases that do not change distances on the graph.

The achieved speed-ups vary greatly between the instances. This is mainly due to the different structure of the underlying graphs, which results in greatly differing expected speed-ups. It is interesting to see that in nearly all cases the best actual speed-ups are close to the expected speed-ups or even higher. This, in combination with the small absolute runtimes in the range of microseconds, makes us expect that there is not much space for further improvement for the single-edge update case.

5.5   Experiments on Batch Updates

Multiple Randomly Chosen Edges. In this experiment we chose 25 edges uniformly at random. For each edge, we chose uniformly at random a value from the interval $(0, 2)$ and multiplied the weight of the edge with that value. For each graph there is hardly any difference between $\Delta(G, U)$ and $\delta(G, U)$. Therefore, the single-edge updates did only interfere marginally with each other. Hence, not much news is to be expected by this setting regarding the comparison of the algorithms. This has been confirmed by the experiments.

However, we ran the batch-algorithms (NARVÁEZ and TUNED SWSF) twice. One time with processing the edges in batch as stated in the description and one time with iteratively processing the edges one after another. Nearly no runtime differences were observed between the iterative and the batch variants, which indicates a low overhead with batch updates.

Node Failure and Recovery. This update class uses the two parameters deg_{min} and deg_{max}. First, a node v with degree between deg_{min} and deg_{max} is chosen uniformly at random. The update consists of two steps. In the first step, v *fails*, i.e. the weights of all edges adjacent to v are set to infinity. In the second step, v *recovers*, i.e. the weights of all edges adjacent to v are reset to their original values. The results can be found in Tables 2 and 3.

We now take a look at the INTERNET instances. The most remarkable result is the bad performance of the Narvaez-framework, which clearly is the inferior

Table 2. Speed-ups of experiments with node failure and recovery updates on INTERNET-instances

degree	AS-HOP			AS-RAN			CAIDA		
	1–10	10–100	100–500	1–10	10–100	100–500	1–10	10–100	100–500
FMN	784	173	23	1368	129	3	7824	2284	185
ittun SWSF-FP	912	228	26	1320	235	11	**12874**	4212	382
SWSF-FP	273	68	8	389	28	1	9651	2203	128
tun SWSF-FP	967	250	24	1417	252	15	**14042**	4693	405
tun SWSF-NAR	407	92	9	410	50	4	9785	2187	122
tun SWSF-RR	1272	**528**	**130**	**2475**	**433**	21	12395	**6839**	969
RR	**1438**	**576**	**142**	**2623**	**490**	17	13915	**7075**	**1163**
Nar-1st Heap	53	21	9	86	59	16	4315	761	75
itNar-1st Heap	52	18	6	71	30	8	4060	573	63
$\delta(G, U)$	1.26	12.16	82.54	1.47	45.01	1365.85	1.97	7.4	90.99
$\Delta(G, U)$	1.07	8.59	71.28	1.1	34.6	712.3	1.45	5.7	85.73
expected speedup	27909	3489	393	27909	821	86	190914	38183	2246

Table 3. Speed-ups of experiments with node failure and recovery updates on UNIT DISK-instances

metric average degree	hop			euclidean			energy		
	7	10	15	7	10	15	7	10	15
FMN	30	40	55	27	21	24	12	14	20
ittun SWSF-FP	238	398	485	116	95	98	56	66	91
SWSF-FP	128	214	236	60	32	36	28	24	22
tun SWSF-FP	260	462	561	**158**	**115**	**141**	75	**86**	**110**
tun SWSF-NAR	106	116	147	101	77	97	57	61	67
tun SWSF-RR	223	395	527	105	75	89	49	54	67
RR	**289**	**504**	**628**	111	91	106	55	63	84
Nar-1st Heap	70	87	131	84	62	111	52	62	74
itNar-1st Heap	55	71	100	64	50	66	36	46	52
$\delta(G, U)$	19	8	6	86	107	99	194	174	132
$\Delta(G, U)$	18	7	5	54	79	55	128	119	98
expected speedup	833	2500	3750	283	190	273	117	126	153

algorithm for that testset. One main reason for that is, that on this testset the edge-weight propagation in the initialization phase creates useless extra effort which gets overwritten later on. The gap between $\delta(G,U)$ and $\Delta(G,U)$ is small to mid-size, favoring RR with its small overhead, but big enough such that TUNED SWSF RR is nearly as fast. This difference also manifests in the small difference between ITTUNED SWSF and TUNED SWSF.

The situation is similar, but a bit clearer, for UNIT DISK graphs. When applying hop distance, $\delta(G,U)$ and $\Delta(G,U)$ are still quite near to each other, TUNED SWSF and RR are the best-performing algorithms (with RR being slightly better). When applying Euclidean or energy edge weights updates, the difference between $\delta(G,U)$ and $\Delta(G,U)$ is much bigger, and TUNED SWSF clearly is the superior algorithm. We also observe the advantage of TUNED SWSF against SWSF-FP being between 2 and 15 times faster.

Traffic Jams. This update class models real-world traffic jams. It derives from the observation that traffic jams often occur along shortest paths. The number k of updated edges is given as a parameter. Initially, a node v is chosen uniformly at random. Then a shortest path SP ending at v and containing exactly k edges is chosen uniformly at random. The update consists of two steps: in the first step, the weights of edges in SP are multiplied by 10. In the second step, the edge weights are reset to their original values. The results can be found in Table 4.

Table 4. Speed-ups of experiments with traffic jam updates

edges	GRID			LUX			NLD			DEU		
	10	20	30	5	10	20	10	20	30	10	20	30
FMN	3	2	1	4	2	1	11	5	2	185	30	7
ittun SWSF-FP	15	9	5	15	7	2	39	17	6	755	100	23
SWSF-FP	13	10	6	15	9	5	75	32	12	873	173	40
tun SWSF-FP	23	16	9	20	12	6	107	44	17	1210	235	55
tun SWSF-NAR	22	16	9	22	13	7	107	41	17	1402	342	79
tun SWSF-RR	16	12	7	15	9	5	72	31	12	957	181	42
RR	17	10	5	20	9	3	43	18	6	924	149	36
Nar-1st Heap	16	9	5	20	10	4	37	15	5	1120	196	35
itNar-1st Heap	19	12	6	24	12	4	57	24	8	1231	219	54
$\delta(G,U)$	4367	7909	15552	1178	3052	8616	12910	32088	93725	7885	39260	142191
$\Delta(G,U)$	2591	3564	6412	821	1366	2567	4134	10899	26153	3884	13701	51252
expected speed-up	35	25	14	37	22	12	229	87	36	1127	320	85

We observe that this update class consists of strongly interfering single-edge updates: there is a big difference between $\delta(G,U)$ and $\Delta(G,U)$. TUNED SWSF and TUNED SWSF NAR are the best-performing algorithms for this testset. This is because pure batch algorithms avoid processing nodes many times. With an increasing number of edges, the interference between the updated edges increases and the advantage of these two algorithms grows.

For a small number of edges in the jam, the NARVÁEZ-framework is comparable to TUNED SWSF. The framework slows down with a growing number of updated edges. This is because the initialization phase processes many nodes one time for each updated edge. It is astonishing to see that the NARVÁEZ-framework is not able to take advantage of the batch-character of the update. This can be seen through a comparison with ITNARVAEZ. The iterative variant is even faster than the batch one, which could be a hint at space for improvement. Again,

FMN is much slower than the other algorithms, as its overhead does not pay of on these instances.

6 Conclusion

In this work we focused on the single-source shortest-path problem with non-negative weights. We gave the first experimental study evaluating the performance for single-edge updates that contains all current algorithms and incorporates a broad set of instance classes. It turned out that the algorithms perform quite differently, but within the same order of magnitude. Furthermore, the achieved speed-ups varied greatly between different instances. This can be explained by measuring the impact of the updates on the graphs.

Moreover, we presented the first experimental study at all for the case of multiple edge changes at a time. One experiment was to choose a set of edges uniformly at random. It turned out that this way the single-edge updates did almost not interfere. Therefore, the results deviated not much from the single-edge case. We also used two more realistic types of batch updates. One is the simulation of node failure and recovery, which affects all incident edges. The single-edge updates interfered for that class, but not very strongly. For internet instances, the best performing algorithms were RR and TUNED SWSF RR with RR being slightly faster. For UNIT DISK graphs, TUNED SWSF was the best algorithm with RR being slightly faster for hop distance. The other update class modelled traffic jams. The single-edge updates interfered greatly, TUNED SWSF and TUNED SWSF NAR were the superior algorithms there.

Furthermore, we presented tuned variants for the SWSF-FP-algorithm and evaluated their performance. For the tuned variants we observed speed-ups compared to SWSF-FP of up to 15. Finally, we gave a simple methodology (based only on Dijkstra's algorithm) to decide if one should try a single-edge or a batch-update algorithm for a given instance class. We compared the 'impact' of the update when processed in batch with the 'impact' when processed iteratively. For updates with a big gap between both values, the algorithms TUNED SWSF or TUNED SWSF RR usually performed best. With a small gap, there was usually a better-performing iterative algorithm.

Concluding, we gave a first experimental overview on the different approaches for the problem, which can be used as a base for further research. The most important information that can be extracted from our experiments is the astonishing level of data dependency within the problem. It turned out that a proper assessment of an algorithm's running time is not possible without full knowledge of the application it is used in. Further, a great amount of experiments is required to get the big picture of an algorithm's efficiency.

References

1. Bauer, R., Wagner, D.: Batch Dynamic Single-Source Shortest-Path Algorithms: An Experimental Study. Technical Report 2009, 6, ITI Wagner, Faculty of Informatics, Universität Karlsruhe, TH (2009),
   http://digbib.ubka.uni-karlsruhe.de/volltexte/1000010926

2. Narváez, P., Siu, K.Y., Tzeng, H.Y.: New Dynamic Algorithms for Shortest Path Tree Computation. IEEE/ACM Transactions on Networking 8, 734–746 (2000)
3. Bruera, F., Cicerone, S., D'Angelo, G., Stefano, G.D., Frigioni, D.: Dynamic Multi-level Overlay Graphs for Shortest Paths. Mathematics in Computer Science (2008) (to appear)
4. Delling, D., Wagner, D.: Landmark-Based Routing in Dynamic Graphs. In: Demetrescu, C. (ed.) WEA 2007. LNCS, vol. 4525, pp. 52–65. Springer, Heidelberg (2007)
5. Wagner, D., Wattenhofer, R. (eds.): Algorithms for Sensor and Ad Hoc Networks. LNCS, vol. 4621. Springer, Heidelberg (2007)
6. Reps, T., Ramalingam, G.: An Incremental Algorithm for a Generalization of the Shortest-Path Problem. Journal of Algorithms 21 (1996)
7. Ramalingam, G., Reps, T.: On the computational complexity of dynamic graph problems. Theoretical Computer Science 158 (1996)
8. Frigioni, D., Marchetti-Spaccamela, A., Nanni, U.: Fully Dynamic Algorithms for Maintaining Shortest Paths trees. Journal of Algorithms 34 (2000)
9. Buriol, L., Resende, M., Thorup, M.: Speeding Up Dynamic Shortest-Path Algorithms. Informs. Journal on Computing 20 (2008)
10. King, V., Thorup, M.: A space saving trick for directed dynamic transitive closure and shortest path algorithms. In: Wang, J. (ed.) COCOON 2001. LNCS, vol. 2108, pp. 268–277. Springer, Heidelberg (2001)
11. Frigioni, D., Marchetti-Spaccamela, A., Nanni, U.: Fully dynamic shortest paths in digraphs with arbitrary arc weights. Journal of Algorithms 49, 86–113 (2003)
12. Demetrescu, C.: Fully Dynamic Algorithms for Path Problems on Directed Graphs. Ph.D thesis, Department of Computer and Systems Science (2001)
13. Demetrescu, C., Italiano, G.F.: Dynamic shortest paths and transitive closure: Algorithmic techniques and data structures. Journal of Discrete Algorithms 4 (2006)
14. Frigioni, D., Ioffreda, M., Nanni, U., Pasqualone, G.: Experimental Analysis of Dynamic Algorithms for the single Source Shortest Path Problem. ACM Journal of Experimental Algorithmics 3 (1998)
15. Taoka, S., Takafuji, D., Iguchi, T., Watanabe, T.: Performance Comparison of Algorithms for the Dynamic Shortest Path Problem. IEICE Transactions on Fundamentals of Electronics, Communications and Computer Sciences E90-A (2007)
16. Cherkassky, B.V., Goldberg, A.V., Radzik, T.: Shortest paths algorithms. Mathematical Programming, Series A 73, 129–174 (1996)
17. HaCon - Ingenieurgesellschaft mbH (2008), http://www.hacon.de
18. University of Oregon Routeviews Project (2008), http://www.routeviews.org/
19. CAIDA: The Cooperative Association for Internet Data Analysis (2008), http://www.caida.org/
20. PTV AG - Planung Transport Verkehr (2008), http://www.ptv.de

Rotated-Box Trees:
A Lightweight c-Oriented Bounding-Volume Hierarchy[*]

Mark de Berg[1] and Peter Hachenberger[2]

[1] Department of Computing Science, TU Eindhoven, P.O. Box 513, 5600 MB
Eindhoven, the Netherlands
[2] MADALGO[**], Department of Computer Science, University of Aarhus, Aarhus,
Denmark

Abstract. We introduce a new type of bounding-volume hierarchy, the
c-oriented rotated-box tree, or *c-RB-tree* for short. A c-RB-tree uses boxes
as bounding volumes, whose orientations come from a fixed set of predefined box orientations. We theoretically and experimentally compare
our new c-RB-tree to two existing bounding-volumes hierarchies, namely
c-DOP-trees and box-trees.

1 Introduction

The range-searching problem—preprocess a set S of objects into a data structure such that the objects from S intersecting a query range can be reported
efficiently—is one of the most fundamental problems in computational geometry. The problem comes in many variants, depending on the type of objects to
be stored (points, segments, etc.), the type of query ranges (boxes, discs, sim-
plices), the dimension of the underlying space, and so on. For many of these
variants, dedicated data structures exist with theoretical performance guarantees that match (or almost match) the theoretical lower bounds—see one of the
surveys [1,3] for an overview. In many applications, however, one would prefer a
more versatile data structure: a data structure that can store different types of
objects and can answer range queries with different types of ranges. Indeed, such
multi-functional geometric structures are used in most practical applications.

Space-partitioning structures form one category of multi-functional data structures. These structures are based on a (usually hierarchical) subdivision of the
space into cells, where with each cell the objects intersecting it are stored. A
query can then be answered by finding the cells intersecting the query range Q

[*] Both authors were supported by the Netherlands' Organisation for Scientific Re-
search (NWO) under project no. 639.023.301. P.H. was also supported in part by
a NABIIT grant from the Danish Strategic Research Council and by the Danish
National Research Foundation.
[**] Center for Massive Data Algorithms, a Center of the Danish National Research
Foundation.

J. Vahrenhold (Ed.): SEA 2009, LNCS 5526, pp. 63–75, 2009.

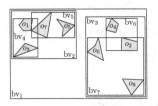 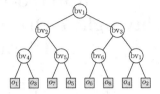

Fig. 1. Example of a bounding-volume hierarchy (here: a box-tree) in the plane

and testing the objects stored with these cells for intersection with Q. Examples of space-partitioning structures are quadtrees and BSP trees. A disadvantage of these structures is that objects may intersect multiple cells and, hence, may have to be stored multiple times. As a result, the storage can be super-linear. An alternative is to use a *bounding-volume hierarchy*, or BVH for short. BVHs are the topic of our paper.

Bounding-volume hierarchies. A BVH for a set S of n objects in $\mathbb{R}^d$ is a tree $\mathcal{T}$ with n leaves, each storing (a pointer to) a distinct object in S. Moreover, each node ν of $\mathcal{T}$ stores a *bounding volume* bv(ν) for the set $S(\nu)$ of objects stored in the leaves of $\mathcal{T}_\nu$, the subtree rooted at ν. Fig. 1 shows an example of a BVH, where the bounding volumes are axis-aligned boxes. A BVH has n leaves by definition and so it uses $O(n)$ storage, assuming the bounding volumes are constant-complexity shapes.

Querying a BVH $\mathcal{T}$ with a range Q is done by traversing $\mathcal{T}$ in a top-down manner, as follows. When a node ν is visited, one tests if bv(ν) $\cap$ $Q = \emptyset$. If so, the subtree $\mathcal{T}_\nu$ need not be explored further, otherwise the children of ν are visited. When a leaf is reached, the object stored with it is tested for intersection (and possibly reported). The efficiency of the query procedure depends on the number of visited nodes, and on the cost of the intersection test that has to be done at each visited node. There are two factors influencing this. One factor is how the objects are distributed over the tree. In Fig. 1, for example, exchanging o_3 and o_8 in the tree would result in much larger bounding volumes which may cause more nodes to be visited. The second factor is the type of bounding volume being used. This is what we will focus on.

Many different types of bounding volumes have been proposed—see the thesis of Haverkort [8] for an overview—the most popular of which is the bounding box: the smallest axis-aligned box containing the objects. A BVH using bounding boxes is called a *box-tree*. Bounding boxes allow for very efficient intersection tests, especially when the query range Q is also an axis-aligned box, as is often the case.[1] An additional advantage is that bounding boxes need very little storage. On the other hand, bounding boxes do not always fit the data well, which can have a negative impact on the number of visited nodes. Agarwal *et al.* [2]

---

[1] When Q is not a box, one usually replaces Q with its bounding box to profit from the fast box–box intersections. Testing the original range Q is then only done at the leaf level—see also Section 2.

have obtained theoretical performance guarantees for orthogonal range searching. More precisely, they showed how to construct, for any given set S of n input boxes in $\mathbb{R}^2$, a box-tree that can answer orthogonal range queries in $O(\sqrt{n} + k)$ time, where k is the number of reported boxes; the result generalizes to $\mathbb{R}^d$, where the bound becomes $O(n^{1-1/d} + k)$. For other types of ranges, it is not possible to obtain sublinear worst-case bounds using box-trees.

How tightly fitting a bounding box is, may depend on the choice of the coordinate system. Hence, bounding boxes that are oriented in such a way that they fit the data best—this orientation can be different at each node—have been investigated as well. The resulting BVHs are called OBB-trees [7]. Unfortunately, intersection tests with arbitrarily oriented boxes are significantly more expensive than with axis-aligned boxes. As a result, OBB-trees do not seem the best solution for range-searching applications. (When using BVHs to perform collision checking between two complex polyhedral objects, OBB-trees have been reported to perform well in certain cases where the objects are very close to each other.)

Another option is to use a so-called c-discretely-oriented polytope (c-DOP) as bounding volume [5,6,9,10]: a convex polytope in $\mathbb{R}^d$ whose facets are orthogonal to c predefined directions, for some parameter $c \geqslant d$. (Thus an axis-aligned bounding box in $\mathbb{R}^d$ is a 2-DOP.) The resulting BVH is sometimes called a c-DOP-*tree*. The larger c, the more tightly fitting the bounding c-DOPs will be, but the more costly the intersection test. Moreover, the larger c, the more storage a c-DOP needs. Because a c-DOP-tree may use facets in more orientations than a box-tree, one can obtain performance guarantees for wider class of query ranges. Indeed, De Berg *et al.*[5] showed how to construct, for any given input set of n c-DOPs, a c-DOP-tree that can answer range queries with c-DOP ranges in $O(n^{1-1/c} + k)$ time. When the input objects are disjoint c-DOPs in the plane, the query time reduces to $O(n^{1/2+\varepsilon} + k)$. This bound is almost as good as the bound for box-trees, while it holds for more general query ranges.

Our contribution. We introduce a new type of BVH, the c-*rotated box-tree*, or c-RB-*tree* for short. In a c-RB-tree in $\mathbb{R}^d$, the bounding volumes are boxes that have one of c/d predefined orientations, where c is a multiple of d. (This implies that the total number of orientations for the facets is c.) Which orientation is used at a node depends on its level. In a 4-RB-tree in $\mathbb{R}^2$, for instance, the bounding boxes of nodes at even levels are axis-parallel, while the bounding boxes at odd levels are oriented at a $45°$ angle. A c-RB-tree uses the same amount of storage per node as a box-tree, independent of the value of c. Thus it uses less storage than a c-DOP-tree (for $c > d$). Moreover, the intersection tests are still simple box–box intersection tests, if one precomputes the bounding box of the query range Q in each of the c/d orientations. Compared to box-trees, c-RB-trees are less sensitive to the choice of the coordinate system.

In Section 2 we investigate the theoretical properties of c-RB-trees: we show that with c-RB-trees, it is possible to obtain the same asymptotic bounds on range queries with c-DOP ranges as with c-DOP-trees. Thus, a collection of n c-DOPs can be stored in a c-RB-tree such that c-DOP queries can be answered in $O(n^{1-1/c} + k)$ time, and when the input consists of disjoint c-DOPs in the plane then the query time reduces to $O(n^{1/2+\varepsilon} + k)$. So from a theoretical point of view, c-RB-trees have an advantage over box-trees in the sense that one can obtain performance guarantees for a wider class of queries.

We also experimentally investigate our new c-RB-tree in the plane, and compare it to box-trees and c-DOP-trees. We start in Section 3 by discussing the test setup. In Section 4 we then describe the experiments and analyze the results.

2   The c-Oriented Rotated-Box Tree

The structure. Let $\mathcal{L} = \{\ell_1, \ldots, \ell_c\}$ be a fixed collection of c non-parallel lines through the origin, where c is a (positive) multiple of d. For $0 \leqslant j < c/d$, let $\mathcal{L}_j = \{\ell_{jd+1}, \ldots, \ell_{(j+1)d}\}$. We define a j-*oriented box*, or j-*box* for short, to be a polytope with $2d$ facets, such that for any line in $\mathcal{L}_j$ there are two facets orthogonal to it. Note that a j-box is simply an axis-aligned box in the coordinate system defined by $\mathcal{L}_j$. A c-RB-tree is now defined as a BVH $\mathcal{T}$ such that the bounding volume $\mathrm{bv}(\nu)$ of a node ν in $\mathcal{T}$ is a j-box, where $j = \mathrm{level}(\nu) \bmod (c/d)$. Thus the bounding boxes of nodes at the same level in the tree all have the same orientation, and as we descend down the tree we repeatedly cycle through the c/d different box orientations. Like in a c-DOP-tree one would usually choose the orientations of the lines in $\mathcal{L}$ as evenly spread as possible. Moreover, one may want to choose the lines in each $\mathcal{L}_j$ to be pairwise orthogonal, but this is not necessary.

Answering queries. A range query with a range Q in a c-RB-tree $\mathcal{T}$ recursively traverses $\mathcal{T}$, starting at the root. When a node ν is visited, the algorithm tests if $\mathrm{bv}(\nu) \cap Q = \emptyset$. If this is the case then the search is terminated. Otherwise, when ν is an internal node then its two children are visited recursively, and when ν is a leaf then the object stored at ν is tested for intersection with Q. We call this procedure the *standard query*.

An alternative is to not use the query range Q itself for the intersection tests with the bounding volumes, but to use a bounding volume for Q as well. In a c-RB-tree this means one has to compute c/d different bounding boxes for Q, namely a bounding j-box $B_j(Q)$ for $j = 0, \ldots, (c/d) - 1$. Then the intersection test at a node ν storing a j-box is the test $\mathrm{bv}(\nu) \cap B_j(Q) = \emptyset$. This is, of course, just a normal box–box intersection test in the coordinate system defined by $\mathcal{L}_j$. The range Q itself is only used at the leaf level, when an input object is tested for intersection. We call this procedure the *intersection-efficient query*.

Note that intersection-efficient queries can also be (and in fact, usually are) used in box-trees and c-DOP-trees; then the intersection tests with the bounding volumes of visited nodes are done with the bounding box resp. bounding c-DOP of Q. Intersection-efficient queries have faster intersection tests than standard queries, but they may result in more nodes being visited.

A theoretical comparison of c-RB-trees with box-trees and c-DOP-trees. Let S be a set of n objects, and let $\mathcal{T}$ be a BVH for S. We define the *base tree* of the BVH to be the tree $\mathcal{T}$ without its bounding volumes. Thus the base tree is a tree with n leaves, each storing a distinct object from S.

It is not difficult to see that the asymptotic query time for a c-RB-tree is at least as good as for the box-tree that uses the same base tree. This is formalized in the following theorem; its proof is similar to the proof of Theorem 2 below. We omit the proof for lack of space.

Theorem 1. *Let $\mathcal{T}_{\text{box}}$ be a box-tree on a set S of objects in $\mathbb{R}^d$. There is a c-RB-tree $\mathcal{T}_{\text{rb}}$ for S such that, for any query range Q, the number of visited nodes in $\mathcal{T}_{\text{rb}}$ is less than $2^{c/d}$ times the number of visited nodes in $\mathcal{T}_{\text{box}}$.*

With a box-tree, one can only guarantee sublinear query times for range searching with axis-aligned boxes. With a c-DOP-tree, on the other hand, one can obtain sublinear query times for range searching with c-DOPs. The next theorem shows that this is also possible for c-RB-trees.

Recall that a c-DOP is a polytope with at most $2c$ facets, each orthogonal to one of c predefined directions, and a c-DOP-tree is a BVH whose bounding volumes are c-DOPs. We will compare c-DOP-trees to c-RB-trees when the orientations of their bounding volumes are defined by the same set $\mathcal{L}$ of lines. The following result holds for intersection-efficient queries as well as for standard queries.

Theorem 2. *Let $\mathcal{T}_{\text{dop}}$ be a c-DOP-tree on a set S of objects in $\mathbb{R}^d$. There is a c-RB-tree $\mathcal{T}_{\text{rb}}$ for S such that, for any query range Q that is a c-DOP, the number of visited nodes in $\mathcal{T}_{\text{rb}}$ is less than $2^{c/d}$ times the number of visited nodes in $\mathcal{T}_{\text{dop}}$.*

Proof. Let $\mathcal{T}_{\text{rb}}$ be the c-RB-tree with the same base tree as $\mathcal{T}_{\text{dop}}$. Define $\mathcal{T}_{\text{dop}}(Q)$ to be the subtree of $\mathcal{T}_{\text{dop}}$ consisting of the nodes that are visited when querying with Q. The subtree $\mathcal{T}_{\text{dop}}(Q)$ is rooted at the root of $\mathcal{T}_{\text{dop}}$, and its leaves are either leaves of $\mathcal{T}_{\text{dop}}$ or internal nodes of $\mathcal{T}_{\text{dop}}$ where the search gets terminated.

Consider a node ν of the latter category, and let $\text{bv}_{\text{dop}}(\nu)$ be its bounding volume in $\mathcal{T}_{\text{dop}}$. Since the search in $\mathcal{T}_{\text{dop}}$ is terminated at ν, we have $\text{bv}_{\text{dop}}(\nu) \cap Q = \emptyset$. Because both $\text{bv}_{\text{dop}}(\nu)$ and Q are c-DOPs, this means there is a line parallel to a line $\ell \in \mathcal{L}$ separating $\text{bv}_{\text{dop}}(\nu)$ from Q. Let j be such that $\ell \in \mathcal{L}_j$. Then the search in $\mathcal{T}_{\text{rb}}$ will be terminated as soon as a descendent μ of ν is reached whose bounding volume is a j-box (possibly $\mu = \nu$). Thus the search in $\mathcal{T}_{\text{rb}}$ does not visit any descendent of ν that is more than $c/d - 1$ levels away from ν. Hence, the number of nodes that are visited in $\mathcal{T}_{\text{rb}}$ below ν is at most $2^{c/d} - 2$. Let m be the number of nodes in $\mathcal{T}_{\text{dop}}(Q)$. Then $\lceil m/2 \rceil$ of those nodes are leaves of $\mathcal{T}_{\text{dop}}(Q)$. The number of visited nodes in $\mathcal{T}_{\text{rb}}$ can thus be bounded by $m + \lceil m/2 \rceil(2^{c/d} - 2)$, which is less than $m \cdot 2^{c/d}$, since $m \geqslant 1$. □

Remark. It may seem that Theorem 1 is implied by Theorem 2. This is, however, not the case since Theorem 1 works for any type of query range, while Theorem 2 is only stated for queries with c-DOPs.

3   Experimental Setup

In our experiments we will compare the performance of c-RB-trees, box-trees, and c-DOP-trees for sets of objects in the plane ($d = 2$). Below we discuss our experimental setup in more detail; the next section then discusses the actual experiments.

Construction of the base trees. In all our experiments we take the directions of the lines in $\mathcal{L}$ evenly distributed, and we take the lines in the sets $\mathcal{L}_j$ orthogonal to each other. Thus $\mathcal{L} = \mathcal{L}_0 \cup \cdots \cup \mathcal{L}_{(c/2)-1}$, where $\mathcal{L}_j$ consists of lines whose counterclockwise angles with the positive x-axis are $j \cdot (\pi/c)$ and $\pi/2 + j \cdot (\pi/c)$, respectively. The value of c will be a parameter in our experiments.

Our general strategy is to first construct a base tree, and then generate our BVHs from that. This way we can compare the effect of the choice of bounding volume, while keeping the base tree the same. On the other hand, different bounding volumes may work best for different base trees, so we try several strategies to construct the base trees.

When the input consists of point data, the base tree is obtained in one the following two ways [5].

Our first base-tree construction uses so-called c-kd-trees, as explained next. In a kd-tree [4] the point set is recursively subdivided in equal-sized subsets, where the partition line alternates between being horizontal and being vertical. In other words, at even levels one uses the horizontal line whose y-coordinate is the median y-coordinate of the current point set, while at odd levels one uses the vertical line whose x-coordinate has the median x-coordinate. A c-kd-tree works in the same way; the only difference is that the orientations of the splitting lines cycle through the orientations of each of the c lines in $\mathcal{L}$, rather than only through the horizontal and vertical direction. More precisely, we use the lines in the order $\ell_1, \ell_{c/2}, \ell_2, \ell_{(c/2)+1}, \ldots$, so that the orientation of consecutive splitting lines is not too similar.

Our second base-tree construction uses the so-called c-grid BSP. Here the plane is subdivided as follows. Fix a parameter $s \geqslant 2$. Then for each line $\ell_i \in \mathcal{L}$, one constructs s slabs parallel to ℓ_i such that each slab contains n/s points. The overlay of these c collections of slabs gives a subdivision with $\Theta(s^2 c^2)$ cells. One then constructs a balanced BSP tree whose leaf regions correspond exactly to these cells. Finally, the algorithm is applied recursively within each cell. This construction is used to obtain the $O(n^{\frac{1}{2}+\varepsilon} + k)$ bound for c-DOP queries in a set of disjoint c-DOPs—see [5] for details. Since we also support the construction of c-kd-trees as base trees, the c-grid BSP should clearly differ from the c-kd-tree. Thus s should not be too small. On the other hand, a large value of s is likely not to be very efficient. In our experiments, we use $s = 16$. Moreover, we do not cycle through the different orientations in the same way as in the c-kd-tree construction, but instead use the same orientation at four consecutive levels.[2]

When the input consists of line segments we proceed as follows. We pick a representative point on each of the segments, and apply one of the construc-

---

[2] There are many variations one could try here. However, the c-kd-tree approach has proven to be superior to the c-grid approach in the experiments by Streppel [13]. Even though his experiments are in a different setting—he considers external-memory data structures—we expect the same to be true in our case. Hence, we have not investigated this issue in depth in our experiments.

tions above to generate a BSP on the representative points. The only difference is that whenever a splitting line ℓ is chosen in the recursive partitioning process, all segments intersecting ℓ are dealt with separately; this separate recursive call uses a similar partitioning strategy, except that the direction of ℓ is no longer used. Thus, a node will have three children: one for the segments to the left of ℓ, one for the segments to the right of ℓ, and one for the segments intersecting ℓ. (If desired, this could be converted to a binary tree.) See [5] for details and a proof that this procedure yields good worst-case query times.

Input Data. As input data we consider points and line segments. In each experiment we generate 500.000 random input objects. The point data is generated inside the unit square according to one of the following three distributions.

Uniform distribution: We generate the points uniformly at random.

Clustered distribution: We take 50 points as cluster centers—these 50 points themselves are generated uniformly at random in the unit square—and for each cluster center we generate a cluster of 10,000 points as follows. We continue generating random points within the unit square, but keep only those that are within a fixed distance of the cluster center.

Line distribution: The points are generated on the line $y = ax$, where a is a parameter in our experiments. More precisely, the points are generated uniformly at random on the part of the line inside the unit square.

The line segment data is generated similarly. First we generate one endpoint of the line segment as described above and then determine the other endpoint by generating a random vector of fixed maximum length, making sure the segment lies within the unit square. We also perform a test on a real-world input set, namely the TIGER/Line data set[3] and, in particular, the road network of Kansas, which consists of 1,135,150 segments.

Queries. We use three different shapes for the query ranges in our test series: line segments, triangles, and boxes. The first two represent arbitrary one- and two-dimensional queries, while the box is a very common query type. The efficiency of a search data structure may vary with the size of the query. Therefore we consider two query sizes for each of the query shapes—a small and a large one. The sizes are fixed such that given n uniformly distributed input objects, a small query would return about $\log n$ answers and a large query about $\sqrt{n}$.

The results will be averaged over 100 randomly generated queries of the shape under consideration. We let the distribution of the location of the query follow a distribution that is similar to the distribution of the input data. For example, when querying an input that follows the line distribution, we also take the query range close to that line. This avoids the situation that the averages are dominated by queries in completely empty regions.

---

[3] Available from: http://www.census.gov/geo/www/tiger/

4   Experiments

The experiments are conducted on a machine with a 2.4 GHz AMD Opteron 250 processor and 4 GB RAM. The code is compiled with g++ 3.2 and compiler option -O3. For each scenario, the values listed are the average values of 100 random queries. We enforce identical input data and query ranges for directly compared runs by manipulation of the random-generator seed.

Intersection-Efficient Queries. First, we are interested in the effectiveness of intersection-efficient queries in comparison to standard queries. Our conjecture is that intersection-efficient queries are faster than standard queries in most scenarios. The size of the queries is an influential parameter in this experiment, since the overhead introduced by approximating a query range by a bounding volume grows with the query range. Therefore we test our conjecture in a scenario that favors the standard query, namely large queries on clustered input data. For this experiment, we fix $c = 4$. We investigate the two different base trees, namely c-kd-tree and c-grid BSPs. The results of the test series are listed in Table 1. From Table 1 we can draw several conclusions.

• Similar to the experiments of Streppel [13] on external-memory c-DOP-trees, the c-kd-tree is superior to the c-grid BSP. We therefore restrict ourselves to c-kd-trees as base trees in the upcoming experiments.

• For segment queries in point data, standard queries are better than intersection-efficient queries. This is not strange, as the extra "query volume" generated by the bounding volume of the range is largest (in the relative sense) for such queries—see the big difference in the number of visited nodes.

Table 1. Comparison of standard queries (top values in each cell) and intersection-efficient queries (bottom values in each cell) by means of the query time in milliseconds (left values in each cell) and the number of visited nodes (right values in each cell). As input we use points and line segments randomly generated according to the clustered distribution. The top half lists runs performed with base tree generated from c-kd-trees. In the lower half they are generated from c-grid BSPs.

query	points			line segments		
	box-tree	c-DOP-tree	c-RB-tree	box-tree	c-DOP-tree	c-RB-tree
line	0.22 170	0.67 143	0.20 138	4.71 4458	8.94 3428	4.08 3509
segment	1.52 2621	0.84 1163	0.81 1229	5.11 6025	3.77 3952	3.98 4241
box	1.65 2770	1.94 2764	2.40 2771	5.29 5835	5.38 5411	6.65 5580
	1.65 2770	1.94 2764	1.89 2842	5.27 5835	5.39 5411	5.68 5686
△	2.72 3286	7.53 3262	2.98 3256	9.63 7610	17.89 6855	9.29 6858
	3.99 6581	3.44 4808	3.33 4890	10.28 9782	8.79 7696	9.05 7992
line	0.50 554	1.60 550	0.53 561	5.04 4914	9.92 4023	4.60 4190
segmet	1.72 2918	1.03 1538	1.05 1636	5.34 6462	4.03 4557	4.39 4910
box	1.90 3099	2.11 3099	2.81 3107	5.81 6497	5.76 6162	7.32 6410
	1.90 3099	2.11 3099	2.24 3195	5.73 6497	5.80 6162	6.27 6518
△	3.08 3772	8.33 3767	3.48 3781	10.03 8235	19.16 7595	10.05 7693
	4.35 6886	3.66 5280	3.84 5423	10.79 10380	9.26 8437	9.84 8820

- For c-DOP-trees, our conjecture holds: intersection-efficient queries are faster than standard queries, except for segment queries in point data and even there the difference is relatively small. This is to be expected, since c-DOPs are already tightly fitting and not much larger than the original range. Note that this effect will be even stronger for larger values of c.
- For box-trees, the conjecture does not hold: standard queries are faster than intersection efficient queries in most settings.
- As expected, the results for c-RB-trees are in between those for box-trees and c-DOP-trees, and not very conclusive. For segment data, the difference between standard and intersection-efficient queries is small. For point data, the differences are bigger, with intersection-efficient queries winning for box queries and standard queries winning for segment and triangle queries.

Based on the outcome of our experiment, we continue to use both standard and intersection-efficient queries for box-trees and c-RB-trees in the upcoming tests, while for c-DOP-trees we confine ourselves to intersection-efficient queries.

Number of orientations. Next we are interested in the impact of the value of c, the number of orientations. We conjecture that for c-DOP-trees, one should choose a relatively small value of c (say $c = 4$ or $c = 6$) since larger values increase the time needed for the basic intersection test, while not providing much more tightly fitting bounding volumes. For the c-RB-tree, it probably does not matter so much which value of c is used: increasing c does not give more tightly fitting volumes, nor does it change the basic intersection test.

To test our conjecture, we conduct a test series that compares the performance of c-DOP-trees and c-RB-trees with $c = 4, 6, 8, 10, 12$. Table 2 lists the average query time and the number of visited nodes of those runs that we performed on line segment data, where the left values in each cell give the query time, and the right values give the number of visited nodes. We first discuss the outcome for uniform and clustered data.

- In the c-DOP-tree, increasing the value of c generally leads to fewer visited nodes, as expected. (Note that this is not the case by definition since the set of directions for $c = 10$, for instance, is not a subset of the set of directions for $c = 8$. Indeed, there are a few cases where the number of visited nodes increases slightly when increasing c.) This does not make up for the more expensive intersection tests, however: The best query time is always achieved for $c = 4$ or $c = 6$.
- Interestingly, the number of visited nodes in the c-RB-tree often goes up when c is increased, also leading to higher query times. The explanation lies probably in the underlying base tree. We use the same value of c for the construction of the c-kd-tree as for the orientations of the bounding volumes. Thus a larger value of c leads to more splitting directions in the c-kd-tree. As a result, the splitting directions at different levels are less "consistent". (Imagine for example a set of points inside an axis-parallel rectangle; splitting them with an axis-parallel line seems to lead to better clustering then splitting with a diagonal line.) Thus the base tree may not be as good for higher values of c. On the other hand, using a different value of c for the c-kd-tree than for the bounding volumes does not seem a good idea either, since then the orientations for the splitting and for the

Table 2. Comparison of the efficiency of c-DOP-trees and c-RB-trees. Each cell features average query times in milliseconds (left) and number of visited nodes (right).

query	c	uniform distribution		clustered distribution		line distribution	
		c-DOP-tree	c-RB-tree	c-DOP-tree	c-RB-tree	c-DOP-tree	c-RB-tree
large line segments	4	0.94 917	1.08 1026	3.77 3952	3.97 4241	9.85 10713	9.33 10472
	6	0.98 853	1.14 1041	3.60 3462	3.84 3988	7.40 7336	7.30 8027
	8	0.98 806	1.19 1071	3.70 3329	3.97 4083	8.37 7956	7.97 8610
	10	1.04 794	1.21 1086	3.90 3214	4.02 4119	7.76 6666	7.42 7941
	12	1.07 780	1.29 1163	3.99 3147	4.20 4297	8.68 7135	8.00 8572
small line segment	4	0.21 190	0.23 227	1.13 1176	1.33 1331	3.68 4002	3.61 4019
	6	0.22 175	0.24 233	1.06 977	1.33 1245	2.11 2046	2.29 2317
	8	0.22 167	0.26 249	1.03 881	1.39 1266	2.66 2452	2.71 2781
	10	0.24 164	0.26 253	1.09 850	1.40 1303	2.08 1683	2.25 2200
	12	0.25 165	0.30 285	1.13 831	1.52 1429	2.65 2108	2.78 2804
small box	4	1.20 1141	1.36 1256	5.39 5411	5.63 5686	17.26 17683	17.25 18024
	6	1.30 1136	1.47 1327	5.80 5340	5.79 5816	18.55 18028	17.76 18632
	8	1.36 1133	1.56 1373	6.10 5323	5.89 5954	19.56 17806	18.00 18655
	10	1.50 1139	1.62 1431	6.45 5332	6.29 6105	21.48 18147	18.76 19206
	12	1.54 1139	1.64 1440	6.83 5319	6.27 6172	22.28 18059	18.92 19292
large box	4	0.32 277	0.43 319	1.99 1994	2.20 2173	5.91 6012	6.03 6246
	6	0.33 265	0.49 331	2.00 1818	2.26 2126	5.43 5166	5.42 5504
	8	0.34 255	0.52 347	2.06 1731	2.33 2157	5.66 5110	5.64 5615
	10	0.37 254	0.51 353	2.20 1706	2.41 2219	6.16 5079	5.83 5735
	12	0.38 254	0.54 380	2.28 1681	2.54 2327	6.34 5045	5.89 5837
large △	4	2.93 2497	3.12 2683	8.79 7696	9.10 7992	23.91 21398	23.96 21597
	6	2.95 2328	3.14 2655	8.69 7075	9.00 7643	23.29 19604	22.69 20428
	8	2.97 2221	3.22 2680	8.99 6977	9.15 7770	23.82 19042	22.63 20104
	10	3.12 2183	3.24 2706	9.48 6872	9.23 7859	25.22 18846	23.08 20414
	12	3.22 2164	3.39 2829	9.70 6754	9.46 8006	25.88 18540	23.25 20468
small △	4	0.44 360	0.57 409	2.97 2580	3.21 2768	7.24 6518	7.33 6643
	6	0.45 338	0.62 416	2.92 2333	3.17 2658	6.14 5105	6.25 5501
	8	0.47 328	0.65 435	2.95 2246	3.25 2709	6.55 5184	6.53 5675
	10	0.50 327	0.65 443	3.11 2200	3.30 2749	6.52 4775	6.44 5532
	12	0.52 326	0.69 473	3.20 2176	3.47 2896	7.01 4937	6.73 5835

bounding volumes are not consistent. The conclusion is that $c = 4$ and $c = 6$ seem the best choices for a c-RB-tree.

The line distribution shows a strange behavior. In particular, the bad performance for $c = 4$ seems unexpected, since then one of the orientations in $\mathcal{L}$ is parallel to the line on which the data are generated. This behavior has already been discussed by Streppel [13]. His tests with external-memory data structures use $c = 2, 3, 4, 5, 6$. In case of his line distribution scenario, which also uses the line $y = x$, the tests show a clear peak in the number of visited nodes for $c = 4$. He argues that using the orientation of the line also for constructing the base tree results in many long and skinny bounding volumes, especially in a c-grid BSP tree. Probably the same effect plays a role here. We also performed tests with the lines $y = 0.5x$ and $y = 0.2x$, which showed no distinguishable effect.

Table 3. Comparison of the query times (in milliseconds) of box trees, c-DOP trees, and c-RB-trees. The latter two are parametrized according to the previous experiment.

distribution	query	points					line segments				
		box-tree std	int-eff	c-DOP tree	c-RB-tree std	int-eff	box-tree std	int-eff	c-DOP tree	c-RB-tree std	int-eff
uniform distribution	line segs	0.15	0.42	0.24	0.13	0.24	1.14	1.35	0.95	1.08	1.08
		0.05	0.04	0.03	0.03	0.02	0.37	0.31	0.21	0.28	0.23
	box	0.35	0.35	0.40	0.52	0.42	1.20	1.20	1.20	1.58	1.36
		0.06	0.06	0.05	0.08	0.05	0.40	0.40	0.32	0.51	0.43
	△	0.91	1.48	1.12	0.96	1.11	3.09	3.94	2.91	3.05	3.12
		0.09	0.10	0.07	0.12	0.09	0.66	0.59	0.45	0.64	0.57
clustered distribution	line segs	0.23	1.55	0.63	0.22	0.60	4.71	5.03	3.65	4.09	3.85
		0.04	0.04	0.04	0.04	0.03	2.10	1.75	1.06	1.59	1.33
	box	1.64	1.64	1.94	2.40	1.90	5.29	5.29	5.45	6.52	5.68
		0.17	0.17	0.20	0.28	0.22	2.56	2.56	2.04	2.60	2.23
	△	2.80	4.07	3.25	3.01	2.94	9.71	10.38	8.71	9.31	8.97
		0.40	0.58	0.45	0.45	0.43	4.26	3.96	2.92	3.59	3.17
line distribution	line segs	0.04	2.65	0.97	0.04	0.83	9.31	9.83	7.38	7.76	7.37
		0.03	0.08	0.05	0.04	0.04	3.93	3.20	2.08	2.72	2.26
	box	7.42	7.42	8.62	10.76	8.39	15.87	15.87	17.32	20.44	17.26
		1.22	1.22	1.55	0.82	1.36	5.23	5.23	5.47	6.16	5.49
	△	7.18	10.60	9.62	7.80	8.48	24.30	26.51	23.40	23.48	22.81
		0.88	1.36	1.18	0.98	1.06	8.15	7.32	6.15	7.07	6.28
tiger/ line data of Kansas	line segs						0.14	0.32	0.18	0.12	0.19
							0.04	0.04	0.03	0.05	0.03
	box						1.20	1.20	1.32	0.96	1.38
							0.10	0.10	0.10	0.17	0.13
	△						0.81	1.09	0.81	0.76	0.82
							0.08	0.09	0.06	0.11	0.09

Efficiency of different BVHs. In our final test series, we compare the efficiency of the different BVHs, with c fixed according to the best results of the previous experiment. Table 3 shows the results. We can draw the following conclusions.

• For segment queries, box-trees perform worse than c-DOP-trees and c-RB-trees; for segment queries in point data, the c-RB-tree is generally best, while for the segment data the c-DOP-tree is the winner.
• For box queries and triangle queries in point data, the box-tree performs best. The only clear exception are small box queries on points on a line, where standard queries on c-RB-trees are clearly the fastest.
• For segment data, box-trees perform best for large box queries, as expected. For other queries on segment data c-DOP-trees are usually best. Thus, c-DOP-trees provide a good alternative to box-trees if the increase in storage it not an issue.
• In contrast to our expectations, the results for the Tiger/Line set of Kansas are not consistent with those of the uniform distribution. Here, the c-RB-tree is best for large queries, and the c-DOP-tree is the winner for small queries.

• Because c-DOP-trees use more storage than box-trees and c-RB-trees (for $c = 4$ the extra storage is a factor 1.57, for $c = 6$ even 2.14), one may want to choose only between box-trees and c-RB-trees. Here we see that, as one would expect, box-trees are the structure of choice for box queries. For other types of queries, however, the c-RB-tree usually beats the box-tree. Thus when most queries are segment or triangle queries, the c-RB-tree might be the structure of choice.

5   Conclusion

We presented a new bounding-volume hierarchy in $\mathbb{R}^d$, the c-RB-tree, and proved that it has the same worst-case asymptotic query bounds as the c-DOP-tree. Moreover, a c-RB-tree uses the same amount of memory as a box-tree, which is less than a c-DOP-tree (for $c > d$). Then we conducted several experiments in which we compared the performance of range queries on c-RB-trees, c-DOP-trees and box-trees. The experiments suggest that a c-DOP-tree could be the structure of choice when the extra memory is not an issue, especially when may queries are not box queries. This interesting conclusion is different than the conclusion by Streppel [13], who compared c-DOP-trees to box-trees when used as an external-memory structure, and found that box-trees are generally better. The reason for this is probably that in an external-memory setting, a node must store a bounding volume for each of its many children. Since a c-DOP uses more storage than a box, this implies that a node can store less bounding c-DOPs than bounding boxes. As a result, the degree of the nodes is smaller, and less objects fit into the leaves, resulting in more nodes being accessed. If one does not want to spend the extra memory for a c-DOP-tree, and most queries are segment or triangle queries, then the c-RB-tree is a good alternative to the c-DOP-tree. This suggest that an c-RB-tree may also be useful in an external-memory setting; it would be interesting to perform experiments to test this.

References

1. Agarwal, P.K.: Range searching. In: Goodman, J., 'O'Rourke, J. (eds.) CRC Handbook of Discrete and Computational Geometry. CRC Press, Boca Raton (2004)
2. Agarwal, P.K., de Berg, M., Gudmundsson, J., Hammar, M., Haverkort, H.J.: Box-trees and R-trees with near-optimal query time. Discr. Comput. Geom. 28, 291–312 (2002)
3. Agarwal, P.K., Erickson, J.: Geometric range searching and its relatives. In: Chazelle, B., Goodman, J., Pollack, R. (eds.) Advances in Discrete and Computational Geometry. Contemporary Mathematics, vol. 223, pp. 1–56. American Mathematical Society, Providence (1998)
4. de Berg, M., Cheong, O., van Kreveld, M., Overmars, M.: Computational Geometry: Algorithms and Applications, 3rd edn. Springer, Heidelberg (2008)
5. de Berg, M., Haverkort, H.J., Streppel, M.: Efficient c-oriented range searching with DOP-trees. Comput. Geom. Theory Appl. 42, 250–267 (2009)
6. Fünfzig, C., Fellner, D.W.: Easy realignment of k-DOP bounding volumes. In: Graphics Interface, pp. 257–264 (2003)

7. Gottschalk, S., Lin, M.C., Manocha, D.: OBB-Tree: a hierarchical structure for rapid interference detection. In: Proc. SIGGRAPH 1996, pp. 171–180 (1996)
8. Haverkort, H.J.: Results on Geometric Networks and Data Structures. Ph.D. Thesis, Utrecht University (2004)
9. Jagadish, H.V.: Spatial search with polyhedra. In: Proc. Int. Conf. Data Engineering (ICDE), pp. 311–319 (1990)
10. Kay, T.: Ray tracing complex scenes. In: Proc. SIGGRAPH 1986, pp. 269–278 (1986)
11. Klosowski, J.T., Held, M., Mitchell, J.S.B., Sowizral, H., Zikan, K.: Efficient collision detection using bounding volume hierarchies of k-DOPs. IEEE Transactions on Visualization and Computer Graphics 4, 21–36 (1998)
12. Nievergelt, J., Widmayer, P.: Spatial data structures: concepts and design choices. In: van Kreveld, M., Nievergelt, J., Roos, T., Widmayer, P. (eds.) CISM School 1996. LNCS, vol. 1340, pp. 153–197. Springer, Heidelberg (1997)
13. Streppel, M.: Multifunctional Geometric Data Structures. Ph.D thesis, TU Eindhoven (2007)
14. Sitzmann, I., Stuckey, P.J.: The O-Tree – A Constraint-Based Index Structure. Technical report, University of Melbourne (1999)
15. Zachmann, G.: Rapid collision detection by dynamically aligned DOP-trees. In: Proc. of IEEE Virtual Reality Annual International Symposium, pp. 90–97 (1998)

psort, Yet Another Fast Stable Sorting Software

Paolo Bertasi, Marco Bressan, and Enoch Peserico

Dipartimento di Ingegneria dell'Informazione, Università degli Studi di Padova, Italy
`psort@dei.unipd.it`

Abstract. *psort* was the fastest sorting software in 2008 according to the Pennysort benchmark, sorting 181GB of data for 0.01\$ of computer time. This paper details its internals, and the careful fitting of its architecture to the structure of modern PCs-class platforms, allowing it to outperform state-of-the-art sorting software such as *GNUsort* or *STXXL*.

1   Introduction

This paper details the internals of *psort*, the fastest sorting software of 2008 according to the Pennysort benchmark [3]. This introduction provides a brief history of Pennysort and related sorting benchmarks (Subsection 1.1), and a simple taxonomy of the "mainstream" sorting techniques for large datasets that helps put our work into perspective (Subsection 1.2), followed a high-level overview of *psort* and of the organization of the rest of this paper (Subsection 1.3).

1.1   Datamation, Pennysort, and Other Sorting Benchmarks

Datamation [4] defined the first [15] public sorting benchmark - sort a million 100 byte records, initially in random order, according to the first 10 bytes *from disk to disk.*

In 1985 the time required to complete the Datamation benchmark was almost 1 hour; but within 10 years it had dropped to a few seconds, and it appeared that it would soon become obsolete.

Therefore, in 1995, [22] proposed two new sorting benchmarks, MinuteSort and PennySort. MinuteSort, aimed at supercomputer-class platforms, requires sorting as many records as possible within 1 minute. Pennysort, aimed at PC-class platforms, requires sorting as many records as possible with 0.01\$ of computing time, assuming that the price of a machine is amortized over 3 years (thus, on a x dollar machine, one is allowed $\frac{0.01}{x} \cdot 3$ years of computing time). In both cases record format coincides with that of the datamation benchmark. Also, in both cases a distinction is made between "Daytona" software, designed for general purpose sorting, and "Indy" software, specifically optimized for the benchmark. Over the years a number of slight refinements have been added to the rules, and new benchmarks such as TeraSort and JouleSort have been introduced. All details can be found at [3].

J. Vahrenhold (Ed.): SEA 2009, LNCS 5526, pp. 76–88, 2009.
© Springer-Verlag Berlin Heidelberg 2009

1.2   A Simple Taxonomy of Sorting

Hundreds of articles and even entire books (e.g. [18]) have been written on sorting. This subsection provides a simple taxonomy of sorting techniques for large data-sets to help put *psort* into perspective compared to existing software.

Virtually all efficient sorting software today is either *distribution*-based, *merge*-based, or a hybrid of the two. Distribution-based sorting distributes the data into two or more bins, in such a way that for each pair of bins all keys of one precede all keys of the other; then it recursively sorts each bin. Merge-based sorting splits the input into two or more runs, sorts each run, and then merges the sorted runs. Distribution and merge-based sorting can obviously be combined. For example, one might use the former to "locally" sort separate runs, that are in turn "globally" merged - the approach of Alphasort [22].

Distribution-based sorting has two major advantages over merge-based sorting. First, it can be easily performed completely in parallel, and thus is virtually the only approach used today - at least at the "global" level - for sorting on large PC-clusters (e.g. [13,10,21]). Second, it can be very efficient in terms of number of key lookups - this makes it a favorite of all past record holders of the Pennysort benchmark (e.g [14,25,19,23,5]) at least for some phases of the sort.

Merge-based sorting is instead intrinsically "comparison based" and thus requires at least $n \lg(n)$ key lookups to sort n keys ([18]) though this disadvantage is more apparent than real, as we shall see. On the other hand, merge-based sorting always results in well-balanced sub-problems of predictable size; this makes it more "robust" and easier to fine tune to a memory hierarchy. This predictability could in theory be achieved by a careful selection of the thresholds between bins in distribution-based sorting, but only at a high cost (see e.g. [18]).

1.3   Our Results

psort is a fast stable external sorting software (available as source, binary and/or library) that can sort collections of records of arbitrary size according to an arbitrary infix. *psort* sorted 181GBs of data for 0.01$ of computer time in 2008, making it the fastest sorting software in 2008 according to the Pennysort benchmark. The careful fitting of its architecture to the structure of modern PC-class platforms (made easier by its pure merge-based nature) allows it to outperform state-of-the-art sorting software such as *GNU sort*[2], *qsort*[1], *C* + + STL *sort* or *STXXL*[12], even for record and key sizes and distributions quite different from those of the Pennysort benchmark.

In order to understand *psort* optimizations, one has to understand the complex architecture of modern PCs to a greater level of detail than that offered by most theoretical models today. Section 2 provides this information.

Section 3 describes *psort* itself and its tuning to a modern PC.

Section 4 describes our experimental results, comparing *psort* with other state of the art sorting software - both under the Pennysort rules (including the selection of the "best" PC to run it) and under a number of different scenarios.

Finally, Section 5 summarizes our results and discusses their significance before concluding with the bibliography.

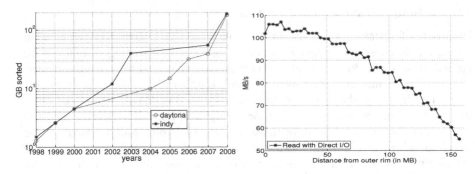

Fig. 1. The Pennysort Benchmark record history (*psort* = 2008)

Fig. 2. Disk bandwidth as a function of distance from outer rim

2   The Anatomy of a PC

Current hardware is extremely complex. While a number of abstract models from the past decades attempt to capture the main aspects of modern architectures - hierarchical memory [6], block transfer[7,9], pipelining[11], parallelism[26] etc. - they are generally insufficient to abandon the ivory tower of big O notation and squeeze out of a machine at least 50% of its the peak performance ([8]). It is impossible to review all the details of the performance of a modern PC, but this section provides a comprehensive overview of those (often disregarded) factors that can be crucial to data-intensive software such as sorting - starting from disks and filesystem (Subsection 2.1), then moving to motherboard and memory (Subsection 2.2), and finally to the processor chip (Subsection 2.3).

2.1   Disks and Filesystem

Modern disks provide the abstraction of a (logical) linear array of data blocks, with access to a sequence of contiguous blocks requiring a fixed *seek time* independent of the amount of data (typically of the order of $10ms$), plus a *transfer time* that is directly proportional to it (typically $10 - 100ms/MB$). Obviously, to approach peak performance, transfer time should dominate seek time (contiguous data transfers indicatively of $1MB$ or more). Transfer time is lower for data logically closer to the beginning of the array, corresponding physically to the area of the disk closer to the outer rim; this should be regarded as the "high performance" portion of the drive (see figure 2).

Multiple disks (usually up to $4 - 8$) can be used in parallel (RAID 0) as a single disk with the same seek time but proportionally larger transfer speed, by splitting data into "stripes" divided, in a round robin fashion, between different disks. This can be done in hardware or by the filesystem/OS - we found the latter approach effective and of minimal CPU cost.

Most software today does not access the disk directly, but through a filesystem. Filesystems offer a lot of functionality over raw disk access, but at a price in terms of performance (in fact, many applications with high performance disk access, such as DBMSs, bypass the filesystem altogether); however, most sorting benchmarks (including Pennysort) enforce the use of a filesystem for disk access. To minimize the CPU and memory system overhead, it is then crucial to access data through asynchronous, direct I/O from the device directly into user-space.

2.2 Motherboard and Main Memory

Disks (or rather their on-board caches) communicate with the motherboard through an interface generally designed to have higher throughput (e.g. $300MB/s$ for SATA2) than the drives themselves, to ensure interface longevity. Motherboards typically support 4 to 6 (rarely 8) of these interfaces, all connected to a chip known as the *south bridge*. The south bridge aggregates the traffic of disks and other devices (e.g. keyboard, network card) and usually directs it to another chip, the *north bridge*, that controls traffic to/from/between the main memory, the processor chip and graphics hardware (in some architectures at least part of the north bridge is incorporated in the processor).

Modern south and north bridges can manage transfers of large data aggregates directly between memory and disk with only minimal CPU involvement ("Direct Memory Access"). Bandwidth between bridges is typically lower than the sum of the disk interface bandwidths supported by the motherboard, but is rarely a bottleneck for all but the fastest and largest same-generation disk RAIDs (typical values are $350-800MB/s$). The bandwidth between the other hardware connected to the north bridge is higher still, typically several GB/s.

There are a few more issues to consider in terms of processor/memory communication. First, over a hundred processor cycles can typically pass between a request for a datum in main memory and its availability on the processor chip. Second, data is transferred to the processor chip only in multiples relatively large *cache lines* (typically 128 to 512 bytes); and if 2 bytes of a datum belong to 2 different cache lines, both lines are accessed. Third, virtual memory addresses have to be translated to physical addresses; while the most recent translations are cached in the processor chip, very sparse memory accesses will force additional accesses to the translation tables [24]. This problem can be minimized using large memory pages, and thus smaller translation tables for the same space.

2.3 The Processor Chip

The basic components of the processor chip are its CPU(s) and its cache(s). Most modern processors have 2 levels of cache, the first smaller (up to a few hundred KB vs. one or more MB), but faster to access (typically a few cycles vs. $10-30$). Each memory block can only be placed in a small number of cache locations, the *associativity* of the cache (typically $4-16$ for L1 caches, $8-32$ for L2 caches). According to [17] associativity above 8 provides few benefits, but we have found this to be true only if data layout is very carefully planned [20]. All decisions on which data to keep in the cache are made by the processor, typically replacing data used furthest in the past with data more recently accessed. This can result in undesired behaviours: for example large streams of read-once data can "pollute" the cache, evicting data used repeatedly but somewhat infrequently. The programmer can sometimes attempt to influence cache replacement by issuing extra memory requests to "nail" or prefetch data into the cache.

In general modern processors employ sophisticated circuitry to reorganize instructions, "guess" yet unavailable data, and backtrack from incorrect guesses. This makes it extremely difficult to understand whether, and how, a code snippet can be redesigned to improve efficiency: "optimizations" that increase performance on one system often decrease it on others. Fine tuning through experimentation on the target architecture can net substantial performance gains.

3   The Anatomy of *psort*

psort is a fast, stable sorting software designed for large data sets on PC-class platforms. *psort* is a simple, merge-based sorter that first sorts individual data runs approximately the size of main memory, and then merges them into a single sorted output. In fact, *psort*'s high-level simplicity is probably the source of its performance, allowing careful, low-level tuning to the complex structure of today's PCs. Subsection 3.1 provides a "global overview" of *psort*. Subsections 3.2 and 3.3 provide the details of the first and second phase of the sort.

3.1   A Theorist's View of *psort*

psort is a merge-based sorter tuned to the memory hierarchy. For each pair of consecutive hierarchy layers (e.g. main memory and disk) the structure of *psort* depends on the size S_0 of the smaller and on the size S_1 of the larger.

In a nutshell *psort* implements a *p-pass merge* between the two layers as follows. Choose a *block size* B such that $(\frac{S_0}{B})^p \approx \frac{S_1}{B}$. To implement the 1st pass, split the data in the larger layer into $\approx \frac{S_1}{S_0}$ data runs of size $\approx S_0$ and sort each run in the smaller layer. Data are sorted in the smallest layer, the L1 cache, using a simple mergesort. To implement the 2nd pass, split the sorted runs into sets of $w = \frac{S_0}{B}$ runs each, and merge each set into a sorted run of size wS_0 using a w-way selection tree merger [18], where each way is assigned a buffer of size B. The entire tree then always resides in cache, so that, during the pass, each item is brought into, and evicted from, the cache only once - and each transfer is amortized over a block of size B. Similarly, to implement the ith pass, merge sets of w runs from the $(i-1)^{th}$ pass into sorted runs of size $w^{i-1}S_0$.

The minimum block size that still allows an efficient transfer places an upper bound on the number of ways of a merge, and thus a lower bound on the number of passes. It is chosen to minimize $(p-1)(t_B + t_{peak})$ where t_B is the amortized read time per bit using blocks of size B, and t_{peak} is the amortized write time per bit at the peak transfer rate using blocks of size comparable to S_0 (since during a merge phase, the w input buffers must be small, but there is only one write buffer, that can be large). Note that the first term of the product is $(p-1)$: in the first pass, one can essentially perform both reads and writes of a size comparable to S_0. In practice, on modern computing platforms, p almost always equals 2 or (more rarely) 3 at all layers of the memory hierarchy, with the possible exception of the lowest two (memory and disk), where data sets small enough to fit in memory can be sorted with a single read/write pass of the disk(s). This is not the case for the Pennysort benchmark, where data is usually one to two orders of magnitude larger than memory - making $p = 2$ and thus entailing 2 "large writes" passes, 1 "large reads" pass (the first), and 1 "small reads" pass (the second). In this case, B is chosen to match the size of a few disk tracks (i.e. a few MBs - to minimize the overhead of seek time) times the number of disks in the RAID.

Two things are worth noting. First, the first pass may always be performed in place, with at most a "slack" of a single run; this is generally impossible for other passes. Second, as noted e.g. in [12], utilizing RAID at the disk layer is asymptotically suboptimal (as the number of disks grows to infinity) compared to (randomized) strategies that control disks independently. However, such strategies have the potential of actually being asymptotically slower than pure RAID on pathological inputs. Perhaps more importantly, their "sweet spot" requires larger disk arrays and sorts than those encountered in practice when dealing with today's PC-class platforms, where they can actually end up being slower even on "average" inputs due to their small (in fact *asymptotically* negligible) fluctuations in disk utilization (see Subsection 4.3) and their reduced ability to exploit the faster zones of the disk (see Subsection 4.2).

3.2   The First Phase

The first phase of *psort* essentially involves reading a data run approximately the size of the main memory from disk, sorting it in memory, and writing it back to disk. The devil is in the details.

For I/O efficiency, *psort* makes use of direct asynchronous I/O that transfers data between the disks and a set of userspace buffers (dimensioned so as to achieve near peak transfer rate without consuming too much memory): this requires minimal CPU involvement (even with software RAID, less than 4% of CPU time) and bypasses the space and time costs of moving data through kernelspace buffers. With two buffers one can guarantee that the disks never fall idle: while the CPU operates (reading or writing data) on one, disks exchange data with the other, thus completely overlapping I/O with computation.

As soon as a read buffer is filled *psort* must transfer data to its main memory space. Since this requires moving the data through the processor's cache, it piggybacks some computation onto the transfer, acting on data "microruns" of size roughly equivalent to the L2 cache. More precisely, if keys are sufficiently small compared to the records' remaining payload, it separates keys and payloads, attaching to each key a pointer to the corresponding payload. *psort* at this stage also offers the possibility of restructuring keys (e.g. from big to small endian) to make later comparisons more efficient. *psort* then sorts those keys (and eventually reshuffles the payloads) and finally writes the whole "microrun" to a new memory area. This involves, for each datum, at most one write and one read in memory that would be performed anyway to empty the read buffers.

The sorting algorithm used in L1 cache for the microruns is a simple merge-sort, with some tweaks. First, a single pass on the data (the same that possibly detaches keys and records) sorts sets of 4 consecutive elements using a simple selection sort. Second, careful placement of data ensures that only 5/4 of the total space s occupied by the records (or by the detached keys) is used, rather than the "common" factor 2 of mergesort: this is achieved by first sorting half of a microrun in space $\frac{s}{2} + \frac{1}{2} \cdot \frac{s}{2}$ (see [18]), then the other half (for a total space of $\frac{s}{2} + \frac{s}{2} + \frac{1}{2} \cdot \frac{s}{2} = \frac{5s}{4}$), and finally piggybacking the merge of the two halves over the transfer out of the read buffer. Third, *psort* offers the opportunity to minimize program branches. As a first option, it can use bitwise — rather than logical — ANDs and ORs when comparing multi-word keys. As a second option, it can use the result of a comparison (and its inverse) as an offset to a pointer to the new positions of the keys, saving a branch at the cost of a few extra operations. The data sorted in L1 are then merged in L2.

Sorted microruns are then merged in a second pass (and possibly a third, depending on the ratio between the size of the memory and of the L2 cache, and the associativity of the latter). Two important potential hurdles at this stage are associativity misses (in early experiments these reduced performance by as much as 20%) and "stream pollution" of the cache (see Section 2). *psort* employs a careful data layout to minimize the former, and offers against the latter the option of a periodic cache refresh through dummy reads.

The output of the final pass is directly written to the I/O buffers, potentially recombining keys and payloads, and possibly (if there is no second phase) in-verting the initial restructuring of the keys. Again, double buffering allows full overlap between I/O and computation.

3.3   The Second Phase

The second phase (which only takes place if the dataset does not fit in main memory) is much simpler: w sorted runs at a time are streamed from disk and merged (with the same code that merges microruns in the first phase), and the output is streamed back to disk. Recalling Subsection 3.1, it is reasonable to use a single merge pass if the number of data runs (i.e. the ratio between data and memory size) does not exceed the ratio between the size of the memory and that of one "efficient" read from disk.

Data is read with direct asynchronous I/O into (userspace) dynamically sized buffers, one per run. When the amount of data in a buffer falls below a threshold the buffer is "refilled" from the appropriate run. In theory, if data were consumed uniformly from all w runs, one could divide the total available buffer space B in such a way that a newly refilled buffer held $\approx \frac{w}{w^2+w} 2B \approx 2B/w$ bytes of data, the previously refilled one $\frac{w-1}{w^2+w} 2B$, and so on. This would allow reads of about twice the size achievable with static buffers of size B/n. This can be highly ineffective, however, if data are not consumed uniformly, and in particular if they are consumed more rapidly from recently refilled buffers. For this reason, *psort* allows the user to specify buffer geometry, choosing a tradeoff between safety and optimization.

4   *psort* vs. the Competition

This section compares *psort* to its competitors, in terms of the Pennysort benchmark (Subsections 4.1 and 4.2) and also in a wider variety of sorting scenarios (Subsection 4.3).

4.1   Choosing and Configuring the Hardware

To test *psort* we looked for a hardware platform delivering maximum performance at the minimum cost, to maximize our results under the Pennysort benchmark, but also to understand the bottlenecks in today's PC architectures for data-intensive tasks.

Our choice for a motherboard was an ASRock ALive NF6P-VSTA with an Nvidia nForce 430 Southbridge, an inexpensive but high performance "Linux friendly" motherboard supporting 4 SATA2 channels, for a maximum aggregated traffic of $500MB/s$. We paired it with 4 Western Digital WD1600AAJS drives, which can individually deliver a whopping $100MB/s$ peak read/write rate. We configured them with GNU/Linux (Gentoo) "vanilla" software RAID. As a filesystem, we tested SGI's XFS, IBM's JFS, ReiserFS and ext2fs. The best performers where XFS and JFS, with JFS slightly better overall but XFS outperforming it very slightly for the read and write sizes of interest to us (see figures 3 and 4). In both cases CPU usage to saturate the disk transfer rate was negligible - less than 2%. Thus, we finally settled for XFS. The best performance was achieved with a stripe size of $128KB$. Note that this is a very "disk heavy" PC, with about half the total cost being taken by the 4 disks (see figure 8).

RAM choice must take into account three parameters: size, speed and price. A RAM that is twice as large more than doubles the size of runs in the first pass. This, in turn, allows reads that are over 4 times longer during the second pass. It turned out that the best compromise was $2GB$. RAM speed is another

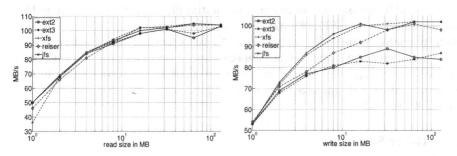

Fig. 3. Filesystem read speed as a function of read size

Fig. 4. Filesystem write speed as a function of write size

important parameter. PC4200 RAM has a *theoretical* "peak" transfer rate of 4.2GB/s - an order of magnitude faster than the southern bridge. In practice, we found that accessing RAM can have a large number of "hidden" costs - e.g. due TLB lookups and to the fact that it is accessed in whole "cache lines". Even just two or three read+write passes can consume the majority of the available RAM bandwidth, and it is extremely difficult to coax the compiler to overlap RAM to cache transfers with processor operations. In practice, it turned out that even using 2 banks of OCZ 800MHz PC6400 RAM with CAS 4 latency almost half the "CPU" time during the first pass was spent accessing RAM.

The choice of the actual processor strongly depends on that of the other components - probably more than on the "number crunching" power of the CPU itself. We chose a cheap, single core Athlon 1620LE running at $2.4GHz$, with $128KB$ of L1 cache and $1MB$ of L2 cache. The total cost of the hardware at NewEgg.com on May 19th 2008 was 357.78\$. Under the Pennysort formula, adding the mandatory 35\$ "assembly fee", this allowed us a total time budget of slightly more than 2408.6766 seconds.

4.2   Pennysort Results

We tested *psort* on the hardware described in Section 4.1. We compiled it with:
`-march=k8 -O3 -funroll-loops -funsafe-loop-optimizations`
`-B /usr/share/libhugetlbfs/ -Wl,--hugetlbfs-link=B`.
We positioned the input file into an appropriately sized partition on the outer rim of the disks, overwrote it with the output of the first pass, and had the second pass create the output file in a second partition. This guaranteed 3 of the 4 passes took place on the fastest partition of the disk, and only 1 on a slower partition. Note that, had we been using independent disks for the intermediate files (RAID would still have been necessary for the initial input and final output, since the rules of the Pennysort benchmark enforce a single file for each), at most 2 passes could have taken place on the faster partition.

The first pass was slightly limited by the CPU, or, more correctly, by the combination of CPU and RAM. More expensive CPUs did not yield sufficient increases in performance to justify their use. The second pass was entirely limited by the disks. *psort* (using 2^{16} record cache merge and a 2^8-way merger-tree, $50MB$ read/write buffers, overwriting the initial file with the intermediate file) sorted $108 \cdot 2^{24} = 1,811,939,328$ records taking less than 2405 seconds. We then manually "retooled" *psort* into an "Indy" version adapted solely for the Pennysort benchmark (eliminating unnecessary "general purpose sorting" code, manually unrolling loops etc.). This yielded a small, but observable gain in performance. *psort* Indy managed to sort $113 \cdot 2^{24} = 1,895,825,408$ records in less than 2407 seconds.

4.3   Other Scenarios

In order to evaluate *psort* outside the Pennysort context, we compared it to some state-of-the-art sorting libraries under different scenarios. We had *psort*

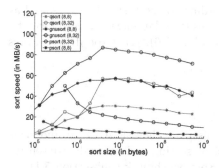
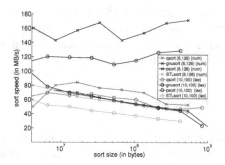

Fig. 5. *psort* vs *qsort* and *GNU sort* for small record sizes

Fig. 6. *psort* vs *qsort*, *STL sort* and *GNU sort* for large record sizes

compete against the high performance (non-stable) external sorting *STXXL* library [12] on sorting 128, 32 and 8 byte records according to the first 8 bytes for a variety of input sizes, from $10MB$ to $100GB$ (we only managed to run STXXL with power-of-2 key size, which seems in line with [12]). *psort* speed was always 20% higher or more for 128 byte records, and either higher (for small sort sizes) or essentially identical for 32 byte and 8 byte records (see figure 7). It should be noted that STXXL has a large number of tuning parameters, that we tweaked trying to achieve optimal performance. We found the most crucial one to be block size - in our scenario $32MB$ (vs. the default value of $2MB$) offered the best performance. Interestingly, and contrary to what one might expect from [12], using independent disks never increased performance significantly, and sometimes even slightly decreased it, due to the slight fluctuations it introduces in disk usage. Apparently the theoretical advantage of using independent disks translates into a practical advantage only for significantly larger sorts.

We also had *psort* compete *in main memory* against both *qsort* [1], *GNU sort* [2] and the C++ STL *sort* (which do not support external memory sorting),

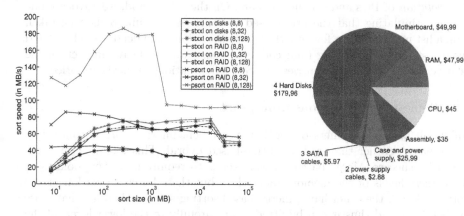

Fig. 7. *psort* vs *STXXL*

Fig. 8. Budget of the testbed machine

on sorting 128, 32 and 8 byte records according to the first 8 bytes, as well as 100 byte records according to the first 10 bytes (the datamation record format), again for a variety of input sizes (see figures 5 and 6). *psort* speed was always higher, from 20% to 300%.

5   Conclusions

This section briefly summarizes our results, discusses their significance, and looks at future directions both in terms of efficient sorting and other data-intensive software (Subsection 5.2) and of the Pennysort benchmark (Subsection 5.1).

5.1   10 Years of Pennysort

It is interesting to compare our results with the prediction of 10 years ago by Gray et al. [16] that price-performance would double yearly for the next 10 years, yielding $1.50TB$ for 1 penny by 2008. Instead, price-performance has "only" increased by an average factor of about 1.6/year (see figure 1). This almost exactly matches Moore's Law - but looking at the 1998 winners it is easy to see that improvement is not due solely to better hardware. We sorted more than 120 times the data of Gray et al. [16] using 3 times the time budget, a set of disks with about 15 times the peak (total) bandwidth, and a $2.4GHz$ Athlon 1620LE vs. a $266MHz$ Pentium II - the latter having a lesser comparative gap to the memory. Software engineering advances are then responsible for at least a factor $2 - 3$ of improvement (note that *psort*'s basic algorithms are decades old).

After 10 years Pennysort is still an excellent benchmark for the lower levels of the memory hierarchy - less so for the processor. As reflected by our budget (see figure 8) our machine had superb disks, an excellent motherboard, good memory, and one of the cheapest processors of the market. While the Pennysort "spirit" has remained the same over the years, the rules keep changing slightly every year. We believe this reduces the value of Pennysort as a benchmark for the *evolution* of PCs and sorting software. On the other hand, we advocate one change: stipulating that the prices used to compute the time budget be taken from a list made public a few months before the submission deadline. This would avoid last minute shopping (and coding) frenzies and/or heavy impact on the relative results of different entries caused by fluctuations of hardware prices.

5.2   Some (Ugly?) Lessons from *psort*

Unlike all previous winners of the Pennysort benchmark, *psort* is completely merge-based, rather than a distribution-based hybrid. This might be one reason of its success. Merge-based software tends to require more key lookups at the highest levels of the memory hierarchy - but these levels are no longer the bottleneck. On the other hand, merge-based sorting software is somewhat more predictable - and thus it can be fitted more carefully to the lower levels of the memory hierarchy, avoiding performance losses where they count.

psort exploits many simple tricks that can be expected to boost the performance of any data-intensive software. Perhaps the ultimate lesson of *psort* is that a lot of ugly work is necessary to transform any simple, elegant algorithm into a software that preserves at least 50% of the performance potential of today's PCs. This explains why sorting, despite its relative simplicity, its long history, and its great practical importance, can still see non-trivial improvements through simple algorithm engineering (rather than algorithmic breakthroughs).

References

1. The GNU C library - Array Sort Function, http://www.gnu.org/
2. GNU Coreutils - sort, http://www.gnu.org/
3. Sort Benchmark Home Page, http://www.hpl.hp.com/hosted/sortbenchmark/
4. A measure of transaction processing power. Datamation 31(7), 112–118 (1985)
5. Aaron Darling, A.M.: DMSort: A PennySort and Performance/Price Sort, http://www.hpl.hp.com/hosted/sortbenchmark/DMsort.pdf
6. Aggarwal, A., Alpern, B., Chandra, A., Snir, M.: A model for hierarchical memory. In: Proc. of ACM STOC 1987, pp. 305–314 (1987)
7. Aggarwal, A., Chandra, A.K., Snir, M.: Hierarchical memory with block transfer. In: Proc. of IEEE FOCS 1987, pp. 204–216 (1987)
8. Ailamaki, A., Dewitt, D.J., Hill, M.D., Wood, D.A.: Dbmss on a modern processor: Where does time go. In: Proc. of VLBD (1999)
9. Alpern, B., Carter, L., Feig, E., Selker, T.: The uniform memory hierarchy model of computation. Algorithmica 12, 72–109 (1994)
10. Arpaci-Dusseau, A.C., Arpaci-Dusseau, R.H., Culler, D.E., Hellerstein, J.M., Patterson, D.A.: High-performance sorting on networks of workstations. In: Proc. of ACM SIGMOD 1997, vol. 26(2), pp. 243–254 (1997)
11. Bilardi, G., Ekanadham, K., Pattnaik, P.: Optimal organizations for pipelined hierarchical memories. In: Proc. of ACM SPAA 2002, pp. 109–116 (2002)
12. Dementiev, R., Sanders, P.: Asynchronous parallel disk sorting. In: Proc. of ACM SPAA 2003, pp. 138–148 (2003)
13. Dewitt, D.J., Naughton, J.F., Schneider, D.A.: Parallel sorting on a shared-nothing architecture using probabilistic splitting. In: Proc. of PDIS 1991 (1991)
14. Govindaraju, N.K., Gray, J., Kumar, R., Manocha, D.: Gputerasort: High performance graphics coprocessor sorting for large database management. In: Proc. of ACM SIGMOD ICMD 2006 (2006)
15. Gray, J.: A measure of transaction processing 20 years later. CoRR, abs/cs/0701162 (2007)
16. Gray, J., Coates, J., Nyberg, C.: Price/performance sort and 1998 pennysort performance/price sort and pennysort (1998)
17. Hennessy, J., Hennessy, J.L., Goldberg, D., Patterson, D.A.: Computer Architecture: A Quantitative Approach, 1st edn. Morgan Kaufmann Publishers, San Francisco
18. Knuth, D.E.: Art of Computer Programming, 2nd edn. Sorting and Searching, vol. 3. Addison-Wesley, Reading (1998)
19. Lei Yang, Z.W., Huang, H., Song, T.: SheenkSort, Performance/Price Sort and PennySort (2003), http://www.hpl.hp.com/hosted/sortbenchmark/SheenkSort.pdf
20. Mehlhorn, K., Sanders, P.: Scanning multiple sequences via cache memory. Algorithmica 1(35), 75–93 (2003)

21. Nodine, M.H., Vitter, J.S.: Greed sort: optimal deterministic sorting on parallel disks. J. ACM 42(4), 919–933 (1995)
22. Nyberg, C., Barclay, T., Cvetanovic, Z., Gray, J., Lomet, D.B.: Alphasort: A cache-sensitive parallel external sort. VLDB J. 4(4), 603–627 (1995)
23. Liu, P., Shi, Y., Zhang, L.: 2002 Performance / Price Sort and PennySort (2002), http://www.hpl.hp.com/hosted/sortbenchmark/THsort.pdf
24. Rahman, N., Cole, R., Raman, R.: Optimised predecessor data structures for internal memory. In: Brodal, G.S., Frigioni, D., Marchetti-Spaccamela, A. (eds.) WAE 2001. LNCS, vol. 2141, pp. 67–78. Springer, Heidelberg (2001)
25. Ramey, R.: Postman's Sort, http://www.rrsd.com/
26. Valiant, L.: A bridging model for parallel computation. Comm. ACM 33(8), 103–111 (1990)

A Heuristic for Fair Correlation-Aware Resource Placement

Raouf Boutaba, Martin Karsten, and Maxwell Young[*]

David R. Cheriton School of Computer Science, University of Waterloo, ON, Canada
rboutaba@uwaterloo.ca, mkarsten@uwaterloo.ca,
m22young@uwaterloo.ca

Abstract. The configuration of network resources greatly impacts the communication overhead for data intensive tasks and constitutes a critical problem in the design and maintenance of networks. To address the issue of resource placement, we analyze and implement a semidefinite programming-based heuristic for solving a known NP-complete graph optimization problem called MAXIMUM SIZE BOUNDED CAPACITY CUT. Experimental results for our heuristic demonstrate promising performance on both synthetic and real world data. Next our heuristic is used as a sub-routine to solve another known NP-complete problem called MIN-MAX MULTIWAY CUT whose traits we adapt to yield a resource placement scheme that exploits correlations between network resources. Our experimental results show that the resulting placement scheme achieves a significant savings in communication overhead.

1 Introduction

While a measure of cost in a network is often dependent upon the given scenario, cost is generally coupled with the volume of communication between network entities. In turn, the communication between network entities is dependent upon the allocation of resources in the network. There are several scenarios characterized by frequent access to limited and multiple resources such as complex queries in distributed databases, which often require the aggregation of several objects, or the use of keyword indices by search engines in order to efficiently resolve user queries, or peer-to-peer (P2P) file sharing, where latency and bandwidth usage are highly dependent upon where files are stored.

In such environments, the location of resources has an impact on the efficiency of the system. For example, if resources A and B are often requested together, but stored at different locations, the communication overhead incurred by queries for these resources can be significantly greater than if A and B are colocated. In this work, we consider the communication costs between network entities and model the task of resource placement using a known graph optimization problem. Under this model, the goal is to distribute resources to locations in a network such that the maximum cost between any given pair of partitions is minimized; this aspect lends an important notion of *"fairness"* to our resource placement scheme.

We also demonstrate that correlation is crucial for motivating a notion of cost in our placement scheme. Correlation values are calculated using data from prior network

---

[*] Corresponding author. This research was partially supported by NSERC.

transactions. Therefore, it is assumed that knowledge of communication trends is available either as traffic engineering matrices, snapshots of past communications, or design considerations. For instance, Internet Service Providers routinely trace traffic in P2P networks and this information can be used to improve network performance [12]. Naturally, this approach raises the question of whether such correlations are stable enough over a sufficient period of time to warrant the additional computational costs of a correlation-aware placement scheme. Indeed, it has been demonstrated that resource correlations *do* remain stable over at least *month-long periods* [22].

From a practical perspective, the resource placement scheme we propose can be managed by an authority in a distributed setting, such as an ISP; there exist proposals for cooperation between ISPs and P2P networks [5,20]. Alternatively, in a centralized scenario, such a scheme would be useful within single-administrative domains, such as allocation in data-centers. We treat the details of such a setup as outside the scope of this paper, and instead focus on the analysis of our scheme.

1.1 Our Contributions

We propose a resource placement scheme that exploits correlation information between network resources; we call this a *correlation-aware resource placement scheme*. In formulating our scheme, two known NP-complete problems are dealt with: MAXIMUM SIZE BOUNDED CAPACITY CUT (MAXSBCC) and MIN-MAX MULTIWAY CUT (MMMC). We mathematically analyze and implement a heuristic for MAXSBCC based on the technique of semidefinite programming (SDP). Our heuristic is then employed as a subroutine for solving MMMC. To the best of our knowledge, our work provides the first empirical evaluation of these two problems.

We then identify two challenges to employing MMMC as a model of resource placement. First, naively employing a cost metric can lead to insensible solutions. Second, privacy issues are absent from the model. We address both challenges by showing how correlation adequately motivates important cost metrics, and we mathematically extend the model to account for privacy constraints. Finally, an experimental evaluation of our correlation-aware resource placement scheme is conducted with real-world data.

1.2 Related Work

Our work differs from a number of previous treatments on data placement where cost metrics are not motivated by correlation and, moreover, cost is measured as an aggregate [14] or average [6] across the entire network. The most relevant related work is [22] where the authors address the use of correlation in placing data items in a network. However, again, their work aims to minimize an *aggregate* notion of cost in contrast to the substantially different min-max approach used here.

Our correlation-aware resource placement scheme is based on the MIN-MAX MULTIWAY CUT problem which was introduced by Tardos and Svitkina [18]. The MIN-MAX MULTIWAY CUT problem is NP-complete and the best known approximation algorithm relies on obtaining an efficient solution to a sub-problem called the MAXIMUM SIZE BOUNDED CAPACITY CUT problem. In [18], the authors cite an algorithm developed in [9] to achieve a polylogarithmic approximation. However, the algorithm of [9] is extremely intricate and, several years after the theoretical result, no implementation exists.[1] Moreover, due to the large number of constraints, this algorithm is likely to be extremely computationally intensive, even for very small problem instances.

In contrast, our work incorporates an efficient heuristic based on the technique of semidefinite programming (SDP). SDP has figured prominently in the development of heuristics for problems in the areas of phylogenic reconstruction [16], machine learning [21], sensor network layout [8], bioinformatics [13] and graph partitioning [10]; these results demonstrate that heuristics can benefit greatly from this optimization technique.

2   Our Heuristic and Analysis

To solve the MIN-MAX MULTIWAY CUT problem, we solve a subproblem known as the MAXIMUM-SIZE BOUNDED-CAPACITY CUT (MAXSBCC) problem introduced in [18]. The input to MAXSBCC is an undirected graph $G = (V, E)$ with weights on the vertices $w(v)$, capacities on the edges $c(e)$, source and sink vertices $v_s, v_t \in V$, and an integer B. Given a partition of V into S and T, denote by $\delta(S)$ the total weight of the cut edges and denote by $w(S)$ the total weight of the vertices in S. The MAXSBCC problem is to find an s-t cut (S, T) such that $\delta(S) \leq B$, and $w(S) = \sum_{v \in S} w(v)$ is maximized. In [18], the authors focus on a (α, β)-bicriteria approximation algorithm for MAXSBCC. That is, given an instance of MAXSBCC with an optimal solution $(S_\star, T_\star)$, returns in polynomial time a solution (S', T') such that $\delta(S') \leq \alpha B$ and $w(S') \geq \beta w(S_\star)$ where $\alpha \geq 1$ and $0 < \beta \leq 1$. Therefore, solutions *may exceed the budget* and this also turns out to be true for our heuristic.

Consider the quadratic program specified by Equations (1)-(4). Variable x_i corresponds to $v_i \in V$, w_i is the weight of vertex v_i, and w_{ij} is the edge weight of (v_i, v_j) which is zero if no such edge exists.

$$\max \sum_{i=1}^{n} \frac{1 + x_s x_i}{2} w_i \tag{1}$$

$$s.t. \sum_{i<j} x_i x_j \, w_{ij} \geq M \tag{2}$$

$$x_s x_t = -1 \tag{3}$$

$$x_i \in \{-1, 1\} \tag{4}$$

$$\max \sum_{i=1}^{n} \frac{1 + y_{si}}{2} w_i \tag{5}$$

$$s.t. \; \mathbf{Y} = (y_{ij}) \succeq 0 \tag{6}$$

$$\sum_{i<j} y_{ij} \, w_{ij} \geq M \tag{7}$$

$$y_{ii} = 1 \tag{8}$$

$$y_{st} = -1 \tag{9}$$

Equation (1) counts the cumulative weight of the vertices in S. Let τ denote the cumulative weight of the edges internal to the set T and let σ denote the cumulative weight of the edges internal to the set S. Equation (2) counts $\sigma + \tau - \delta(S)$. Equation (3) states that the source and sink nodes must be in separate partitions. Equation (4) guarantees that each vertex belongs to one and only one partition. Treating each vertex variable x_i as a vector $\mathbf{v_i}$, and letting $y_{ij} = \mathbf{v_i} \cdot \mathbf{v_j}$, the SDP specified by Equations (5)-(9) is obtained where '$\succeq$' denotes that Y is positive semidefinite. We now analyze the semidefinite program to motivate its suitability for MAXSBCC. Due to space constraints, proofs of the following results are omitted and can be found in the full version of this paper [7]. Let $W^* \geq M$ denote the value of Equation (7) given by solving the

---

[1] To the best of our knowledge, no such implementation exists. Furthermore, via email correspondence, Professor R. Krauthgamer (one of the authors of [9]) stated he was unaware of any such implementation.

semidefinite program and let $W \leq W^*$ denote its value after applying the standard technique of hyperplane rounding.

Lemma 1. *Hyperplane rounding of the SDP provides a W such that $E[W] \geq 0.945W^*$.*

Let S^* denote the value of Equation (5) given by the solution to the semidefinite program and let S denote the value of this quantity after hyperplane rounding. We establish the following critical result:

Lemma 2. *Let ϵ be a small positive constant. With probability at least $1 - \frac{1}{n}$, hyperplane rounding need only be applied $\lceil 2 \ln n/\epsilon \rceil$ times before a rounded solution to the SDP is obtained such that $W \geq (1.823 - \lambda)W^*$ and $w(S) \geq \beta'w(S^*)$ where $1 \geq \lambda \geq 0.823$.*

Finally, we address the quality of our solution:

Theorem 1. *With probability at least $1 - \frac{1}{n}$, for $\lambda \in [0.823, 1]$, the above algorithm achieves a cut (S, T) such that:*

$$\delta(S) \leq \lambda + \left(\frac{1 - \lambda}{2}\right) \frac{w(E)}{B}$$

$$w(S) \geq (1.823 - \lambda) \cdot w(S^*)$$

As λ approaches 1, the quality of our approximation increases correspondingly. Therefore, the quality of our solution depends heavily on λ which, in turn, depends on the value M that we set in our program. This relationship suggests a heuristic approach whereby we attempt to improve our solution quality by modifying M. Throughout, we refer to the above algorithm by $\text{ALG}(M)$ to reflect that the performance depends on M.

2.1   Our Full Heuristic: MaxSBCC Solver

We seek a solution to MAXSBCC by performing multiple iterations of ALG(M) and modifying the value of M in our semidefinite program formulation at each iteration. Denote the output of ALG by the tuple (v, S, T, B_{actual}) where:

- v: is a boolean variable with value `true` if the solution returned is an *s-t* cut; `false` otherwise.
- S: is the set of nodes on the source side of the cut.
- T: is the set of nodes on the sink side of the cut.
- B_{actual}: is the bound resulting from our setting of M.

Due to space constraints, we outline our heuristic which we call MAXSBCC SOLVER; the pseudocode is given in the full paper. The input into MAXSBCC SOLVER is the graph $G = (V, E)$, the desired input bound B_0 and the number of iterations r of ALG(M). The core idea is to modify M at each iteration of ALG until we achieve a B value close, or equal, to our original desired bound B_0. M is modified by essentially performing a binary search through the possible values of the input bound B.

MAXSBCC SOLVER begins by storing our original input bound $B = B_0$ and $M = W(E) - 2B$. Once executed, the solution is checked for validity by inspecting the boolean v variable. If MAXSBCC SOLVER failed to find an *s-t* cut, the next iteration proceeds with the input bound B doubled. Once a valid solution is returned, B_{actual} is examined; if it is larger than our desired B_0, B decreases by $\lceil B/2^j \rceil$ where j denotes the number of iterations where a valid solution has been achieved. Conversely, if B_{actual} is smaller or equal to B_0, B increases by $\lceil B/2^j \rceil$. Note that such changes in the value of B are equivalent to modifying M, since $M = W(E) - 2B$ in our SDP.

At the end of r iterations of this process, S is returned. If a valid solution was found, the solution that gave a bound closest to B_0 (ie. the iteration where $|B_0 - B_{actual}|$ was smallest) is returned. Otherwise, if no valid solution was found, S will be empty; however, if there exists a cut of size B_0 or less, then our algorithm will return a valid cut with high probability and with the attributes demonstrated by our analysis above.

2.2   MaxSBCC Solver: Experimental Results

Our experiments are performed on systems with up to 1300 nodes; there are two reasons for this system size. First, although this situation is improving, solving semidefinite programs is still computationally expensive. The area of semidefinite programming is relatively new and available software for solving such programs is consequently limited. Many solvers scale as $O(n^3)$ where n is the dimension of the semidefinite matrix. However, despite its current computational costs, semidefinite programming is a standard technique for solving many challenging problems. Moreover, a number of recent results address the issue of scalability such as parallelized implementations [17] as well as devising problem formulations that can be computed more efficiently [15]. We believe such techniques can allow our algorithm to scale to much larger system sizes but are outside the scope of this paper.[2]   The second reason is that, for our experiments with MAXSBCC, we wish to compare against the optimum solution. This allows a particularly unforgiving comparison in judging the performance of our algorithm. We formulate an integer linear program (ILP) for each of the experimental problem instances. Using the ILP solver CPLEX [1], an optimal solution can be achieved for the purposes of comparison. For our experiments we use the SDP solver 'SemiDefinite Programming Algorithm in Matlab [2].

Our first data set consists of three unweighted (Table 1) and three weighted (Table 2) Barabási-Albert scale free graphs created using the BRITE topology generator [3]. Each graph is connected[3], consists of 300 vertices and, for the weighted case, edge capacities are exponentially distributed in the range $\{1, ..., 1024\}$ while node weights are chosen uniformly at random in the range $\{1, 1024\}$. Our second data set consists of three unweighted (Table 3) and three weighted (Table 4) connected Waxman graphs [19] of 300 nodes, with the same capacity and node weight distributions. Finally, we use a real-world data set collected in [11] consisting of a trace of peer-to-peer (P2P) traffic and containing information on data objects both advertised and queried over the course of two months; we restrict our use of this data to mp3 files. Correlation values are derived for each object by examining how often file x and file y were colocated at a peer. Edges with correlation values less than 0.25, using a p-value of 0.1, are discarded yielding a connected component of 358 nodes which we used. Each correlation value is multiplied by a factor of 100 and rounded to the nearest integer in order to provide integer input values for the SDP program. We then select a node of

---

[2] In terms of our later application to resource placement, previous results of [22] have shown that only a small fraction of the total system need be optimized to achieve substantial savings.

[3] If the graph is disconnected, our algorithm can be used on each component individually. For simplicity, we used only connected graphs in our experiments.

Tables 1 & 2. Results of the MAXSBCC heuristic on unweighted and weighted Barabási-Albert graphs, respectively

Trial	B	S	δ	S_*	δ_*	$S/S_*, \delta/\delta_*$
	2	1	2	1	2	1.000, 1.000
	10	5	10	9	10	0.556, 1.000
	18	9	18	20	18	0.450, 1.000
1	26	15	27	29	24	0.517, 1.125
	34	16	30	38	34	0.421, 0.882
	42	25	43	49	42	0.510, 1.024
	46	299	46	299	46	1.000, 1.000
	2	1	2	1	2	1.000, 1.000
	7	4	8	5	7	0.800, 1.143
	12	6	12	10	12	0.600, 1.000
2	17	10	22	15	17	0.667, 1.294
	22	12	23	21	22	0.571, 1.045
	27	299	31	27	27	11.074, 1.148
	31	299	31	299	31	1.000, 1.000
	2	1	2	1	2	1.000, 1.000
	9	4	8	7	9	0.571, 0.889
	16	9	19	15	16	0.600, 1.188
3	23	12	23	23	23	0.522, 1.000
	30	15	31	30	27	0.500, 1.148
	37	17	38	38	37	0.447, 1.027
	41	299	41	299	41	1.000, 1.000

Trial	B	S	δ	S_*	δ_*	$S/S_*, \delta/\delta_*$
	6	973	6	973	6	1.000, 1.000
	1096	13409	1360	20362	1096	0.659, 1.241
	2186	20951	2590	31666	2186	0.662, 1.185
1	3276	34038	3943	41133	3271	0.828, 1.205
	4366	151828	4367	152244	4362	0.997, 1.001
	5456	156249	5460	156243	5423	1.000, 1.007
	5460	156249	5460	156249	5460	1.000, 1.000
	14	369	14	369	14	1.000, 1.000
	751	10592	959	13222	743	0.801, 1.291
	1488	13000	1496	23478	1487	0.554, 1.006
2	2225	22865	2332	31874	2223	0.717, 1.049
	2962	151079	2976	151003	2940	1.001, 1.012
	3699	155300	3702	155284	3656	1.000, 1.013
	3702	155300	3702	155300	3702	1.000, 1.000
	3	917	3	917	3	1.000, 1.000
	770	14476	1073	14958	770	0.968, 1.393
	1537	17647	1572	24342	1532	0.725, 1.026
3	2304	22645	2317	33561	2304	0.675, 1.006
	3071	147415	3065	147992	3064	0.996, 1.000
	3838	151589	3838	151589	3838	1.000, 1.000

Tables 3 & 4. Results of the MAXSBCC heuristic on unweighted and weighted Waxman graphs, respectively

Trial	B	S	δ	S_*	δ_*	$S/S_*, \delta/\delta_*$
	2	1	2	1	2	1.000, 1.000
	5	2	6	3	5	0.667, 1.200
	8	3	11	7	8	0.428, 1.375
1	11	5	12	10	11	0.500, 1.091
	14	5	12	14	14	0.357, 0.857
	17	299	17	299	17	1.000, 1.000
	2	1	2	1	2	1.000, 1.000
	4	3	5	2	4	1.500, 1.250
	6	5	8	5	6	1.000, 1.333
2	8	5	8	7	8	0.714, 1.000
	10	7	11	9	10	0.778, 1.100
	12	7	11	11	12	0.636, 0.917
	14	299	16	13	14	23.000, 1.143
	16	299	16	299	16	1.000, 1.000
	2	1	2	1	2	1.000, 1.000
	5	4	6	3	5	1.333, 1.200
	8	5	12	6	7	0.833, 1.714
3	11	5	12	10	11	0.500, 1.091
	14	9	16	12	14	0.750, 1.143
	17	299	17	299	17	1.000, 1.000

Trial	B	S	δ	S_*	δ_*	$S/S_*, \delta/\delta_*$
	13	916	13	916	13	1.000, 1.000
	344	6047	289	9049	342	0.668, 0.845
	675	8987	642	15041	675	0.598, 0.951
1	1006	149667	1125	149377	991	1.002, 1.135
	1337	150765	1366	150501	1256	1.001, 1.088
	1668	151290	1668	151290	1668	1.000, 1.000
	11	78	11	78	11	1.000, 1.000
	343	5844	459	6705	336	0.872, 1.366
	675	157520	606	156514	606	1.006, 1.000
2	1007	157520	1051	157146	987	1.002, 1.065
	1339	158254	1672	157520	1051	1.005, 1.591
	1672	158254	1672	158254	1672	1.000, 1.000
	15	631	15	631	15	1.000, 1.000
	387	8372	367	9895	385	0.846, 0.953
	759	12554	734	15957	756	0.787, 0.971
3	1131	159297	1173	157260	1106	1.013, 1.061
	1503	160081	1474	160115	1471	0.999, 1.002
	1876	161490	1876	161490	1876	1.000, 1.000

Table 5. Results of the MAXSBCC heuristic on real world data

B	S	δ	$S_\star$	$\delta_\star$	$S/S_\star, \delta/\delta_\star$
25	159975	25	159975	25	1.000, 1.000
115	160957	127	163526	113	0.984, 1.124
205	166445	228	165908	202	1.003, 1.129
295	167438	278	167108	278	1.002, 1.000
385	167825	354	168476	381	0.996, 0.929
476	176567	476	176567	476	1.000, 1.000

Table 6. Results of testing RANDOM, GREEDY, MMMC SOLVER on MMMC test cases

Trial	c	N	τ	RANDOM	GREEDY	MMMC SOLVER
1	0.30	111	4	1290	831	34
				1475	888	66
				1222	875	39
2	0.28	379	5	2708	2273	163
				2495	2190	300
				2480	1962	136
3	0.26	813	6	3756	2990	152
				3567	3179	238
				3514	3114	219
4	0.24	1256	7	5246	4145	288
				5208	4060	256
				5297	4556	538

minimum degree, d_{min}, to be the source and a node of maximum degree, d_{max}, to be the sink. Budget values are selected as even increments in the range $[d_{min}, d_{max}]$. We set $r = 5$ in MAXSBCC SOLVER which we found yielded good solutions.

Tables 1-5 provide the results of MAXSBCC SOLVER in the S and δ columns, while the optimum solution is given in the S^* and δ^* columns. For both the weighted and unweighted synthetic data sets, the worst source side approximation is 0.421 and the worst cut approximation is 1.393 and, generally, the approximations are even significantly better than these worst cases. Moreover, for our experiment using the real-world data, we observe in Table 5 that our heuristic yields *very high quality solutions*. We also note that MAXSBCC SOLVER performs relatively quickly, completing within no more than 4 hours on a machine utilizing a single 1.3 GHz Intel Itanium2 CPU running SuSE Linux. In comparison, the trials with CPLEX frequently required up to 48 hours on the same system. Overall, the performance of MAXSBCC SOLVER is promising and, with our heuristic in hand, we now move onto MMMC and resource placement.

3   Towards Resource Placement: Min-Max Multiway Cut

As a basic abstract model of resource placement, we use the MIN-MAX MULTIWAY CUT (MMMC) problem as defined in [18]. Given an undirected graph G = (V,E) with weighted edges and a subset of the vertices $T = \{t_1, ..., t_k\}$ which are called terminals, a multiway cut is a partition of V into disjoint sets $S_1, ..., S_k$, such that S_i contains t_i for $i = 1, ..., k$. The goal in MMMC is to partition V such that the maximum cut between any two partitions is minimized. If cut size is related to the cost of communication between two partitions, then MMMC seeks to *fairly* distribute this cost over all partitions. Therefore, we deal with a setting where local communication is relatively cheap, while communication between individual machines or domains is costly.

The 'fairness' aspect of MMMC distinguishes this work from a number of other resource placement models. In contrast, by minimizing the aggregate or average notion of cost, it is possible for some network participants to suffer a disproportionate amount of traffic. Here, we are not concerned with the number of resources allocated to a network entity, but with the cost of inter-partition traffic incurred by holding such items. With abundant availability of disk-space, we treat bandwidth and latency as principal aspects in our model.

3.1 Experimental Results

In [18], the authors show how a (α, β)-approximation algorithm for MAXSBCC can be used to achieve an $(\alpha \log_\beta n)$-approximation for MIN-MAX MULTIWAY CUT. Here, we employ their result using our heuristic for MAXSBCC; we call this algorithm MMMC SOLVER. Due to space constraints, we refer the reader [18] for further details.

Using the data set from [11], the most widely held 2000 mp3 files are extracted. Pair-wise correlations are computed between all data objects and we discard edges with a correlation value below a cutoff point c which differs per trial. Edge capacities were then multiplied by a factor of 100 and rounded to the nearest integer. For each trial, τ terminal nodes are chosen uniformly at random from the total number of nodes N and the input bound was chosen to be an arbitrary value of 1000. MMMC SOLVER is then run on the graph problem. Since no ILP formulation for MMMC is known, we compare against two other algorithms in order to evaluate the quality of our solution. The first is a simple random placement of nodes to partitions which models the behavior we would expect from employing a secure hash function; we denote this algorithm by RANDOM. The second is a greedy algorithm, denoted by GREEDY, that begins with a random assignment to partitions and then attempts to reduce the size of the maximum cut by greedily reassigning vertices. In particular, each non-terminal node involved in a maximum cut is tested in all other partitions. If such a relocation reduces the maximum cut, the new assignment is immediately kept; otherwise, the node remains at its original location. A solution is returned when no further reduction can be obtained.

Table 6 provides the results of our experiments. The longest running experiment consisted of 1256 nodes with 7 terminals and the running time of MMMC SOLVER was under 8 hours. Over all four trials, MMMC SOLVER demonstrates superior performance in the size of the maximum cut. The discrepancy between RANDOM and MMMC SOLVER is significant although expected. More strikingly, the difference between GREEDY and MMMC SOLVER is substantial suggesting that the greedy approach becomes trapped in local optima which MMMC SOLVER is able to avoid. In particular, the maximum cut between any two partitions yielded by MMMC SOLVER is *never more than* 14% of that yielded by GREEDY.

4   Correlation-Aware Resource Placement: Extending MMMC

Initially, the MIN-MAX MULTIWAY CUT problem appears to model almost any resource placement problem. However, this is not the case for two reasons:

1. *Cost Dependencies:* consider modelling a P2P network as discussed in the original work [18] where the terminals are peers and the remaining non-terminal vertices are data items. A data item belonging to S_i is stored at peer t_i and edge capacities reflect expected communication patterns. The goal is to allocate data items among the peers so as to minimize the expected communication cost. *However, costs will likely depend on where items are placed in the network.* Here, MMMC requires input for the edges and yet this input will be defined by the very solution we seek. Consequently, cost metrics need to be carefully motivated.

2. *Privacy:* there are often constraints on where resources can be placed in the network and, by itself, the MIN-MAX MULTIWAY CUT problem does not address this issue. This may be due to privacy issues where sensitive data can only be allocated to secure locations. Conversely, a server may refuse to maintain a particular resource given legal concerns or quality of service constraints. Resources may even be physically restricted to certain locations.

We now show how to solve these two problems and arrive at our correlation-aware resource placement scheme.

4.1 Cost Metrics Motivated by Correlation

In this section, we demonstrate how the use of correlation in our model avoids the problem of cost dependencies and motivates two important cost metrics. Overall, the main benefit of correlation information is that it is independent of location. Throughout, assume we are provided with positive correlation values between nodes in the network.[4]

Latency as a Cost Metric. Consider resources d_1 and d_2, which are strongly correlated. They may be colocated in order to reduce the communication overhead involved in obtaining them both. For instance, in response to a query involving d_1, both resources d_1 and d_2 may be fetched in anticipation of a follow-up request for d_2; alternatively, less inter-machine communication may be required if both resources are located on one, or even a small number of machines, if substantial inter-machine communication is required for a query. Therefore, under scenarios where the size of resources is relatively small, the correlation values on our input graph to the MIN-MAX MULTIWAY CUT give rise to latency as a plausible cost metric.

This problem domain is suited to a number of applications. For instance, text search engines typically utilize inverted indices in order to be efficient. Primarily, an inverted index stores information matching a keyword to documents that contain it. A query with K terms often requires that the inverted indices of all K terms be accessed. For distributed search systems, these indices are placed on many different machines. Consequently, the communication overhead between machines storing the indices required for resolving the same query presents a critical factor in supporting fast search [22].

Bandwidth as a Cost Metric. For sizable data items, bandwidth becomes the dominating cost, not latency. Consider two large files d_1 and d_2 that are highly correlated in the sense that if a user obtains one, he is likely to obtain the other. As a simple example, d_1 and d_2 might be two jpeg files by a user's favorite artist. In this case, it does not matter whether d_1 and d_2 are colocated since our cost metric is dominated by bandwidth consumption which does not necessarily bear any relationship to the correlation value on the edge (d_1, d_2) in our input graph.

The situation, however, is quite different for queries involving the collection of more than one data source. For instance, a user may wish to compute a function over the aggregation of d_1 and d_2. There are a number of settings where such complex queries are useful for allowing richer search capabilities. A range query might require a join operation on d_1 and d_2. Here, it makes sense to have d_1 and d_2 colocated; the query can be resolved without downloading of *at least one of* d_1 *or* d_2. Such complex queries motivate a meaningful relationship between correlation and bandwidth consumption.

---

[4] Our approach can incorporate negative correlations; however, for simplicity we restrict our attention to positive correlation in the context of our work.

4.2 Adding Privacy Constraints

We enforce privacy constraints by embedding them into the MAXSBCC sub-problem. The following primal semidefinite form for our SDP of Section 2 can be obtained from Equations (5)-(9) by standard methods:

$$\max \ C \cdot X \ s.t. \ (1/2)A \bullet X = W$$
$$E_{ii} \bullet X = 1, \text{ for } 1 \leq i \leq n$$
$$(1/2)E_{st} \bullet X = -1$$
$$X \succeq 0$$

where $P \bullet Q$ denotes the standard $\sum_i \sum_j P_{ij} Q_{ij}$. B' is such that entry $b'_{11} = 1$ and all other entries in the first row and first column have value $1/2$; the rest of the entries are 0. Then $C = (1/2)I + (1/2)B'$, $X_{ij} = \mathbf{v_i v_j}$, and A is the capacity matrix for G. E_{ij} is an $n \times n$ matrix with a 1 in the ij^{th} and ji^{th} entries and zeros everywhere else.

Using the primal form, privacy constraints are added in the following fashion. Assuming feasibility, if we wish to constrain the location of a particular resource b to a terminal v, we include $(1/2)E_{v,b} \bullet X = 1$; alternatively, $(1/2)E_{v,b} \bullet X = -1$ ensures that b will not be stored at v. We can also force resources a and b to be colocated or separated by setting $(1/2)E_{a,b} \bullet X = 1$ or $(1/2)E_{a,b} \bullet X = -1$, respectively. The mathematical analysis of Section 2 changes little and the results remain unchanged.

4.3 Experimental Results

Our experimental work is in the context of Section 4.1. Assume a homogenous multi-administrative network where users in the network are issuing text queries and the cost of resolving a query *within the domain of the particular issuer of the query* is inexpensive, while communication between administrative domains is costly. We consider the problem of placing inverted indices such that (1) the communication overhead between domains during query resolutions is reduced and (2) no domain is involved in an excessive number of transactions involving multiple domains.

We utilized the query data[5] of [11] which totals 5462 queries, each consisting of several terms, by users in the network. Using the SMART 'stopword' list [4], queries were pruned to remove trivial terms.[6] From this data set, the K most prevalent terms were extracted and correlation values between each pair were calculated. The most prevalent terms are not necessarily correlated with one another; therefore, we extracted the largest group of terms that did share positive correlations. Represented as a connected component where nodes are terms and edges are weighted by correlation values, this graph was used as our input. 10 randomly chosen terminals were chosen to correspond to domains. There were 12 trials in total, consisting of 53, 144, 238, 336, 423, 523, 630, 734, 847, 956, 1064 and 1157 nodes, respectively. In each trial, all 5462 pruned queries were executed. For each term in a particular query that matched a top key word in our trial, data was collected; otherwise, the term was ignored.

Distinct Domains Accessed per Query. If terms within a single query require accessing multiple domains, then the number of unique domains accessed provides a measure

---

[5] All query data was used, not just queries related to mp3 files.

[6] Pruning did not remove queries, but trivial terms were removed from a large number of queries.

Tables 7. Number of unique domain accesses per query aggregated over all queries

Trial	GREEDY	MMMC SOLVER
1	1218	1111
2	2247	1996
3	2942	2409
4	3439	2333
5	3808	2493
6	4191	2693
7	4561	2909
8	4886	3024
9	5285	3159
10	5555	3291
11	5858	3471
12	5997	3485

Table 8. Total percentage of queries that can be resolved within a single domain

Trial	GREEDY	MMMC SOLVER
1	83.0%	87.7%
2	73.6%	81.4%
3	68.3%	83.9.%
4	64.8%	96.8%
5	62.4%	94.7%
6	59.4%	97.6%
7	58.3%	96.8%
8	56.8%	97.8%
9	55.2%	98.8%
10	54.4%	98.8%
11	53.6%	97.8%
12	53.1%	99.4%

Table 9. Of the remaining queries that require accessing two or more domains to resolve, the percentage attributed to the domain involved in the most number of such transactions

Trial	GREEDY	MMMC SOLVER
1	7.0%	10.0%
2	9.7%	12.1%
3	11.5%	13.7%
4	11.7%	3.2%
5	14.5%	5.3%
6	13.7%	2.4%
7	14.6%	3.2%
8	14.7%	2.2%
9	16.1%	1.3%
10	15.7%	1.2%
11	15.9%	2.1%
12	16.2%	5.8%

of communication overhead per query. Table 7 illustrates the sum of such access data over all queries. In comparison with the placement scheme given by GREEDY, MMMC SOLVER achieves substantially fewer unique domain accesses over the course of executing all 5462 queries. In particular, for Trial 4 and above, MMMC SOLVER incurs *only 68% down to 58% of the unique domain accesses* performed by GREEDY.

Queries Involving Multiple Domains. The number of queries requiring communication between multiple domains concerns both the amount of communication overhead and also the aspect of fairness. Table 8 depicts data for both MMMC SOLVER and GREEDY on the number of queries that were resolved through a single domain only and the number of queries that required two or more domain accesses. For MMMC SOLVER, *at least 81% of all queries could be resolved at a single domain.* Moreover, for Trial 4 and above, this value grew to be 95% or more. In contrast, a significantly smaller percentage of queries were resolved at a single domain using GREEDY. In terms of fairness, with MMMC SOLVER, no domain participated in transactions with other machines for more than 19% of the queries. This value can be dissected further by examining how much of this 19% is attributed to each domain. Table 9 gives this information; no domain is ever forced to participate in more than 14% of these transactions involving more than one domain. Moreover, this value decreases substantially as the number of key words is increased, *dropping below 6% after Trial 4.*

5 Conclusion

In this work, we proposed a novel correlation-aware resource placement scheme. A heuristic was developed for solving the MAXSBCC problem. This heuristic is used as a critical sub-routine for solving MMMC which, after extensions to address cost metrics and privacy constraints, yields our correlation-aware resource placement scheme. The results of our experiments were encouraging and demonstrated that our scheme can yield substantial savings in communication overhead. Interesting future work includes analyzing the performance benefits of using negative correlation information and parallelized implementations of our algorithms.

Acknowledgements. We gratefully thank Jared Saia for his helpful discussions.

References

1. ILOG CPLEX, http://www.ilog.com/products/cplex/
2. SDPA, http://homepage.mac.com/klabtitech/sdpa-homepage/
3. BRITE, http://www.cs.bu.edu/brite/
4. SMART, ftp://ftp.cs.cornell.edu/pub/smart/
5. Aggarwal, V., Feldmann, A., Scheideler, C.: Can ISPs and P2P users cooperate for improved performance. Computer Communication Review 37(3), 29–40 (2007)
6. Baev, I.D., Rajaraman, R.: Approximation algorithms for data placement in arbitrary networks. In: 12th Annual ACM-SIAM Symposium on Discrete algorithms, pp. 661–670 (2001)
7. Boutaba, R., Karsten, M., Young, M.: A heuristic for fair correlation-aware resource placement. Technical Report CS-2009-04, University of Waterloo Technical Report (January 2009)
8. Carter, M., Jin, H., Saunders, M., Ye, Y.: Spaseloc: An adaptive subproblem algorithm for scalable wireless sensor network localization. SIAM Journal on Optimization 17(4), 1102–1128 (2006)
9. Feige, U., Krauthgamer, R.: A polylogarithmic approximation of the minimum bisection. In: IEEE Symposium on Foundations of Computer Science, pp. 105–115 (2000)
10. Ghaddar, B., Anjos, M., Liers, F.: A branch-and-cut algorithm based on semidefinite programming for the minimum k-partition problem (2007) (unpub. manuscript)
11. Goh, S.-T., Kalnis, P., Bakiras, S., Tan, K.-L.: Real datasets for file-sharing peer-to-peer systems. In: Zhou, L.-z., Ooi, B.-C., Meng, X. (eds.) DASFAA 2005. LNCS, vol. 3453, pp. 201–213. Springer, Heidelberg (2005)
12. P. Heywood. Caching in on p2p, www.lightreading.com/document.asp?doc_id=34399
13. Kalpakis, K., Namjoshi, P.: Haplotype phasing using semidefinite programming. In: Proc. 5th IEEE Symposium on Bioinformatics and Bioengineering, pp. 145–152 (2005)
14. Krick, C., Racke, H., Westermann, M.: Approximation algorithms for data management in networks. In: SPAA 2001, pp. 237–246 (2001)
15. Kulis, B., Surendran, A.C., Platt, J.C.: Fast low-rank semidefinite programming for embedding and clustering. In: 11th Int. Conf. on A.I. and Statistics, pp. 512–521 (2007)
16. Moran, S., Rao, S., Snir, S.: Using semi-definite programming to enhance supertree resolvability. In: Casadio, R., Myers, G. (eds.) WABI 2005. LNCS (LNBI), vol. 3692, pp. 89–103. Springer, Heidelberg (2005)
17. Nakata, K., Yamashita, M., Fujisawa, K., Kojima, M.: A parallel primal-dual interior-point method for semidefinite programs using positive definite matrix completion. Parallel Computing 32(1), 24–43 (2006)
18. Svitkina, Z., Tardos, É.: Min-max multiway cut. In: Jansen, K., Khanna, S., Rolim, J.D.P., Ron, D. (eds.) RANDOM 2004 and APPROX 2004. LNCS, vol. 3122, pp. 207–218. Springer, Heidelberg (2004)
19. Waxman, B.M.: Routing of multipoint connections. IEEE Journal on Selected Areas in Communications 6(9), 1617–1622 (1988)
20. Xie, H., Yang, Y.R., Krishnamurthy, A., Liu, Y., Silberschatz, A.: P4P: Provider portal for applications. In: SIGCOMM, pp. 351–362 (2008)
21. Zhang, Y., Burer, S., Nick Street, W.: Ensemble pruning via semi-definite programming. Journal of Machine Learning Research 7, 1315–1338 (2006)
22. Zhong, M., Shen, K., Seiferas, J.: Correlation-aware object placement for multi-object operations. In: 28th International Conference on Distributed Computing Systems (ICDCS), pp. 512–521 (2008)

Measuring the Similarity of Geometric Graphs

Otfried Cheong[1,*], Joachim Gudmundsson[2], Hyo-Sil Kim[1,*],
Daria Schymura[3,**], and Fabian Stehn[3,**]

[1] KAIST, Korea
{otfried,hyosil}@tclab.kaist.ac.kr
[2] NICTA,[***] Australia
joachim.gudmundsson@nicta.com.au
[3] FU Berlin, Germany
{schymura,stehn}@inf.fu-berlin.de

Abstract. What does it mean for two geometric graphs to be similar?
We propose a distance for geometric graphs that we show to be a metric,
and that can be computed by solving an integer linear program. We also
present experiments using a heuristic distance function.

1 Introduction

Computational geometry has studied the matching and analysis of geometric
shapes from a theoretical perspective and developed efficient algorithms measuring the *similarity* of geometric objects. Two objects are similar if they do
not differ much geometrically. A survey by Alt and Guibas [1] describes the significant body of results obtained by researchers in computational geometry in
this area.

This paradigm fits a number of practical shape matching problems quite well,
such as the recognition of symmetries in molecules, or the self-alignment of a
satellite based on star patterns. Other pattern recognition problems, however,
seem to require a different definition of "matching." For instance, recognizing
logos, Egyptian hieroglyphics, Chinese characters, or electronic components in
a circuit diagram are typical examples where this is the case. The same "pattern" can appear in a variety of shapes that differ geometrically. What remains
invariant, however, is the "combinatorial" structure of the pattern.

We propose to consider such patterns as geometric graphs, that is, planar
graphs embedded into the plane with straight edges. Two geometric graphs can

---

* This work was supported by the Korea Science & Engineering Foundation through
the Joint Research Program (F01-2006-000-10257-0).

** This research was partially funded by Deutsche Forschungsgemeinschaft within the
Research Training Group (Graduiertenkolleg) "Methods for Discrete Structures",
the DFG-projects AL253/6-1, KN591/2-2 and through the Joint Research Project
446 KOR 113/211/0-1.

*** NICTA is funded by the Australian Government as represented by the Department
of Broadband, Communications and the Digital Economy and the Australian Research Council through the ICT Centre of Excellence program.

be considered similar if both the underlying graph and the geometry of the planar embedding are "similar." The distance measures considered in computational geometry, such as the Hausdorff distance, Fréchet distance, or the symmetric difference, do not seem to apply to geometric graphs.

Pattern recognition systems that combine a combinatorial component with a geometric component are already used in practice—in fact, *syntactic* or *structural* pattern recognition is based on exactly this idea: A syntactic recognizer decomposes the pattern into geometric primitives and makes conclusions based on the appearance and relative position of these primitives [2,7]. While attractive from a theoretical point of view, syntactic recognizers have not been able to compete with numerical or AI techniques for character recognition [6]. In general, the pattern recognition community may be said to consider graph representations as expressive, but too time-consuming, as subgraph isomorphism in general is known to be intractable.

An established measure of similarity between (labeled) graphs is the *edit distance*. The idea of an edit distance is very intuitive: To measure the difference between two objects, measure how much one object has to be changed to be transformed into the other object. To define an edit distance, one therefore defines a set of allowed operations, each associated with a cost. An edit sequence from object A to object B is a finite sequence of allowed operations that transforms A into B. The distance between A and B is the minimal cost of an edit sequence from A to B.

The edit distance originally stems from string matching where the allowed operations are insertion, deletion and substitution of characters. The edit distance of strings can be computed efficiently, and the string edit distance is used widely, for instance in computational biology.

Justice and Hero [5] defined an edit distance for vertex-labeled graphs that additionally allows relabeling of vertices, and give an integer linear programming formulation of the edit distance. The edit operations are insertion and deletion of vertices, insertion and deletion of edges, and a change of a vertex label.

It is natural to try to define an edit distance for geometric graphs as well. Simply considering a geometric graph as a graph whose vertices are labeled with their coordinates is not sufficient, as the cost of inserting and deleting an edge should also be dependent on the length of the edge. This leads to the following operations: Insertions and deletions of vertices, translations of vertices, and insertions and deletions of edges. However, it is difficult to give bounds on the length of an edit sequence: vertices can move several times to make insertions and deletions cheaper. We give some examples in the following section.

This leads us to define another graph distance function in Section 3. It is not an edit distance, and so we need to prove explicitly that it is a metric. We also give an integer linear programming formulation that allows us to compute our distance for small graphs with an ILP solver. Unfortunately, we do not know how to compute or even approximate our graph distance for larger graphs. In fact, we give two reductions from NP-hard problems, but both result in non-"practical" instances of the problem.

We therefore turn our attention to a heuristic. We define the *landmark distance* of two geometric graphs, and present pattern retrieval experiments on a database of 25056 graphs created from glyphs of Chinese characters. The idea of the landmark distance is to represent a geometric graph on n vertices as a set of n points in $\mathbb{R}^6$. The landmark distance between two geometric graphs is then the Earth Mover's distance between the point sets representing the graphs.

2   Why Not an Edit Distance?

A graph edit distance for geometric graphs needs to support at least the following primitive operations: insertion and deletion of vertices and edges, and translation of vertices.

Throughout this paper, we assume that geometric graphs are given with absolute coordinates in the plane. In other words, a translated, rotated, or scaled copy of a graph drawing is not necessarily similar to the original graph drawing. If a similarity measure that is invariant under some motions is needed, this can always be achieved by minimizing over all motions of interest.

Let us first assume that insertions and deletions of edges have cost identical to the length of the edge, and the cost of a vertex translation is the distance by which the vertex was translated. This leads to very artificial edit sequences, where vertices "hop around" several times. For instance, in the example shown in Figure 1, the cheapest edit sequence transforming the graph (a) into graph (b) (where both graphs are meant to be at the same location in the plane) is to translate x first close to a, to insert edge xa, then to translate x close to b, to insert edge xb and so on. The costs are roughly $d(x, a) + d(a, g) + d(x, g)$, which is much less than the sum of the lengths of the edges. In addition there is actually no optimal edit sequence—to reach the optimum, one would have to insert a vertex *on top* of an existing vertex—so the graph distance would have to be defined as the infimum over all edit sequences.

To fix this problem, we can change the cost of a vertex translation to account for the change in length of all the incident edges. Unfortunately, there is then no bound on the length of an optimal edit sequence, and we again have to define the graph distance as the infimum over all edit sequences. For instance, in the simple example of Figure 1 (c), where we want to measure the distance between the graph consisting of the single edge uu' and the graph consisting of the single

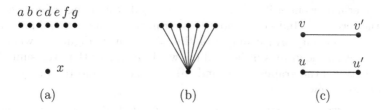

Fig. 1. Bad examples for edit distances

edge vv', an optimal edit sequence would be to move both vertices alternatingly by an infinitesimal amount so as to minimize the change in edge length incurred.

3   Geometric Graph Distance

Our distance is inspired by the graph edit distance in that it is based on the primitive operations above. However, we do not allow arbitrary sequences of the operations. Instead the edit operations must be performed in this order:

1. Edge deletions
2. Vertex deletions
3. Vertex translations
4. Vertex insertions
5. Edge insertions

Only isolated vertices can be inserted or deleted, and this operation is free. Insertion or deletion of an edge e of length $|e|$ has cost $C_e|e|$. Translating a vertex has cost C_v times the distance of the translation plus, for each incident edge, C_e times the *change in the length* of the edge.

Note that *we measure the change in edge length between the two graphs*, and not for individual operations!

The ordering of the five types of operations is really the only ordering suitable for this definition: It has to be symmetric and so vertex translations have to appear in the middle; since only isolated vertices can be deleted, edges have to be deleted before vertices; and allowing deletions after insertions is never useful.

A different way of looking at the distance is the following: Let $A = (V_A, E_A)$ and $B = (V_B, E_B)$ be the two graphs. We chose a subset $V^* \subseteq V_A$ and an injection $\sigma : V^* \to V_B$, and associate with them the following cost:

(i) we pay $C_e|uv|$ for any edge $(u, v) \in E_A$ such that not both $u, v \in V^*$ or $\sigma u \sigma v \notin E_B$;

(ii) we pay $C_e|uv|$ for any edge $(u, v) \in E_B$ such that not both $u, v \in \sigma V^*$ or $\sigma^{-1}u\sigma^{-1}v \notin E_A$;

(iii) we pay $C_e||uv| - |\sigma u \sigma v||$ for each edge $(u, v) \in E_A$ with both $u, v \in V^*$ and $\sigma u \sigma v \in E_B$;

(iv) we pay $C_v|u\sigma u|$ for each $u \in V^*$.

The geometric graph distance $ggd(A, B)$ is the minimum of this cost over all choices of V^* and σ.

The reader may wonder if it is necessary to include the change of edge length, that is, the term (iii). Indeed, without this term, the distance would not satisfy the triangle inequality. An example of this is shown in Figure 2, where the distance between the graphs in (a) and (d) would be larger than the sum of the distances between the graphs in (a) and (b), (b) and (c), and (c) and (d).

We can now show:

Theorem 1. *The geometric graph distance defined above is a metric on the set of geometric graphs without isolated vertices for positive C_v and C_e.*

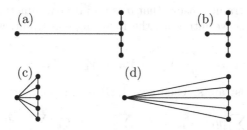

Fig. 2. The triangle inequality does not hold without accounting for the change in edge length

Proof. Let A, B and C be geometric graphs without isolated vertices. It follows from the definition that $ggd(A, B) \geq 0$ and that $ggd(A, B) = ggd(B, A)$. We clearly have $ggd(A, A) = 0$. Assume now that $ggd(A, B) = 0$. Then we must have $V^* = V_A$ and $\sigma V^* = V_B$, and for each $u \in V_A$ we must have $u = \sigma u$, and so $A = B$.

It remains to show the triangle inequality. Let $V_1^* \subset V_A$ and $\sigma_1 : V_1^* \to V_B$ be as in the definition of $ggd(A, B)$, and let $V_2^* \subset V_B$ and $\sigma_2 : V_2^* \to V_C$ be as in the definition of $ggd(B, C)$.

Let $V^* := \sigma_1^{-1}(\sigma_1 V_1^* \cap V_2^*)$, and let $\sigma : V^* \to V_C$ be defined as $\sigma u = \sigma_2 \sigma_1 u$. We evaluate the cost of V^* and σ. The cost of terms (iii) and (iv) is bounded by the sum of the corresponding terms in $ggd(A, B)$ and $ggd(B, C)$. Edges $(u, v) \in E_A$ such that not both of $u, v \in V_1^*$ are accounted for in $ggd(A, B)$, and edges $(u, v) \in E_C$ such that not both of $u, v \in \sigma_2 V_2^*$ are accounted for in $ggd(B, C)$.

For an edge $(u, v) \in E_A$ with $u \in V_1^* \setminus V^*$ and $v \in V_1^*$, we have that $\sigma_1 u \notin V_2^*$. It follows that $(\sigma_1 u, \sigma_1 v) \in E_B$ falls into case (i) for $ggd(B, C)$, and so the cost of deleting this edge is bounded by the change in edge length in $ggd(A, B)$ and the deletion cost in $ggd(B, C)$.

A symmetric argument holds for edges $(u, v) \in E_C$ with $u \in \sigma_2 V_2^* \setminus \sigma V^*$ and $v \in \sigma_2 V_2^*$. □

The geometric graph distance can be formulated as an ILP as follows: For each pair of vertices $u \in V_A$ and $v \in V_B$, we introduce a binary variable V_{uv}. We will have $V_{uv} = 1$ if $u \in V^*$ and $\sigma u = v$, and 0 otherwise. Similarly, for each pair of edges $e \in E_A$ and $e' \in E_B$, we introduce a binary variable $E_{ee'}$. This variable will be 1 if and only if both endpoints u, v of e are in V^*, and $\sigma u, \sigma v$ are the endpoints of e'.

The constraints are as follows: For each $u \in V_A$, we have

$$\sum_{v \in V_B} V_{uv} \leq 1,$$

and for each $v \in V_B$ we have

$$\sum_{u \in V_A} V_{uv} \leq 1$$

Together, these constraints ensure that σ is an injective function. Furthermore, for each variable $E_{ee'}$ we introduce the following constraint, where $e = (u,v)$ and $e' = (u',v')$:

$$E_{ee'} \leq \frac{1}{2} \cdot (V_{uu'} + V_{vv'} + V_{uv'} + V_{vu'})$$

These are all the constraints. Then $ggd(A,B)$ is the minimum of the function:

$$C_v \sum_{u,v} |uv| \cdot V_{uv} + C_e \sum_{e \in E_A} |e| + C_e \sum_{e' \in E_B} |e'|$$
$$-C_e \sum_{e,e'} \Big(|e| + |e'| - \big||e| - |e'|\big| \Big) \cdot E_{ee'}.$$

The first term is the cost of translating vertices (the remaining vertices are deleted and inserted, which is free). The second and third terms are the total cost of deleting all edges of A, and inserting all edges of B. The only way to avoid deleting and inserting an edge is by moving it from A to B. In that case, $E_{ee'} = 1$, and the cost is the difference in edge length, which is modeled by the fourth term, that is, we subtract $|e|$ and $|e'|$ and add $\big||e| - |e'|\big|$ instead.

We were able to compute distances between some small graphs with less than ten vertices using the ILP solver of the Gnu linear programming toolkit (GLPK). Unfortunately, we do not know how to compute the distance for larger graphs, and indeed this is a hard problem.

Theorem 2. *The problem of computing the geometric graph distance as defined above is NP-hard.*

We give two proofs. However, the first proof assumes that graphs can be non-planar, the second proof assumes that we can choose $C_v \ll C_e$. Thus, both results are for non-"practical" instances of the problem.

Proof (assuming graphs can be non-planar). We reduce 3DMATCHING to the problem. Remember that the input for 3DMATCHING consists of three disjoint copies X, Y, Z of $\{1, \ldots, n\}$, and a set T of m triples from $X \times Y \times Z$. The problem is to determine whether there is a subset S of T of exactly n triples that cover X, Y, Z completely.

We can reduce 3DMATCHING to graph matching as follows: Pick four points t_0, x_0, y_0, z_0 in the plane as the corners of a unit square. Pick n points $x_1, x_2, \ldots, x_n$ very close to x_0, and do the same for $y_1, \ldots, y_n$ and $z_1, \ldots, z_n$. Finally, pick m points $t_1, \ldots, t_m$ very close to t_0.

The graph G_0 consists of $4n$ vertices and edges, forming the n disjoint loops $t_i x_i y_i z_i t_i$.

The graph G_1 consists of $3n+m$ vertices as follows: Let triple i in T be (j,k,l). Then G_1 includes the loop $t_i x_j y_k z_l t_i$. Let $M \leq 4m$ be the number of edges of G_1.

Clearly we have to insert $M - 4n$ edges to go from G_0 to G_1. If there is a subset S of n triples covering X, Y, Z, then we can map the loops of G_0 to the loops corresponding to the triples in S, and the total edit cost is very close to $M - 4n$. On the other hand, if there is no such subset, then at least one edge has to be deleted from G_0, and the total cost is at least close to $M - 4n + 1$. And so 3DMATCHING can be decided by computing the geometric graph distance for a highly non-planar graph. □

Proof (assuming C_e and C_v are part of the input and $C_v \ll C_e$ is allowed). Consider the decision version of our problem, that is, given two geometric graphs G_0 and G_1 and three positive real constants C_v, C_e and K, is the geometric graph edit distance between G_0 and G_1 less than K?

The reduction is done by using the well-known Hamiltonian path problem restricted to grid graphs, which is known to be NP-complete [4]. An instance of the Hamiltonian path problem is a grid graph G with n vertices and m edges.

The reduction is as follows. Set $C_e = 1$, $C_v = 1/(n \cdot \text{diam}(G_0 \cup G_1))$ and $K = m - n + 2$. Let G_0 be a grid path on n vertices, where each edge has length 1, and let G_1 be the grid graph defined by G on n vertices. We claim that G_1 contains a Hamiltonian path if and only if the geometric graph edit distance between G_0 and G_1 is at most K.

Consider an optimal transformation of G_0. We will argue that the optimal solution will always translate a maximum number of edges in G_0, and that all the $n - 1$ edges in G_0 can only be used if there is a Hamiltonian path in G_1.

Since G_1 is a grid graph all edges have length 1, thus, translating an edge in G_0 does not change its length. Due to the choice of C_v and C_e this implies that moving an existing edge is always cheaper than inserting a new edge.

Consider the cost of an optimal solution. Since G_0 only contains $n - 1$ edges and G_1 contains m edges one has to insert at least $m - n + 1$ new edges, with a total cost of at least $m - n + 1$. The cost of translating the vertices of G_0 does not exceed 1 in total. The $n - 1$ edges of G_0 can only be reused if G_1 contains a Hamiltonian path. Otherwise, for each deleted and inserted edge the additional cost is 2. Thus, if all the edges in G_0 can be used then the total cost is at most $K = m - n + 2$ and there exists a Hamiltonian path in G, otherwise not. □

4 Landmark Distance

Since we do not know how to compute our geometric graph distance efficiently, we turned to a heuristic distance measure, the *landmark distance*. The idea of the landmark distance is to designate a few vertices of the graph as its *landmarks* and to represent the vertices of the geometric graph by their distances to the landmarks.

Formally, let $G = (V, E)$ be a geometric graph, and let K_G be the complete graph on the vertex set V. An edge (u, v) of K_G is given a weight as follows: if $(u, v) \in E$, then its weight is $|uv|$, otherwise its weight is $p \cdot |uv|$, where $p > 1$ is a fixed *penalty* value. The distance $d_G(u, w)$ between a vertex $u \in V$ and a landmark $w \in V$ is then defined as the length of the shortest path between u and v in K_G. After testing different values for the penalty p we chose $p = 1.6$ for our experiments.

Let $L = w_1, \ldots, w_k$ be k vertices of G called *landmarks* (in our experiments $k = 4$). For a vertex $v \in V$ with coordinates (x, y), we define the *L-vector* L_v of v

as the $(k+2)$-dimensional vector containing the distances of v to the k landmarks as well as the coordinates of v:

$$L_v = (d_G(v, w_1), \ldots, d_G(v, w_k), x, y).$$

The *landmark representation* $R(G)$ of the graph G is now simply the set of L-vectors of all vertices: $R(G) := \{L_v \mid v \in V\}$.

We define the *landmark distance* $d_L(G_0, G_1)$ between two geometric graphs G_0 and G_1 (both with given landmarks) as the *normalized earth mover's distance* between the point sets $R(G_0)$ and $R(G_1)$.

The normalized Earth Mover's Distance (nEMD) is a distance measure defined on weighted point sets [8]. Let P and Q be two weighted point sets with $\sum_{p \in P} w(p) = \sum_{q \in Q} w(q) = 1$, where $w(u) \geq 0$ is the weight of a point u.

Intuitively, a point $p \in P$ can be seen as a pile of earth of size $w(p)$, while a point $q \in Q$ can be seen as a hole in the ground of size $w(q)$. The Earth Mover's Distance is then defined as the *cheapest* way to move the earth into the holes, where piles can be split and the cost of transporting s units of earth from a pile to a hole is equal to s times the distance between the pile and the hole.

Formally, nEMD can be defined by the following linear program. Let f_{pq} denote the flow of earth from $p \in P$ to $q \in Q$. Then

$$\mathrm{nEMD}(P, Q) = \min \sum_{p \in P} \sum_{q \in Q} d(p, q) f_{pq},$$

where $d(p, q)$ is the distance between p and q in some underlying metric, subject to the constraints:

$$\begin{aligned}
\forall p \in P \,\, \forall q \in Q & \quad f_{pq} \geq 0 \\
\forall p \in P & \quad \sum_{q \in Q} f_{pq} = w(p) \\
\forall q \in Q & \quad \sum_{p \in P} f_{pq} = w(q)
\end{aligned}$$

In our definition of the landmark distance, we give all points of $R(G)$ equal weight, and we use the ℓ_1-metric[1] in $\mathbb{R}^{k+2}$ as the underlying distance for the nEMD.

The landmark distance is "nearly" a metric:

Theorem 3. *The landmark distance on graphs with given landmarks has the following properties:*
(i) $d_L(G_0, G_0) = 0$ *and* $d_L(G_0, G_1) \geq 0$,
(ii) $d_L(G_0, G_1) = d_L(G_1, G_0)$,
(iii) $d_L(G_0, G_2) \leq d_L(G_0, G_1) + d_L(G_1, G_2)$.

Proof. All the properties in the theorem follow directly from the fact that the normalized EMD is a metric [8]. □

However, it is easy to see that $d_L(G_0, G_1) = 0$ does not imply $G_0 = G_1$.

---

[1] Whether the ℓ_1- or ℓ_2-metric is used makes little difference in the experiments. The advantage of the ℓ_1-metric is that it does not need floating point arithmetic.

5 Experimental Results

We performed pattern retrieval experiments on a database of graphs generated from Chinese character glyphs using the landmark distance. The motivation here was not to build a Chinese character recognition system—a lot of research has been done in this area, and it is not our intention to compete with these finely tuned results of years of research. We turned to Chinese characters because we found them to be a source of a large number of geometric graphs with known semantics.

We selected six different fonts, including two Korean, two Japanese, and two Chinese fonts. We picked a set of 4176 Chinese characters (or, more precisely, Unicode code points) that exist in all six fonts, and generated graphs for each of these $6 \times 4176 = 25056$ glyphs as follows: We draw the glyph and compute its medial axis. We prune away small features, and then simplify each chain of degree-two vertices using the Imai-Iri algorithm [3]. Figure 3 shows three examples of glyphs and the corresponding graphs. For the distance computations, all graphs were then linearly scaled to fill a unit square (not shown in the figure).

As explained, our database consists of 4176 sets of six graphs that represent the same abstract Chinese character. In principle, the graphs representing the same character should be similar, and we have assumed this as the ground truth of our database. In reality, there can be considerable variation between graphs generated from glyphs with the same semantics, due to a different glyph style, or actual variations in character shape.

The experiment. We selected one of the six fonts (the Korean *dotum* font) as our reference font, and built a database of 4176 *model* graphs from this font.

We then considered each of the remaining 20880 graphs as a *pattern*, and computed its distance to each model, using the EMD implementation by Rubner et al. [8]. The models were then sorted in order of distance from the pattern. Ideally, the nearest model should be a graph for the same Chinese character, so

Fig. 3. Three glyphs and generated graphs for the Chinese character U+6f11 "slowly flowing water"

we determined the index of the occurrence of the same Chinese character in the ranked list of models.

As a control experiment, we first tried this experiment using the Hausdorff distance of the graph vertices. The results are given in the first line of Table 1: For 31.8% of the 20880 patterns the best (most similar) model belonged to the same Chinese character, for 50.9% of the patterns, one of the first (most similar) ten models belongs to the same character.

Table 1. A summary of our experimental studies of Chinese character retrieval

Graph distance		Index in ranked model list							
		1	2	3	4	5	6–10	11–20	21–200
Hausdorff-distance	# patterns	6647	1272	705	496	387	1115	1257	4283
of vertices	accum %	31.8	37.9	41.3	43.7	45.5	50.9	56.9	77.4
EMD	# patterns	16844	1646	519	258	184	286	193	200
of vertices	accum %	80.7	88.6	91.0	92.3	93.2	94.5	95.4	96.4
Landmark	# patterns	17814	1298	454	260	152	317	195	221
distance	accum %	85.3	91.5	93.7	95.0	95.7	97.2	98.1	99.2

It is clear that the Hausdorff-distance does not capture the problem well enough (even though we ignore edge information here, we do not believe that using the edge information would actually help).

The Earth Mover's distance, however, does much better, even when we ignore all information about the edges of the graph. The second line of our table shows this experiment, where we ranked the models by nEMD of the graph vertices only (that is, a graph drawing is interpreted as a set of points in the plane). The results are surprisingly good considering that the edge information of the graphs is not used at all: for 94.5% of the patterns, the correct Chinese character is found in the top ten.

Landmark selection. We fixed four landmarks in each model graph, by choosing the four vertices that are extreme in the four diagonal directions, that is, the vertices extreme in the directions $(1, 1)$, $(1, -1)$, $(-1, -1)$, and $(-1, 1)$.

Ideally, we would apply the same approach to the pattern graph, but very often this does not find the "right" vertex, as for example in Figure 4. In such a case, the distance between the graphs is far larger than it should be—selecting the right landmarks in each graph is critical to the success of the landmark distance.

We actually try all plausible choices of landmarks for a given pattern, and then use the landmarks that result in the smallest distance to a given model (so we could select different landmarks for different models). We consider a vertex to be a plausible landmark if it is the corner of a quadrant that contains no vertices.

Speeding up the computation. It turns out that computing the EMD is a rather expensive operation, and computing 4176 × 20880 EMD distances in every experiment is very time-consuming. For instance, for U+6f01, whose graph has

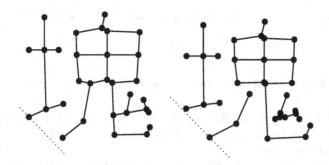

Fig. 4. Extremal vertices are not robust

33 vertices, computing the EMD for each model in the entire model database takes 31 seconds.

We therefore used a heuristic to speed up the computation: Instead of the EMD, we compute a *simplified landmark distance* $d'_L(G_0, G_1)$ which is defined as follows: For each point u in $R(G_0)$, find the nearest point $nn(u) \in R(G_1)$. Then

$$d'_L(G_0, G_1) := \sum_{u \in R(G_0)} ||u - nn(u)||_1.$$

Note that this "distance" is not symmetric—we compute it with G_0 being the pattern, G_1 being one of the models. It is also by itself not a very good distance measure, and ranks the models much more poorly than the EMD.

So what we do instead is to first rank all models using the simplified landmark distance. We then look at the top 200 models (that is, the 200 models with the smallest simplified landmark distance), and recompute the distance from the pattern to these 200 models using the landmark distance (that is, using the EMD computation).

The heuristic greatly speeds up the computation: Comparing the character U+610f mentioned above against the entire model database now takes only 1.8 seconds, a speed-up of around 17. This speed-up comes at nearly no cost—the quality of the recognition is nearly identical to using the full EMD computation.

The result. The result of the our landmark distance experiment is given in the last line of Table 1. For 85.3% of the patterns our approach finds the correct Chinese character in the top position, and in 97.2% of the cases it is in the top ten. This is a small but real improvement over the EMD for this dataset.

6   Conclusions and Open Problems

We believe that we have only scratched the surface of this problem. We gave some evidence that our geometric graph distance is hard to compute, but we lack a formal proof that it is NP-hard for planar graphs with realistic values of C_e and C_v.

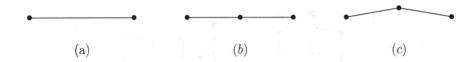

(a)   (b)   (c)

Fig. 5. Our distance does not handle bending edges well

Is there a PTAS for our distance, or at least a constant-factor approximation?

If we do not want insertions and deletions of vertices to be free, can we incorporate that into our distance?

Finally, a major problem of our metric is that it does not allow us to cheaply "bend" an edge, that is, to insert a degree-two vertex into an edge and then to move that vertex slightly. The three graphs shown in Figure 5 are all quite similar, but our distance is large between (a) and (b) and small between (b) and (c). Can we define a metric that allows this operation while still being computable at least through integer linear programming?[2]

As for our experiments, perhaps the graphs in our database are "too easy" in the sense that even ignoring the edge information, the EMD does pretty well. It would be interesting to try harder graph sets, or to consider arbitrary rigid motions.

References

1. Alt, H., Guibas, L.J.: Discrete geometric shapes: Matching, interpolation, and approximation. In: Handbook of Computational Geometry, pp. 121–153. Elsevier B.V., Amsterdam (2000)
2. Fu, K.S.: Syntactic Pattern Recognition and Applications. Prentice-Hall, Englewood Cliffs (1982)
3. Imai, H., Iri, M.: Polygonal approximations of a curve - formulations and algorithms. In: Toussaint, G.T. (ed.) Computational Morphology, pp. 71–86. Elsevier B.V., Amsterdam (1988)
4. Itai, A., Papadimitriou, C.H., Szwarcfiter, J.L.: Hamilton paths in grid graphs. SIAM Journal on Computing 11(4), 676–686 (1982)
5. Justice, D., Hero, A.: A binary linear programming formulation of the graph edit distance. IEEE Transactions on Pattern Analysis and Machine Intelligence 28(8), 1200–1214 (2006)
6. Lucas, S., Vidal, E., Amiri, A., Hanlon, S., Amengual, J.C.: A comparison of syntactic and statistical techniques for off-line OCR. In: 2nd International Colloquium Grammatical Inference and Applications, pp. 168–179. Springer, Heidelberg (1994)
7. Pavlidis, T.: Structural Pattern Recognition. Springer, New York (1977)
8. Rubner, Y., Tomasi, C., Guibas, L.J.: The earth mover's distance as a metric for image retrieval. International Journal of Computer Vision 40(2) (2000), http://robotics.stanford.edu/~rubner

---

[2] A natural approach would be to allow for free insertion of vertices on edges before all other operations, and deletion of degree-two vertices where the incident edges form an angle of 180° after all other operations. This, however, defines a measure that is not a metric.

A Heuristic Strong Connectivity Algorithm for Large Graphs

Adan Cosgaya-Lozano* and Norbert Zeh**

Faculty of Computer Science, Dalhousie University, Halifax, Nova Scotia, Canada
{acosgaya,nzeh}@cs.dal.ca

Abstract. We present a contraction-based algorithm for computing the strongly connected components of large graphs. While the worst-case complexity of the algorithm can be terrible (essentially the cost of running a DFS-based internal-memory algorithm on the entire graph), our experiments confirm that the algorithm performs remarkably well in practice. The strongest competitor is the algorithm by Sibeyn et al. [17], which is based on a semi-external DFS algorithm developed in the same paper. Our algorithm substantially outperforms the algorithm of [17] on most of the graphs used in our experiments and never performs worse. It thus demonstrates that graph contraction, which is the most important technique for solving connectivity problems on *undirected* graphs I/O-efficiently, can be used to solve such problems also on *directed* graphs, at least as a heuristic.

1 Introduction

Driven by the availability of massive amounts of data in a wide range of application areas, tremendous efforts have been made over the last two decades to develop algorithms that can process data sets beyond the size of a computer's main memory efficiently. Traditional algorithms perform poorly on such inputs, as most of these algorithms exhibit little or no access locality and cause a disk access for almost every computation step, which results in a slow-down by a factor of about 10^6 compared to processing the data in memory. *I/O-efficient algorithms*, on the other hand, are designed to access data sequentially or in large blocks, in order to reduce the number of disk accesses to the point where massive data sets can be processed efficiently.

In the algorithms community, much work has focused on developing provably I/O-efficient algorithms for a wide range of fundamental problems, particularly for geometric and graph problems. See [4, 19] for surveys. For graph problems,

---

* Supported by scholarships funded by the National Council for Science and Technology of Mexico, the Natural Sciences and Engineering Research Council of Canada, and Dalhousie University.
** Supported in part by the Natural Sciences and Engineering Research Council of Canada, the Canadian Foundation for Innovation, and the Canada Research Chairs programme.

J. Vahrenhold (Ed.): SEA 2009, LNCS 5526, pp. 113–124, 2009.

much progress has been made on *undirected* graphs and special graph classes. In contrast, no provably efficient algorithms are known for general *directed* graphs. This lack of theoretical results motivates the study of *heuristic* techniques for processing directed graphs I/O-efficiently. The most successful effort so far is the depth-first search (DFS) algorithm by Sibeyn et al. [17], which is a *semi-external* algorithm; that is, it can process the edges of the graph I/O-efficiently if the vertices fit in memory. Since DFS is a central building block used in many classical graph algorithms, the algorithm of [17] provides a general tool for solving problems on directed graphs efficiently if the vertices fit in memory. If, on the other hand, the size of the vertex set exceeds the memory size, the performance of the algorithm deteriorates to that of an internal-memory DFS algorithm. Sibeyn et al. demonstrated the effectiveness of their approach in the semi-external case by using it to compute the strongly connected components of a variety of directed graphs. A directed graph is *strongly connected* if, for every vertex pair (x, y), there exists a directed path from x to y. The *strongly connected components* (SCC's) of a graph are its maximal strongly connected subgraphs (SCSG's).

While Sibeyn et al. used strong connectivity merely as an example to demonstrate the efficiency of their DFS procedure, we propose a heuristic specifically for computing SCC's in this paper. The aim is (a) to achieve a better performance than [17] on graphs whose vertex sets fit in memory and (b) to process graphs whose vertex sets do not fit in memory, which, our experiments confirm, the algorithm of [17] cannot do in a reasonable amount of time. Our algorithm achieves both goals on a variety of input graphs, outperforming the semi-external algorithm by a factor of 2–4 on most of the test graphs whose vertices fit in memory, and being able to efficiently process graphs well beyond the reach of the semi-external algorithm.

Our algorithm is based on graph contraction: it identifies and contracts strongly connected subgraphs until the graph fits in memory, and then computes the SCC's in internal memory. Thus, given its good performance, our algorithm demonstrates that, at least as a heuristic, graph contraction is useful for solving connectivity problems on directed graphs. This is interesting because this technique is the most important tool for solving connectivity problems on *undirected* graphs I/O-efficiently, both in theory and in practice.

The remainder of this paper is organized as follows. Section 2 reviews previous work on implementing I/O-efficient graph algorithms; Section 3 describes the algorithm; Section 4 discusses implementation details; Section 5 discusses experimental results; and Section 6 offers concluding remarks.

2   Previous Work

While much theoretical work has focused on developing I/O-efficient graph algorithms, much less is known about their practical efficiency. The main reason is their algorithmic complexity. Most of these algorithms build on a number of widely used primitives—list ranking, Euler tour construction, etc.—in addition to internal-memory algorithms that are used to process the parts of the graph loaded into memory. No good implementations of these primitives are publicly available, which makes implementing any I/O-efficient graph algorithm a formidable task, as it requires the implementation not only of the actual algorithm but also of a number of more elementary, yet non-trivial, building blocks.

In spite of these challenges, a number of experimental results have been obtained for *undirected* graphs. Dementiev et al. [10] provided a carefully engineered implementation of a minimum spanning tree (MST) algorithm based on ideas from [18]. Their algorithm is theoretically inferior to the MST algorithms of [1,5,8] but performs extremely well in practice. Ajwani et al. [2,3] provided implementations of the undirected breadth-first search algorithm by Mehlhorn and Meyer [14] and obtained excellent results on a wide range of graph classes. The semi-external DFS algorithm by Sibeyn et al. [17] seems to be the only work that focused specifically on solving fundamental problems on directed graphs.

Other related work includes a large body of work on preprocessing large graphs, particularly road networks, for fast shortest path queries. The most recent results in this area include [6,12,13,15].

3   A Contraction-Based Strong Connectivity Algorithm

This section describes a simple contraction-based SCC algorithm referred to as EM-SCC throughout this paper. Section 4 discusses its implementation.

The algorithm consists of two phases: a *preprocessing phase* and a *contraction phase*. The contraction phase looks for SCSG's in the input graph G and contracts each into a single vertex, thereby reducing the size of G without altering its connectivity. This process continues until the graph fits in memory, at which point the algorithm loads it into memory and computes its SCC's using an internal-memory algorithm. In this sense, EM-SCC resembles the connectivity algorithm for undirected graphs by Chiang et al. [8]. In the undirected case, however, the graph is guaranteed to fit in memory after a logarithmic number of contraction steps, while, in the directed case, the algorithm succeeds only if each round finds sufficiently many and large SCSG's to contract.

The contraction phase searches for SCSG's by loading memory-sized subgraphs of G into memory and computing their SCC's. The preprocessing phase tries to group the vertices and edges of G so that the chance of finding non-trivial SCC's in these subgraphs is maximized.

Next we discuss these two phases in detail. Throughout this discussion, we assume the input graph is connected. It is not hard, however, to extend the algorithm to disconnected graphs with little or no impact on its performance.

3.1   Preprocessing Phase

The preprocessing phase of EM-SCC is conceptually simple. It arranges the vertices of G in a list V_0 in the order of their first occurrences along an Euler tour of a spanning tree T of G. It stores the edges in a list E_0, which is the concatenation of "one-sided" adjacency lists of the vertices in V_0 arranged in the same order as the corresponding vertices in V_0. The adjacency lists are one-sided in the sense that an edge xy is stored in the adjacency list E_x of x if $x > y$, and in E_y otherwise; vertices are compared by their positions in V_0.

The contraction phase discussed in Section 3.2 below sweeps the two lists V_0 and E_0 in tandem and processes maximal groups of consecutive vertices in V_0 that induce memory-sized subgraphs of G. Intuitively, the ordering of the vertices in V_0 produced by the preprocessing phase should ensure that the processed subgraphs are connected or have few connected components (in the undirected sense). Assuming sufficiently random edge directions and sufficiently many non-tree edges, this should lead to non-trivial SCC's in the processed subgraphs.

To compute lists V_0 and E_0, the algorithm has to compute the tree T, its Euler tour, and a ranking of the Euler tour. To compute the spanning tree, we use the MST algorithm by Dementiev et al. [10] (setting all edge weights to 1). Sorting and scanning the edge set of T suffices to compute an Euler tour of T. To rank this tour, we use the list ranking algorithm by Sibeyn [18].

Give the ranked tour, the algorithm finds the first occurrence of every vertex of G in the tour by sorting and scanning the node list of the tour, numbers the vertices of G in the order of these occurrences, and places them into V_0 in order. The edge list E_0 is constructed by sorting and scanning the edges of G three times: twice to label each edge with the numbers of its endpoints, and once more to arrange the edges in the order described above.

Before trying the simple preprocessing strategy discussed here, we experimented with a more sophisticated hierarchical clustering approach, which clustered vertices based on their degrees. The contraction phase then considered (contracted versions of) clusters of increasing size and decreasing density in its search for SCSG's. The intuition was that, assuming the edge directions are sufficiently random, dense graphs are more likely to contain large SCSG's, so that this degree clustering approach should lead to a rapid reduction of the graph size early on in the contraction phase. The cost of computing this clustering, however, was prohibitive, and the speed-up of the contraction phase compared to the simple preprocessing described here was insignificant.

3.2   Contraction Phase

The contraction phase of EM-SCC proceeds in *rounds*. Each round produces a more compressed version of G from the previous version by identifying SCSG's and contracting them. Let $G = G_0, G_1, \ldots, G_r$ be the sequence of graphs this produces; that is, round i produces graph G_i from graph G_{i-1}. The algorithm represents each graph G_i using two lists V_i and E_i whose structure is identical to that of V_0 and E_0 described in the previous section.

The ith round partitions V_{i-1} into subsets $V_1', V_2', \ldots, V_k'$ of consecutive vertices such that the graphs $G_j' = G_{i-1}[V_j']$ they induce fit in memory. The algorithm loads these graphs into memory, one at a time, and identifies and contracts their SCC's. In more detail, the ith round scans V_{i-1} and E_{i-1} in tandem, collecting the vertices and edges in the current graph G_j' in memory. Let x be the first vertex in V_{i-1} that belongs to G_j', and let n_j and m_j respectively be the numbers of vertices and edges currently in G'_j. To decide whether to include the next vertex y in V_{i-1} in G'_j, the algorithm scans E_y and counts the edges whose lower endpoints belong to G'_j, that is, are no less than x; let m_y be their number.

If $n_j + 1$ vertices and $m_j + m_y$ edges fit in memory, the algorithm includes y in G'_j and partitions the edges in E_y into two groups: those with lower endpoints no less than x and those with lower endpoints less than x. It loads the former into memory (thereby adding them to G'_j) and appends the latter to an initially empty edge list E''_i to be processed at the end of this round. Then the algorithm proceeds to the next vertex in V_{i-1}.

If adding m_y edges to G'_j would make it exceed the memory size, the algorithm declares vertex y to be the first vertex of G'_{j+1} and appends its entire adjacency list to E''_i. Then it computes the SCC's of G'_j in memory, contracts them, and eliminates parallel edges that result from these contractions. At the end, the vertices in G'_j are labelled with ID's of their SCC's, that is, with the ID's of their corresponding super-vertices in G_i. The algorithm writes this mapping information back to V_{i-1} and appends the sorted list of super-vertices to V_i. The edges of the contracted version of G'_j are appended to an initially empty edge list E'_i. This finishes the processing of G'_j, and the algorithm starts to construct G'_{j+1} with y as its first vertex.

The ith round ends after the last vertex in V_{i-1} has been consumed. At this point, the algorithm discards the edge list E_{i-1}, but not V_{i-1}, as the information stored in V_{i-1} is necessary to compute the final component labelling of the vertices of G. If the algorithm numbers the vertices of G_i in increasing order as it produces them, V_i already contains the sorted vertex list of G_i. To produce E_i, the endpoints of all edges in E''_i have to be replaced with their corresponding super-vertices in G_i. Since the edges in E''_i are already sorted by their upper endpoints in G_{i-1}, a single scan of V_{i-1} and E''_i suffices to replace those endpoints. To replace the lower endpoints, the algorithm sorts the edges in E''_i by these endpoints and scans V_{i-1} and E''_i again. Finally, it concatenates the resulting list with E'_i, and sorts the concatenation primarily by upper endpoints (in V_i) and secondarily by lower endpoints. A single scan now suffices to eliminate duplicates from this list, which produces the edge list E_i of G_i.

3.3 Postprocessing

Let G_r be graph produced by the last round of the contraction phase; that is, G_r fits in memory. Then the algorithm loads G_r into memory and labels every vertex in V_r with the SCC containing it. What remains to be done is to copy these labels back to the vertices in G. This is done by iteratively copying these labels from V_i to V_{i-1}, for $i = r, r - 1, \ldots, 1$.

To copy the labels from V_i to V_{i-1}, the algorithm sorts the vertices in V_{i-1} by their corresponding super-vertices in V_i. Now every vertex in V_{i-1} can be labelled with the label of its corresponding vertex in V_i using a single scan of the two sorted lists. Finally, the algorithm returns the vertices in V_{i-1} to their original order, in preparation for the next iteration.

4   Implementation Details

We implemented algorithm EM-SCC in C++ using the STXXL library [9], which provides I/O-efficient counterparts of the C++ STL containers and algorithms. In particular, we used STXXL vectors to store the vertex and edge lists of graphs, the STXXL sorting procedure to perform all sorting steps in the algorithm, and the STXXL priority queue implementation in the list ranking step of the preprocessing phase. The rest of this section discusses the most important implementation choices made in the different parts of the algorithm.

Graph representation. As already discussed, each graph G_i is represented by a vertex list V_i and an edge list E_i. In our implementation, every vertex in V_i was represented using two integers, one being its own ID, the other one the ID of the corresponding super-vertex in G_{i+1}.

Edges were represented as pairs of vertex ID's, that is, using two integers. The only exception was the addition of an extra integer to represent the edge weight up to and including the MST computation. This could have been avoided by modifying the MST implementation to compute an arbitrary spanning tree of an unweighted graph. We did not do this, as the MST computation did not account for a major part of the running time of our algorithm.

MST algorithm. We used the MST algorithm of [10] to compute the spanning tree T in the preprocessing phase. The implementation was available from [16]. That algorithm is a sweeping algorithm, which iteratively removes vertices by contracting the lightest edge incident to each processed vertex. This strategy can be implemented using an external priority queue or using an I/O-efficient bucket structure. The default implementation uses a bucket structure, as it results in slightly better performance; we had no reason to change this.

Euler tour. To compute the Euler tour of T, we used the standard strategy. We created two copies xy and yx of each spanning tree edge xy and sorted the resulting edge list by their first vertices. Then we scanned the sorted edge list and, for each pair of consecutive edges, xy_1 and xy_2, incident to the same vertex x, we made edge xy_2 the successor of edge y_1x in the Euler tour. This was easily implemented by storing the edges in an STXXL vector and using the STXXL sorting algorithm to implement the sorting step.

List ranking. The list ranking algorithm of [18] is a sweeping algorithm similar to the MST algorithm of [10]. The *down-sweep* removes vertices one by one from the list until only one vertex remains. For each removed vertex v, its two incident edges are replaced with a weighted edge between v's neighbours; the weight equals the length of the sublist between these two neighbours. The *up-sweep* re-inserts the removed vertices in the opposite order and computes the rank of each vertex v from the rank of one of the two vertices that became adjacent as a result of the removal of v in the down-sweep.

As discussed in [18], this algorithm can be implemented using a bucket structure, similar to the one used in the MST algorithm, to pass information between vertices in the two sweeps. An alternative implementation uses a priority queue and two stacks. Since our focus was not on engineering an optimal list ranking algorithm, we opted for the easier implementation using a priority queue.

Internal-memory SCC algorithm. We used the one-pass SCC algorithm described in [11] to compute the SCC's of graphs loaded into memory. The implementation of this algorithm requires two stacks to keep track of partially identified SCC's. In order to maximize the amount of memory available for processing each graph G_j', we implemented them using STXXL stacks. This limited the memory footprint of the stacks to 4 pages.

Internal-memory graph representation. To maximize the size of the subgraphs that can be processed in internal memory in each round of EM-SCC, we used a fairly compact graph representation in internal memory, consisting of two arrays: an edge array and a vertex array. The edge array contained the concatenation of adjacency lists of the vertices. Since the SCC algorithm only needed access to the out-edges of each vertex, only those edges were stored in the adjacency lists. When accessing an adjacency list, it was known to which vertex this adjacency list belonged. Hence, the tail vertex of every edge did not have to be stored explicitly. This allowed us to represent every edge using a single integer storing the head vertex of the edge.

We represented every vertex using a two-integer record in the vertex array. The first integer represented the SCC containing this vertex (once identified), the other the index of the first edge in its adjacency list in the edge array. Vertex ID's did not have to be stored explicitly, as a consecutive numbering of the vertices allowed us to use the position of a vertex in the vertex array as its ID.

Since this representation stores edges in a different order than on disk, it was necessary to sort the edges by their tails to construct the internal-memory representation of a graph G_j' from its external one. This required the use of an initial edge representation using both its endpoints during the construction of the internal-memory graph representation. Once the edges were arranged in the right order, we dropped their tail endpoints, thus halving the memory requirements of the representation. Since the ability of our algorithm to identify SCC's improves with the size of the subgraphs it can process in memory, we decided to process subgraphs that occupied all of the available main memory (minus some buffer blocks for caching used by the STXXL vectors) using the compact representation. As a result, the initial sorting step required to construct this representation used the STXXL external sorting algorithm to sort up to $2M$ data, where M denotes the memory size.

5   Experimental Results

This section discusses our experimental results, comparing the performance of EM-SCC with that of the semi-external SCC algorithm by Sibeyn et al. (called SE-SCC here). First we describe our test environment and the data sets used in our experiments. Then we discuss the results of our experiments.

5.1 Environment and Settings

All experiments were run on a PC with a 3GHz Pentium-4 processor, 1GB of RAM, and one 500GB 7200RPM IDE disk using the XFS file system. The operating system was Fedora Core 6 Linux with a vanilla 2.6.20 Linux kernel. The code was compiled using g++ 4.1.2 and optimization level –O3. All of our timing results refer to wall clock times in minutes.

Since STXXL allows the specification of the block size for data transfers between disk and memory, we experimented with different block sizes between 256KB and 8MB. A block size of 2MB resulted in the best performance, since EM-SCC accesses data in a mostly sequential fashion. This block size was used throughout our experiments. Two additional parameters control the amount of memory allocated to the LRU pager used by STXXL vectors to cache accessed blocks. The first parameter is the page size as a multiple of the block size. Data is swapped one page at a time. The other parameter is the number of pages to be cached. We set both parameters to 2, as the mostly sequential data accesses of EM-SCC did not benefit substantially from a bigger cache, but this would have left less memory for the graphs to be processed in memory.[1]

5.2 Data Sets

We tested both algorithms on synthetic graphs and real web graphs. The synthetic graphs were generated using the same data generator used by Sibeyn et al. [17]. The web graphs were produced by real web crawls of the .uk domain, the .it domain, and from data produced by a more global crawl using the Stanford WebBase crawler. They were obtained from http://webgraph.dsi.unimi.it/, and their characteristics are shown as part of Table 1. Next we give an overview of the types of synthetic graphs used in our experiments.

Random: These graphs were generated according to the $G_{n,m}$ model; that is, m edges were generated, choosing each edge endpoint uniformly at random from a set of n vertices.

Cycle: The vertices were evenly spaced on a ring, and every vertex had out-edges to its $d = m/n$ nearest neighbours.

Geometric 1D: The vertices were evenly spaced on a ring of length n. Edges were generated by choosing their tails uniformly at random. If u was chosen as the tail of an edge, vertex v was chosen to be the head of this edge with probability proportional to α^d, where $\alpha < 1$ and d is the distance between u and v. In our experiments, we chose $\alpha = 0.9$.

---

[1] Using a single disk, a block size of 2MB and a page size of two blocks is equivalent to using a block size of 4MB and a page size of one block. We chose the former option because we also tested our algorithms using two disks, in which case the blocks of each page can be assigned to different disks. Using two disks, our algorithm experienced a speed-up of about 30%. Since the semi-external algorithm wasn't able to take advantage of multiple disks, we do not discuss the timings using two disks in detail here.

Geometric 2D: The vertices were placed on a $\sqrt{n} \times \sqrt{n}$ grid wrapped around at the edges to form a torus. Edges were generated as for geometric 1D graphs, but d was chosen to be the Manhattan distance between u and v in the grid. Here we chose $\alpha = 0.8$.

Out-star: Given a star degree s, this graph was generated in $\lfloor m/s \rfloor$ rounds. In each round, a tail vertex and s head vertices were chosen uniformly at random. Then edges were added from the tail to the chosen head vertices. We chose $s = 1000$ in our experiments.

In-out-star: This construction was similar to the out-star construction, but half of the rounds directed the generated edges towards the centre of the star. Again, we chose $s = 1000$.

Simple web: This construction started with a small complete subgraph and added new vertices by connecting them to the current graph. Afterwards, a small fraction (5% in our case) of random edges were added.

5.3   EM-SCC vs. SE-SCC

Table 1 shows the running times of EM-SCC and SE-SCC on different synthetic inputs and on the three web graphs. For the synthetic graphs with 2^{25} vertices, EM-SCC outperformed SE-SCC by a factor between 2 and 4. The only exception were random graphs and geometric 2D graphs, where SE-SCC took only slightly longer than EM-SCC. For the two smaller web graphs, EM-SCC outperformed SE-SCC by a factor between 3 and 4. As can be observed, the performance of SE-SCC depends strongly on the structure of the input graph, whereas (surprisingly) the performance of EM-SCC is much more immune to these variations. Sibeyn et al. characterized geometric 1D graphs as being among the hardest inputs for their algorithm, and geometric 2D and random graphs as being among the easiest inputs. This is in line with our observations. On the other hand, cycle graphs were mentioned as easy inputs in [17], while this was the synthetic input that took SE-SCC the longest to process in our experiments.

The remaining inputs had at least 2^{26} vertices and were beyond the reach of SE-SCC on our hardware, as the vertex set no longer fit in memory. (See [7] for a discussion of the graph representation used by SE-SCC and approximate vertex numbers it can process without using virtual memory.) We ran SE-SCC on the smallest of these graphs (with 2^{26} vertices and 2^{29} edges), using virtual memory, and terminated each of these test runs after 12h without SE-SCC having produced any result. Since the performance of SE-SCC on the semi-external instances of random and geometric 2D graphs was comparable to that of EM-SCC, we expected that SE-SCC would have the least difficulties to process larger instances of these graph classes, and we let the experiments on these inputs run for 24h. Again, SE-SCC did not finish within this amount of time.

In contrast, EM-SCC was able to process most of the test graphs in under two hours, while none took more than 2 1/2 hours. The exceptions were the out-star graphs and the sparsest of the in-out-star and simple web graphs. The next section discusses possible reasons why EM-SCC could not process these inputs, which sheds some light on its limitations.

Table 1. Experimental results on synthetic data and real web graphs. Dashes indicate inputs that could not be processed by the algorithm. For geometric 1D and 2D graphs, m_r denotes the number of edges requested to be generated. Since the data generator filters duplicate edges for these two graph types, the actual number of edges, m, is less than m_r. The ratio m/n in the table reflects this. Notes: (1) experiment terminated after 12h; (2) experiment terminated after 24h; (3) no further compression after a small number of initial contraction rounds, but graph still beyond memory size.

Cycle						Geometric 1D					
n	m	m/n	EM	SE	SCC's	n	m_r	m/n	EM	SE	SCC's
2^{25}	2^{29}	16	58	208	1	2^{25}	2^{29}	13.2	51	161	11
2^{26}	2^{29}	8	71	—[1]	1	2^{26}	2^{29}	7.2	65	—[1]	45084
2^{27}	2^{29}	4	94	—	1	2^{27}	2^{29}	3.8	90	—	5.2m
2^{26}	2^{30}	16	120	—	1	2^{26}	2^{30}	13.2	103	—	17

Geometric 2D						In-out-star					
n	m_r	m/n	EM	SE	SCC's	n	m	m/n	EM	SE	SCC's
2^{25}	2^{29}	15.6	58	62	7175	2^{25}	2^{29}	16	63	141	22490
2^{26}	2^{29}	7.9	70	—[2]	45060	2^{26}	2^{29}	8	79	—[1]	2.6m
2^{27}	2^{29}	4.0	91	—	5.2m	2^{27}	2^{29}	4	—	—	—
2^{26}	2^{30}	15.6	117	—	18	2^{26}	2^{30}	16	134	—	44800

Out-star						Simple web					
n	m	m/n	EM	SE	SCC's	n	m	m/n	EM	SE	SCC's
2^{25}	2^{29}	16	65	109	33m	2^{25}	2^{29}	16	63	113	1.6m
2^{26}	2^{29}	8	—[3]	—[1]	—	2^{26}	2^{29}	8	86	—[1]	10.6m
2^{27}	2^{29}	4	—[3]	—	—	2^{27}	2^{29}	4	—[3]	—	—
2^{26}	2^{30}	16	—[3]	—	—	2^{26}	2^{30}	16	133	—	3.2m

Random						Real web graphs					
n	m	m/n	EM	SE	SCC's	n	m	m/n	EM	SE	SCC's
2^{25}	2^{29}	16	61	63	12	18.5m	298.1m	16.1	29	104	3.8m
2^{26}	2^{29}	8	77	—[2]	45173	41.3m	1,150.7m	27.9	116	517	6.7m
2^{27}	2^{29}	4	109	—	5.2m	118.1m	1,019.9m	8.6	124	—[1]	38.5m
2^{26}	2^{30}	16	133	—	17						
2^{27}	2^{30}	8	159	—	90279						
2^{28}	2^{30}	4	345	—	10.4m						

5.4 The Effect of Graph Structure

The ability of EM-SCC to process certain graphs is limited by the available amount of main memory. The input graph needs to have few enough SCC's to fit in memory, and the SCC's have to be composed of short enough cycles for EM-SCC to find them as part of the memory-sized subgraphs it processes. The inability of EM-SCC to process all but one of the out-star graphs nor the sparsest of the in-out-star and simple web graphs reflects these limitations.

Since EM-SCC was not able to process these graphs, we can of course only extrapolate from the properties of the graphs in these classes the algorithm *was* able to process. The smallest simple web graph had about 1.6m SCC's, and the smallest out-star graph had about 33m SCC's. Compared to at most a few thousand SCC's in the smallest cycle, geometric 1D and 2D, and random graphs, these graphs have significantly more SCC's. For the bigger and sparser inputs, we suspect that the number of SCC's exploded, preventing EM-SCC from compressing the graph down to memory size.

The smallest in-out-star graph had about 22,000 SCC's, which is more than for cycle, geometric 1D and 2D, and random graphs, but significantly less than for out-star and simple web graphs. Therefore, there are two possible explanations for the inability of EM-SCC to process the sparsest in-out-star graph: either the lower density of the graph again resulted in an explosion of the number of SCC's, or the SCC's consisted of very long cycles, which EM-SCC was not able to find using the amount of main memory available on our test machine.

Another interesting observation we made in our experiments was the lack of a smooth transition between graphs EM-SCC could process efficiently and graphs it could not process at all. More precisely, all the graphs it was able to process required one or two contraction rounds, followed by a final round computing the SCC's in internal memory. On the other hand, for all inputs the algorithm was not able to process, it took only a few contraction rounds to reach a stage where no further contraction took place. For the out-star and in-out-star graphs, contraction stopped after at most 4 rounds. It is possible that the algorithm had found all SCC's at that point, but there simply were too many. For the simple web graph of density 4, it took 18 contraction rounds to reduce the graph by only 33%, and subsequent rounds achieved no further contraction. We suspect that more main memory would have helped in this case to identify and contract SCC's consisting of long cycles.

6   Conclusions

We have presented a contraction-based heuristic algorithm, EM-SCC, for computing the strongly connected components of large graphs. Our algorithm demonstrates that graph contraction is a useful tool for computing SCC's of large graphs, as it was able to process a wide range of input graphs faster than the currently best algorithm by Sibeyn et al., and it was able to process graphs whose vertex sets did not fit in memory.

The main limitation of EM-SCC is that it relies on the graph to have relatively few SCC's, consisting of relatively short cycles. This limitation seems impossible to overcome using graph contraction alone.

An interesting strategy that might speed up EM-SCC on inputs it *can* process is the use of pipelining to pass data between successive contraction rounds without writing this data to disk.

References

1. Abello, J., Buchsbaum, A.L., Westbrook, J.: A functional approach to external graph algorithms. Algorithmica 32(3), 437–458 (2002)
2. Ajwani, D., Dementiev, R., Meyer, U.: A computational study of external-memory BFS algorithms. In: Proceedings of the 17th ACM-SIAM Symposium on Discrete Algorithms, pp. 601–610 (2006)
3. Ajwani, D., Meyer, U., Osipov, V.: Improved external memory BFS implementation. In: Proceedings of the Workshop on Algorithm Engineering and Experiments (2007)
4. Arge, L.: External memory data structures. In: Abello, J., Pardalos, P.M., Resende, M.G.C. (eds.) Handbook of Massive Data Sets, pp. 313–357. Kluwer Academic Publishers, Dordrecht (2002)
5. Arge, L., Brodal, G.S., Toma, L.: On external-memory MST, SSSP and multi-way planar graph separation. Journal of Algorithms 53(2), 186–206 (2004)
6. Bauer, R., Delling, D., Sanders, P., Schieferdecker, D., Schultes, D., Wagner, D.: Combining hierarchical and goal-directed speed-up techniques for Dijkstra's algorithm. In: McGeoch, C.C. (ed.) WEA 2008. LNCS, vol. 5038, pp. 303–318. Springer, Heidelberg (2008)
7. Beckmann, A.: Parallelizing semi-external depth first search. Master's thesis, Martin-Luther-Universität, Halle, Germany (October 2005)
8. Chiang, Y.-J., Goodrich, M.T., Grove, E.F., Tamassia, R., Vengroff, D.E., Vitter, J.S.: External-memory graph algorithms. In: Proceedings of the 6th Annual ACM-SIAM Symposium on Discrete Algorithms, pp. 139–149 (1995)
9. Dementiev, R., Kettner, L., Sanders, P.: STXXL: Standard library for XXL data sets. Software: Practice and Experience 38(6), 589–637 (2007)
10. Dementiev, R., Sanders, P., Schultes, D., Sibeyn, J.F.: Engineering an external memory minimum spanning tree algorithm. In: Proceedings of IFIP TCS: 3rd International Conference on Theoretical Computer Science, pp. 195–208 (2004)
11. Dijkstra, E.W.: A Discipline of Programming. Prentice-Hall, Englewood Cliffs (1976)
12. Geisberger, R., Sanders, P., Schultes, D., Delling, D.: Contration hierarchies: Faster and simpler hierarchical routing in road networks. In: McGeoch, C.C. (ed.) WEA 2008. LNCS, vol. 5038, pp. 319–333. Springer, Heidelberg (2008)
13. Goldberg, A.V., Werneck, R.: Computing point-to-point shortest paths from external memory. In: Proceedings of the 7th Workshop on Algorithm Engineering and Experiments, pp. 26–40 (2005)
14. Mehlhorn, K., Meyer, U.: External-memory breadth-first search with sublinear I/O. In: Möhring, R.H., Raman, R. (eds.) ESA 2002. LNCS, vol. 2461, pp. 723–735. Springer, Heidelberg (2002)
15. Sanders, P., Schultes, D., Vetter, C.: Mobile route planning. In: Halperin, D., Mehlhorn, K. (eds.) ESA 2008. LNCS, vol. 5193, pp. 732–743. Springer, Heidelberg (2008)
16. Schultes, D.: External memory minimum spanning trees (2003), http://algo2.iti.uni-karlsruhe.de/schultes/emmst
17. Sibeyn, J., Abello, J., Meyer, U.: Heuristics for semi-external depth-first search on directed graphs. In: Proceedings of the 14th ACM Symposium on Parallel Algorithms and Architectures, pp. 282–292 (2002)
18. Sibeyn, J.F.: External connected components. In: Hagerup, T., Katajainen, J. (eds.) SWAT 2004. LNCS, vol. 3111, pp. 468–479. Springer, Heidelberg (2004)
19. Vitter, J.S.: Algorithms and data structures for external memory. Foundations and Trends in Theoretical Computer Science 2(4), 305–474 (2006)

Pareto Paths with SHARC*

Daniel Delling and Dorothea Wagner

Universität Karlsruhe (TH), 76128 Karlsruhe, Germany
{delling,wagner}@ira.uka.de

Abstract. Up to now, research on speed-up techniques for DIJKSTRA's algorithm focused on *single-criteria* scenarios. The goal was to find the quickest route within a transportation network. However, the quickest route is often not the best one. A user might be willing to accept slightly longer travel times if the cost of the journey is less. A common approach to cope with such a situation is to find *Pareto-optimal* (concerning other metrics than travel times) routes. Such routes have the property that each route is better than any other route with respect to at least one metric under consideration, e.g., travel costs or number of train changes. In this work, we study multi-criteria search in road networks. On the one hand, we focus on the problem of limiting the number of Pareto paths. On the other hand, we present a multi-criteria variant of our recent SHARC algorithm.

1 Introduction

The computation of quickest paths in graphs is used in many real-world applications like route planning in road networks, timetable information for railways, or scheduling for airplanes. In general, DIJKSTRA's algorithm [1] finds a quickest path between a given source s and target t. Unfortunately, the algorithm is far too slow to be used on huge datasets. Thus, several speed-up techniques have been developed (see [2] for an overview) that can retrieve the quickest path in a road network within less than a millisecond.

However, the quickest route in transportation networks is often not the "best" one. For example, users traveling by car may be willing to accept (slightly) longer travel times if the costs of the journey (toll, fuel consumption) is lower. A possible approach to such better routes is to run a multi-criteria query which incorporates other metrics besides travel times for finding a set of attractive routes from which a user can choose. Unfortunately, all methods developed during the last years only work in single-criteria scenarios. We here present an augmented version of our recently developed SHARC (SHortcuts + ARC-flags) algorithm working in such a multi-criteria scenario.

1.1 Related Work

A lot of speed-up techniques for single-criteria scenarios have been developed during the last years. Due to space limitations, we direct the interested reader to [2], which gives a recent overview over single-criteria routing techniques.

---

* Partially supported by the Future and Emerging Technologies Unit of EC (IST priority – 6th FP), under contract no. FP6-021235-2 (project ARRIVAL) and the DFG (project WA 654/16-1).

J. Vahrenhold (Ed.): SEA 2009, LNCS 5526, pp. 125–136, 2009.

Basics. The straightforward approach to find all Pareto optimal paths is the generalization [3,4] of DIJKSTRA's algorithm: Each node $v \in V$ gets a number of multi-dimensional labels assigned, representing all Pareto paths to v. For the bicriteria case, [3] was the first presenting such a generalization, while [5] describes multi-criteria algorithms in detail. By this generalization, DIJKSTRA loses the label-setting property, i.e., now a node may be visited more than once. It turns out that a crucial problem for multi-criteria routing is the number of labels assigned to the nodes. The more labels are created, the more nodes are reinserted in the priority queue yielding considerably slow-downs compared to the single-criteria setup. In the worst case, the number of labels can be exponential in $|V|$ yielding impractical running times [3]. Hence, [3,6] present an FPAS for the bicriteria shortest path problem.

Speed-up Techniques. Most of the work on speed-up techniques for multi-criteria scenarios was done on networks deriving from timetable information. In such networks, [7] observed that the number of labels is often limited such that the brute force approach for finding *all* Pareto paths is often feasible. Experimental studies finding all Pareto paths in timetable graphs can be found in [8,9,10]. However, to the best of our knowledge, all previous work only uses basic speed-up techniques for accelerating the multi-criteria query. In most cases a special version of A^* is adapted to this scenario. Unfortunately, the resulting speed-ups only reach up to a factor of 5 which is much less than for the (single-criteria) speed-up techniques developed during the last years.

1.2 Our Contribution

In this work, we present the first efficient speed-up technique for multi-criteria routing, namely an augmented version of SHARC [11]. Similar to the time-dependent version of SHARC [12], the key observation is that the basic concept of SHARC stays untouched. By augmenting the main subroutines of SHARC to multi-criteria variants and by changing the intuition when setting Arc-Flags [13,14], we end up in a very efficient multi-criteria speed-up technique.

We start our work on multi-criteria routing with basic definitions in Section 2. We also shortly report how SHARC works in a single-criteria scenario. In Section 3, we show how the main ingredients of SHARC—DIJKSTRA's algorithm, contraction, and arc-flags—can be augmented such that correctness can be guaranteed in a multi-criteria scenario. It turns out that adaption of contraction is straight-forward, while for arc-flags, we have to alter the intuition of a true arc-flag slightly. In Section 4 we assemble our augmented ingredients to present a multi-criteria variant of SHARC. The key observation is that the basic concept of SHARC stays untouched, we only need to additionally augment the last ingredient, i.e., arc-flags refinement. This routine can be generalized by substituting local single-criteria DIJKSTRA-searches by multi-criteria ones.

The experimental evaluation in Section 5 confirms the excellent speed-up achieved by our multi-criteria variant of SHARC: The speed-up over the generalized DIJKSTRA's algorithm is the same as in a single-criteria scenario. However, it turns out that in road networks, multi-criteria searches yield too many possible routes to the target. Hence, we introduce several reasonable constraints how to prune unattractive paths both during preprocessing and queries. Here, the key observation is that we define a main metric (we use travel times) and only allow other paths if they do not yield too long of a delay. Moreover, we also introduce a constraint called *pricing*. Paths with longer travel times are only accepted if they yield significant improvements in other metrics. With these additional constraints we are able to compute reasonable Pareto paths in continental-sized road networks. In addition, we run experiments with similar metrics where we do not need the just mentioned constraints and also present results on synthetic data sets. We conclude our work with a summary and possible future work in Section 6.

2   Preliminaries

The main difference between single- and multi-criteria routing is that the labels assigned to edges contain more than one weight. In this work, we restrict ourselves to vectors in $\mathbb{R}_+^k$. Let $L = (w_1,\ldots,w_k)$ and $L' = (w_1',\ldots,w_k')$ be two labels. We use the following notation and operations in $\mathbb{R}_+^k$: L *dominates* another label L' if $w_i < w_i'$ holds for one $1 \leq i \leq k$ and $w_i \leq w_i'$ holds for each $1 \leq j \leq k$. The sum of L and L' is defined by $L \oplus L' = (w_1 + w_1',\ldots,w_k + w_k')$. We call $\underline{L} = \min_{1 \leq i \leq k} w_i$ the minimum component of L, the maximum component $\overline{L}$ is defined analogously.

We also restrict ourselves to directed graphs $G = (V,E)$ with a length function $len : E \rightarrow \mathbb{R}_+^k$, assigning a k-dimensional label to each edge. Note that we allow multi-edges. The reverse graph $\overleftarrow{G} = (V,\overline{E})$ is the graph obtained from G by substituting each $(u,v) \in E$ by (v,u).

The 2-core of an undirected graph is the maximal node induced subgraph of minimum node degree 2. The 2-core of a directed graph is the 2-core of the corresponding simple, unweighted, undirected graph. All nodes not being part of the 2-core are called 1-shell nodes. Note that connected components within the 1-shell are trees. Since each tree is attached to the 2-core, we call these trees *attached trees*.

A *partition* of V is a family $\mathscr{C} = \{C_0, C_1, \ldots, C_k\}$ of sets $C_i \subseteq V$ such that each node $v \in V$ is contained in exactly one set C_i. An element of a partition is called a *cell*. A *multilevel partition* of V is a family of partitions $\{\mathscr{C}^0, \mathscr{C}^1, \ldots, \mathscr{C}^{L-1}\}$ such that for each $l < L - 1$ and each $C_i^l \in \mathscr{C}^l$ a cell $C_j^{l+1} \in \mathscr{C}^{l+1}$ exists with $C_i^l \subseteq C_j^{l+1}$. In that case the cell C_j^{l+1} is called the *supercell* of C_i^l. The supercell of a level-$L - 1$ cell is V. Note that the number of levels is denoted by L. We denote $c_j(u)$ the level-j cell u is assigned to. The *boundary nodes* B_C of a cell C are all nodes $u \in C$ for which at least one node $v \in V \setminus C$ exists such that $(v,u) \in E$ or $(u,v) \in E$.

In a multi-criteria scenario, the length $d(s,t)$ of an s–t path $P = (e_1,\ldots,e_r)$ is given by $len(e_1) \oplus \ldots \oplus len(e_r)$. In contrast to a single-criteria scenario, many paths exist between two nodes that do *not* dominate each other. In this work, we are interested in the *Pareto-set* $\mathscr{D}(s,t) = \{d_1(s,t)\ldots d_x(s,t)\}$ consisting of all non-dominated path-lengths $d_i(s,t)$ between s and t. We call $|\mathscr{D}(s,t)|$ the *size* of a Pareto-set. Note that by storing a predecessor for each d_i, we can compute all Pareto-paths as well.

SHARC-Routing. The original arc-flag approach [13,14] first computes a partition $\mathscr{C}$ of the graph and then attaches a *bitvector* to each edge e. A bitvector contains, for each cell $C_i \in \mathscr{C}$, a flag $AF_{C_i}(e)$ which is true if a shortest path to a node in C_i starts with e. A modified DIJKSTRA then only considers those edges for which the flag of the target node's cell is true. This idea was extended to a 2-level setup in [15]. Preprocessing of static SHARC [11] is divided into three sections. During the *initialization* phase, we extract the 2-core of the graph and perform a *multi-level* partition of G. Then, an *iterative* process starts. At each step i we first *contract* the graph by *bypassing* unimportant nodes and set the arc-flags *automatically* for each removed edge. In the contracted graph we compute the arc-flags of level i by growing a *partial* centralized shortest-path tree from each cell C_j^i. In the *finalization* phase, we assemble the output-graph, refine arc-flags of edges removed during contraction and finally reattach the 1-shell nodes removed at the beginning.

The query of static SHARC is a multi-level Arc-Flags DIJKSTRA adapted from the two-level Arc-Flags DIJKSTRA presented in [15]. The query is a modified DIJKSTRA that operates on the output graph. The modifications are as follows: When settling a node n, we compute the lowest level i on which n and the target node t are in the same supercell. When relaxing the edges outgoing from n, we consider only those edges having a set arc-flag on level i for the corresponding cell of t.

3   Augmenting Ingredients

From our augmentation of SHARC to a time-dependent scenario [12], we learned that it is sufficient to augment its ingredients, i.e., local DIJKSTRA-searches, arc-flags computation, and contraction. In this section we show how to augment all these ingredients such that correctness is guaranteed even in a multi-criteria scenario.

3.1   Dijkstra

Computing a Pareto set $\mathscr{D}(s,t)$ can be done by a straightforward generalization of DIJKSTRA's algorithm, as presented in [3,4]. For managing the different distance-vectors at each node v, we maintain a list of labels $\mathtt{list}(v)$. The list at the source node s is initialized with a label $d(s,s) = (0,\ldots,0)$, any other list is empty. We insert $d(s,s)$ to a priority queue. Then, in each iteration step, we extract the label with the smallest minimum component. Then for all outgoing edges (u,v) a temporary label $d(s,v) = d(s,u) \oplus len(u,v)$ is created. If $d(s,v)$ is not dominated by any of the labels in $\mathtt{list}(v)$, we add $d(s,v)$ to $\mathtt{list}(v)$, add $d(s,v)$ to the priority queue, and remove all labels from $\mathtt{list}(v)$ that are dominated by $d(s,v)$. We may stop the query as soon as $\mathtt{list}(t) \neq \emptyset$ and all labels in the priority queue are dominated by all labels in $\mathtt{list}(t)$.

Pareto Path Graphs. In the following, we construct *Pareto path graphs (PPG)* by computing $\mathscr{D}(s,u)$ for a given source s and all nodes $u \in V$, with our generalized DIJKSTRA algorithm. We call an edge (u,v) a *PPG-edge* if $L \in \mathtt{list}(u)$ and $L' \in \mathtt{list}(v)$ exist such that $L \oplus len(u,v) = L'$. In other words, (u,v) is a PPG-edge iff it is part of at least one Pareto-optimal path from s to v. Note that by this notion one edge of two parallel ones can be a PPG-edge while the other one is not.

3.2   Arc-Flags

In a single-criteria scenario, an arc-flag $AF_C(e)$ denotes whether e has to be considered for a shortest-path query targeting a node within C. In other words, the flag is set if e is important for (at least one target node) in C. In [12], we adapted Arc-Flags to a time-dependent scenario by setting a flag to true as soon as it is important for at least one departure time. The adaption to a multi-criteria scenario is very similar: we set an arc-flag $AF_C(e)$ to true, if e is important for at least one Pareto path targeting a node in C.

Unlike in the time-dependent scenario—where we needed approximations—we can settle for the straightforward approach for augmenting Arc-Flags. We build a Pareto path graph in $\overleftarrow{G}$ for all boundary nodes $b \in B_C$ of all cells C at level i. We stop the growth as soon as all labels in the priority queue are dominated by all labels $L(v,b)$ assigned to the nodes v in the supercell of C. Then we set $AF_C(u,v) = $ true if (u,v) is a PPG-edge for at least one PPG grown from all boundary nodes $b \in B_C$. Moreover, we set all own-cell flags to true.

Multi-Level Arc-Flags. SHARC is based on multi-level Arc-Flags. Hence, we need to augment the concept of multi-level Arc-Flags to a multi-criteria scenario. The augmentation is similar to the one to time-dependent networks. We describe a two-level setup which can be extended to a multi-level scenario easily.

Preprocessing is done as follows. Arc-flags on the upper level are computed as described above. For the lower flags, we grow a PPG in $\overleftarrow{G}$ for all boundary nodes b on the lower level. We may stop the growth as soon as all labels attached to the nodes in the supercell of C dominate all labels in the priority queue. Then, we set an arc-flag to true if the edge is a PPG edge of at least one Pareto path graph.

3.3 Contraction

Our augmented Pareto contraction routine is very similar to a static one: we first reduce the number of nodes by removing unimportant ones and—in order to preserve Pareto paths between non-removed nodes—add new edges, called shortcuts, to the graph. Then, we apply an edge-reduction step that removes unneeded shortcuts.

Node-Reduction. We iteratively *bypass* nodes until no node is *bypassable* any more. To bypass a node x we first remove x, its incoming edges I and its outgoing edges O from the graph. Then, for each combination of $e_i \in I$ and $e_o \in O$, we introduce a new edge with label $len(e_i) \oplus len(e_o)$. Note that we explicitly allow multi-edges. Also note that contraction gets more expensive in a multi-criteria scenario due to multi-edges. As for static node reduction, we use a heap to determine the next bypassable node. Let #*shortcut* be the number of *new* edges that would be inserted into the graph if x was bypassed and let $\zeta(x) = $#shortcut$/(|I| + |O|)$ be the *expansion* of node x. Furthermore, let $h(x)$ be the hop number of the hop-maximal shortcut. Then we set the key of a node x within the heap to $h(x) + 10 \cdot \zeta(x)$, smaller keys have higher priority. To keep the costs of shortcuts limited we do not bypass a node if its removal results in a hop number greater than h or an expansion greater than c. We say that the nodes that have been bypassed belong to the *component*, while the remaining nodes are called *core-nodes*.

Edge-Reduction. We identify unneeded shortcuts by growing a Pareto path graph from each node u of the core. We stop the growth as soon as all neighbors v of u have their final Pareto-set assigned. Then we may remove all edges from u to v whose label is dominated by at least one of the labels list(v). In order to limit the running time of this procedure, we restrict the number of priority-queue removals to 1 000.

4   Multi-criteria SHARC

With the augmented ingredients at hand, we are ready to augment SHARC. Remarkably, the augmentation is now very similar to time-dependent SHARC [12]. During perprocessing, we apply the augmented routines from Section 3 instead of their single-criteria counterparts, while the query is a modified multi-criteria DIJKSTRA pruning unimportant edges.

Preprocessing runs in several phase, explained in the following. During the *initialization* phase, we extract the 2-core of the graph and perform a multi-level partition of G according to an input parameter P. We can safely extract the 2-core since we can directly assign correct arc-flags to attached trees that are fully contained in a cell: Each edge targeting the 2-core gets all flags assigned true while those directing away from the 2-core only get their own-cell flag set true. By removing 1-shell nodes *before* computing the partition we ensure that an attached tree is fully contained in a cell by assigning all its nodes to the cell of its 2-core root. After the last step of our preprocessing we simply reattach the nodes and edges of the 1-shell to the output graph.

After the initialization, our *iterative* process starts. Each iteration step is divided into two parts: contraction and arc-flag computation. First, we apply a *contraction* step according to Section 3. In order to perserve correctness of multi-criteria SHARC, we have to use *cell-aware* contraction, i.e., a node u is never marked as bypassable if any of its neighboring nodes is *not* in the same cell as u. We have to set *arc-flags* for all edges of our output-graph, including those we remove during contraction. As for static SHARC, we can set arc-flags for all removed edges automatically. We set the arc-flags of the current and all higher levels depending on the tail u of the deleted edge. If u is a core node, we only set the own-cell flag to true (and others to false) because this edge can only be relevant for a query targeting a node in this cell. If u belongs to the component, all arc-flags are set to true as a query has to leave the component in order to reach a node outside this cell. Setting arc-flags of those edges not removed from the graph is more time-consuming since we apply the preprocessing of multi-level Arc-Flags from Section 3.

The *final* phase of our preprocessing-routine assembles the output graph. It contains the original graph, shortcuts added during preprocessing and arc-flags for all edges of the output graph. However, some edges may have no arc-flag set to true. As these edges are never relaxed by our query algorithm, we directly remove such edges from the output graph. Moreover, we improve on those flags set to true during the contraction process. by *Refinement of Arc-Flags*. This is achieved by propagating flags of edges outgoing from high-level nodes to those outgoing from low-level nodes. In a time-independent scenario [11], we grow shortest path trees to find the so called exit nodes of each node, while in a time-dependent scenario [12], we use profile graphs to determine these nodes. In our multi-criteria scenario, we now grow Pareto path graphs from each node. The propergation itself stays untouched, the only difference is that a node might have more than one predecessor, which all have to be examined when identifying the corresponding outgoing edge. Unfortunately, growing Pareto path graphs can get expensive. Hence, we limit the growth to $n \log(n)/|V_l|$, where V_l denotes the nodes in level l, priority-queue removals. In order to preserve correctness, we then may only propagate the flags from the exit nodes to u if the stopping criterion is fulfilled before this number of removals.

Query. Augmenting the SHARC-query is straightforward. For computing a Pareto-set $\mathscr{D}(s,t)$, we use a modified multi-criteria DIJKSTRA (Section 3) that operates on the output graph. The modifications are then the same as for the single-criteria variant of SHARC: When settling a node n, we compute the lowest level i on which n and the target node t are in the same supercell. Moreover, we consider only those edges outgoing from n having a set arc-flag on level i for the corresponding cell of t. In other words, we prune edges that are not important for the current query. The stopping criterion is the same as for a multi-criteria DIJKSTRA.

5   Experiments

In this section, we present our experimental evaluation. Our implementation is written in C++ using solely the STL at some points. As priority queue we use a binary heap. Our tests were executed on one core of an AMD Opteron 2218 running SUSE Linux 10.3. The machine is clocked at 2.6 GHz, has 16 GB of RAM and 2 x 1 MB of L2 cache. The program was compiled with GCC 4.2, using optimization level 3.

Inputs. We use four real world road networks for our experimental evaluation. The first one is the largest strongly connected component of the road network of Western Europe, provided by PTV AG for scientific use. It has approximately 18 million nodes and 42.6 million edges. However, it turns out this input is too big for finding all Pareto routes. Hence, we also use three smaller networks, namely the road network of Luxemburg consisting of 30661 nodes and 71619 edges, a road network of Karlsruhe and surrounding (77740 nodes, 196327 edges), and the road network of the Netherlands (892392 nodes, 2159589 edges). Note that we use the latter network for testing the impact of our rules of label reduction. As metrics we use travel times for fast cars/slow trucks, costs (toll + fuel consumption), travel distances, and unit lengths. Note that the last metric depicts the number of street segments of a route. Hence, it somehow reflects the number of turns of a journey.

Default Setting. For Europe, we use a 6-level partition obtained by SCOTCH [16] with 4 cells per supercell on levels 0 to 3, 8 cells per supercell on level 4, and 104 cells on level 5. A 3-level partition is applied when using Luxemburg and Karlsruhe as input, with 4 cells per supercell on levels 0 and 1, and 56 cells on level 2. For the Netherlands, we apply a 4-level partition, with 4 cells per supercell on levels 0 and 1, 8 cells on level 2, and 112 cells on level 3. We use $c = 2.5$ as maximal expansions during node-reduction and for the all levels. The hop-bound of our contraction is set to $h = 10$. To keep preprocessing times limited, we use an economical variant, i.e., we compute arc-flags only for the topmost level and do not refine arc-flags for the lowest two levels. For static single-criteria SHARC, this reduces preprocessing times by a factor of 3, but query performance increases only be a factor of 2. In the following, we report preprocessing times and the overhead of the preprocessed data in terms of *additional* bytes per node. Moreover, we provide the average number of settled nodes, i.e., the number of nodes taken from the priority queue, and the average query time. For random *s-t* queries, the nodes *s* and *t* are picked uniformly at random. All figures in this paper are based on 1 000 random *s-t* queries and refer to the scenario that only distance labels of the Pareto paths have to be determined, without outputting a complete description of the paths. However, our efficient implementation for unpacking shortcuts due to [17] needs about 4 additional bytes per node of preprocessed data. Then it takes less than 0.5 ms to unpack a shortest path. Since we allow multi-edges we could apply this unpacking routine to our multi-criteria variant of SHARC.

Full Pareto-Setting. Table 1 depicts the performance of multi-criteria SHARC on our Luxemburg instance in a full Pareto bicriteria setting. For comparison, we also report the performance of single-criteria SHARC on all five metrics.

We observe a good performance of multi-criteria SHARC in general. Preprecessing times are less than 15 minutes which is sufficient for most applications. Interestingly, the speed-up over DIJKSTRA's algorithm with respect to query times even increases when switching to multi-criteria SHARC. However, comparing single- and multi-criteria, we observe that query performance highly depends on the size of the Pareto set at the target node. For similar metrics (fast car and slow truck), bicriteria queries are only 3 times slower than a single-criteria queries. This stems from the fact that the average size of the Pareto-set is only 2. If more labels are created, like for fast car + costs, multi-criteria queries are up to 673 times slower. Even worse, this slow-down increases even further when we apply our Karlsruhe network. Here, the queries are up to 3 366 times slower.

Table 1. Performance of single- and multi-criteria SHARC applying different metrics for our Luxemburg and Karlsruhe inputs. Column *prepro* shows the computation time of the preprocessing in hours and minutes and the eventual *additional* bytes per node needed for the preprocessed data. For queries, we report the number of labels created at the target node, the number of nodes removed from the priority queue, execution times in milliseconds, and speed-up over DIJKSTRA's algorithm.

	Luxemburg						Karlsruhe					
	PREPRO		QUERY				PREPRO		QUERY			
	time	space	target	#del.	time	spd	time	space	target	#del.	time	spd
metrics	[h:m]	[B/n]	labels	mins	[ms]	up	[h:m]	[B/n]	labels	mins	[ms]	up
fast car (fc)	< 0:01	12.4	1.0	138	0.03	114	< 0:01	12.4	1.0	206	0.04	188
slow truck (st)	< 0:01	12.6	1.0	142	0.03	111	< 0:01	12.7	1.0	212	0.04	178
costs	< 0:01	12.0	1.0	151	0.03	96	< 0:01	15.4	1.0	244	0.05	129
distances	< 0:01	14.7	1.0	158	0.03	87	< 0:01	15.7	1.0	261	0.06	119
unit	< 0:01	13.7	1.0	149	0.03	96	< 0:01	14.1	1.0	238	0.05	147
fc + st	0:01	14.7	2.0	285	0.09	100	0:01	15.3	1.9	797	0.26	108
fc + costs	0:04	24.1	29.6	4 149	6.49	263	1:30	26.6	52.7	15 912	80.88	184
fc + dist.	0:14	22.3	49.9	8 348	20.21	78	3:58	23.6	99.4	31 279	202.15	153
fc + unit	0:06	23.7	25.7	4 923	5.13	112	0:17	26.6	27.0	11 319	16.04	200
costs + dist.	0:02	20.4	29.6	3 947	4.87	119	1:11	21.9	67.2	19 775	67.75	160

Summarizing, the number of labels created, and thus, the loss in query performance over single-criteria queries, is too high for using a full Pareto-setting for a big input like Western Europe. Hence, we show in the following how to reduce the number of labels such that "unimportant" Pareto-routes are pruned as early as possible.

Reduction of Labels. As observed in Tab. 1, the number of labels assigned to a node increases with growing graph size. In order to efficiently compute Pareto-paths for our European road network, we need to reduce the number of labels both during preprocessing and queries. We achieve this by tightening the definition of dominance. Therefore, we define the travel time metric to be the dominating metric W. Then, our tightened definition of dominance is as follows: Besides the constraints from Section 2, we say a label $L = (W, w_1, \ldots, w_{k-1})$ dominates another label $L' = (W', w_1', \ldots, w_{k-1}')$ if $W \cdot (1 + \varepsilon) < W'$ holds. In other words, we only allow Pareto-paths which are up to ε times longer (with respect to the dominating metric). Note that by this notion, this has to hold for all sub-paths as well.

Table 2 reports the performance of bicriteria SHARC using the tightened definition of dominance (with varying ε) during preprocessing *and* queries. As input, we use three networks: Karlsruhe, the Netherlands, and Europe. We here focus on the probably most important combination of metrics, namely fast car travel time and costs). We observe that our additional constraint works: Preprocessing times decrease and query performance gets much better. However, as expected, very small ε values yield a small subset of the Pareto-set and high ε values yield high preprocessing times. For small and mid-size inputs, i.e., less than 1 million nodes, setting ε to 0.5 yields a reasonable amount of Pareto paths combined with good preprocessing times and good query performance. Unfortunately, for our European input, only $\varepsilon \leq 0.02$ yields practical preprocessing and query times.

Table 2. Performance of bi-criteria SHARC with varying ε using travel times and costs as metrics. The inputs are Karlsruhe, the Netherlands, and Europe.

	Karlsruhe				The Netherlands				Europe		
	PREP	QUERY			PREP	QUERY			PREP	QUERY	
	time	target	#del.	time	time	target	#del.	time	time	target	#del.	time
ε	[h:m]	labels	mins	[ms]	[h:m]	labels	mins	[ms]	[h:m]	labels	mins	[ms]
0.000	< 0:01	1.0	265	0.09	0:01	1.0	452	0.21	0:53	1.0	3 299	2.6
0.001	< 0:01	1.1	271	0.09	0:01	1.1	461	0.21	1:00	1.1	3 644	4.1
0.002	< 0:01	1.1	302	0.10	0:01	1.2	489	0.22	1:03	1.2	4 340	7.1
0.005	< 0:01	1.3	307	0.11	0:01	1.4	517	0.24	1:18	1.4	5 012	11.3
0.010	< 0:01	1.5	322	0.11	0:01	1.7	590	0.27	1:58	2.4	9 861	19.2
0.020	< 0:01	1.9	387	0.13	0:01	2.2	672	0.32	4:10	5.0	24 540	48.1
0.050	< 0:01	2.5	495	0.18	0:02	3.3	1 009	0.51	14:12	23.4	137 092	412.7
0.100	< 0:01	4.2	804	0.33	0:04	4.8	1 405	0.82	>24:00	–	–	–
0.200	0:01	6.4	1,989	1.86	0:09	7.2	2 225	1.67				
0.500	0:02	14.0	3 193	3.61	0:39	12.8	4 227	4.85				
1.000	0:13	24.0	9 072	14.86	3:44	20.0	12 481	26.85				
∞	1:30	52.7	15 912	80.88	>24:00	–	–	–				

Further Reduction. As observable in Tab. 2, our approach for reducing the number of labels is only practical for very small ε if we use Europe as input. As we are interested in paths with bigger ε values as well, we add another constraint, called *pricing*, in order to define dominance. Besides the constraints from Section 2 and from above, we say a label $L = (W, w_1, \ldots, w_{k-1})$ dominates another label $L' = (W', w'_1, \ldots, w'_{k-1})$ if $\sum_i w'_i / \sum_i w_i > W/W' \cdot \gamma$ holds for some constant γ. In other words, we only accept labels with longer travel times if this results in a decrease in the other metrics under consideration. With this further tightened definition of label dominance, we are finally ready to run multi-

Table 3. Performance of bi-criteria SHARC with varying γ using travel times and costs as metrics. ε is fixed to 0.5. The input is Europe.

	PREPRO		QUERY		
	time	space	target	#del.	time
γ	[h:m]	[B/n]	labels	mins	[ms]
1.100	0:58	19.1	1.2	2 538	1.8
1.050	1:07	19.6	1.3	3 089	2.2
1.010	1:40	20.4	1.7	4 268	3.2
1.005	2:04	20.6	1.9	5 766	4.1
1.001	3:30	20.8	2.7	7 785	6.1
1.000	7:12	21.3	5.3	19 234	35.4
0.999	15:43	22.5	15.2	87 144	297.2
0.995	>24:00	–	–	–	–

criteria queries on our European instance. Table 3 shows the performance of multi-criteria SHARC with varying γ in a bicriteria scenario (travel times + costs) for Europe. Note that we *fix* $\varepsilon = 0.5$. It turns out that our additional constraints work. With $\gamma = 1.0$, we create 5.3 labels in 35.42 ms on average at the target node, being sufficient for practical applications. Preprocessing times are still within reasonable times, i.e., less than 8 hours. If we want to generate more labels, we could set $\gamma = 0.999$. However, query times drop to almost 300 ms and preprocessing increases drastically. Summarizing, bicriteria queries for travel times and travel costs are possible if we use $\gamma = 1.0$ and $\varepsilon = 1.5$.

Table 4. Performance of multi-criteria SHARC applying different travel time metrics. The inputs are the Netherlands and Europe.

	The Netherlands						Europe					
	PREPRO		QUERY				PREPRO		QUERY			
	time	space	target	#del.	time	speed	time	space	target	#del.	time	speed
metrics	[h:m]	[B/n]	labels	mins	[ms]	up	[h:m]	[B/n]	labels	mins	[ms]	up
fast car(fc)	0:01	13.7	1.0	364	0.11	1 490	0:25	13.7	1.0	1,457	0.69	7 536
slow car(sc)	0:01	13.8	1.0	359	0.10	1 472	0:24	13.8	1.0	1,367	0.67	7 761
fast truck(ft)	0:01	13.9	1.0	365	0.10	1 332	0:23	13.9	1.0	1,486	0.71	7 324
slow truck(st)	0:01	13.9	1.0	363	0.10	1 306	0:25	13.9	1.0	1,423	0.68	7 647
fc+st	0:05	16.2	2.2	850	0.33	2 532	2:24	18.3	3.8	6 819	4.35	12 009
fc+ft	0:05	16.2	2.0	768	0.29	2 371	1:30	18.3	3.2	5 466	3.91	11 349
fc+sc	0:05	15.5	1.2	520	0.19	1 896	1:08	17.1	2.0	4 265	2.26	10 234
sc+st	0:05	16.2	1.9	742	0.29	2 009	1:53	18.1	3.3	5 301	4.02	10 874
sc+ft	0:05	16.2	1.7	679	0.26	1 850	1:49	16.2	3.2	5 412	3.65	10 663
ft+st	0:05	15.7	1.3	551	0.21	1 692	1:28	17.4	3.0	5 157	3.73	12 818
fc+sc+st	0:06	19.0	2.3	867	0.37	2 580	2:41	20.3	4.5	6 513	5.70	12 741
fc+sc+ft	0:06	18.9	2.0	764	0.32	2 385	2:47	21.5	3.9	5 989	4.87	12 144
sc+ft+st	0:06	19.0	1.9	740	0.30	2 134	2:59	22.0	4.2	6 348	5.12	13 412
fc+sc+ft+st	0:07	21.8	2.5	942	0.43	2 362	4:41	24.5	6.2	12 766	7.85	15 281

Similar Metrics. Our last experiment for road networks deals with the following scenario. We are interested in the quickest route for different types of vehicles. Hence, we perform multi-criteria queries on metrics all based on travel times. More precisely, we use typical average speeds of fast cars, slow cars, fast trucks, and slow trucks. Due to the very limited size of the resulting Pareto-sets, we afford not to use our tightened definition of dominance for this experiment. Tab. 4 shows the performance of multi-criteria SHARC in such a single-, bi- and tri-, and quadro-criteria scenario.

We observe that a full Pareto-setting is feasible if metrics are similar to each other, mainly because the number labels is very limited. Interestingly, the speed-up of multi-criteria SHARC over multi-criteria DIJKSTRA is even *higher* than in a single-criteria scenario. The slow-down in preprocessing times and query performance is quite high but still, especially the latter is fast enough for practical applications. Quadro-criteria queries need less than 8 ms for our European road networks, being sufficient for most applications. A generalized DIJKSTRA needs about 120 seconds on average for finding a Pareto-set in this quadro-criteria scenario. This speed-up of more than 15 000 is achieved by a preprocessing taking less than 5 hours.

6 Conclusion

In this work, we presented the first efficient speed-up technique for computing multi-criteria paths in large-scale networks. By augmenting single-criteria routines to multi-criteria versions, we were able to present a multi-criteria variant of SHARC. Several experiments confirm that speed-ups over a multi-criteria DIJKSTRA are at least the same as in a single-criteria scenario, in many cases the speed-up with respect to query times is even higher. However, if metrics differ strongly, the number of possible Pareto-routes increases drastically making preprocessing and query times impractical for large instances. By tightening the definition of dominance, we are able to prune unimportant Pareto-routes both during preprocessing and queries. As a result, SHARC provides a feasible subset of Pareto-routes in continental sized road network.

Regarding future work, one can think of other ways for pruning the Pareto-set. Maybe other constraints yield better subsets of the Pareto-set computable in reasonable time as well. An open challenging problem is the adaption of multi-criteria SHARC to a fully realistic timetable information system like the ones presented in [9,10]. Another interesting question is how good alternative routes can be found in a single-criteria scenario.

References

1. Dijkstra, E.W.: A Note on Two Problems in Connexion with Graphs. Numerische Mathematik 1, 269–271 (1959)
2. Delling, D., Sanders, P., Schultes, D., Wagner, D.: Engineering Route Planning Algorithms. In: Lerner, J., Wagner, D., Zweig, K.A. (eds.) Algorithmics of Large and Complex Networks. LNCS. Springer, Heidelberg (to appear 2009),
   http://i11www.ira.uka.de/extra/publications/dssw-erpa-09.pdf
3. Hansen, P.: Bricriteria Path Problems. In: Fandel, G., Gal, T. (eds.) Multiple Criteria Decision Making – Theory and Application, pp. 109–127. Springer, Heidelberg (1979)

4. Martins, E.Q.: On a Multicriteria Shortest Path Problem. European Journal of Operational Research 26(3), 236–245 (1984)
5. Theune, D.: Robuste und effiziente Methoden zur Lsung von Wegproblemen. Ph.D thesis (1995)
6. Warburton, A.: Approximation of Pareto Optima in Multiple-Objective Shortest-Path Problems. Operations Research 35(1), 70–79 (1987)
7. Müller-Hannemann, M., Weihe, K.: Pareto Shortest Paths is Often Feasible in Practice. In: Brodal, G.S., Frigioni, D., Marchetti-Spaccamela, A. (eds.) WAE 2001. LNCS, vol. 2141, pp. 185–197. Springer, Heidelberg (2001)
8. Pyrga, E., Schulz, F., Wagner, D., Zaroliagis, C.: Efficient Models for Timetable Information in Public Transportation Systems. ACM Journal of Experimental Algorithmics 12, Article 2.4 (2007)
9. Müller-Hannemann, M., Schnee, M.: Finding All Attractive Train Connections by Multi-Criteria Pareto Search. In: Geraets, F., Kroon, L.G., Schoebel, A., Wagner, D., Zaroliagis, C.D. (eds.) Railway Optimization 2004. LNCS, vol. 4359, pp. 246–263. Springer, Heidelberg (2007)
10. Disser, Y., Müller-Hannemann, M., Schnee, M.: Multi-Criteria Shortest Paths in Time-Dependent Train Networks. In: McGeoch, C.C. (ed.) WEA 2008. LNCS, vol. 5038, pp. 347–361. Springer, Heidelberg (2008)
11. Bauer, R., Delling, D.: SHARC: Fast and Robust Unidirectional Routing. In: Munro, I., Wagner, D. (eds.) Proceedings of the 10th Workshop on Algorithm Engineering and Experiments (ALENEX 2008), pp. 13–26. SIAM, Philadelphia (2008)
12. Delling, D.: Time-Dependent SHARC-Routing. In: Halperin, D., Mehlhorn, K. (eds.) ESA 2008. LNCS, vol. 5193, pp. 332–343. Springer, Heidelberg (2008)
13. Lauther, U.: An Extremely Fast, Exact Algorithm for Finding Shortest Paths in Static Networks with Geographical Background. In: Geoinformation und Mobilität - von der Forschung zur praktischen Anwendung. IfGI prints, vol. 22, pp. 219–230 (2004)
14. Hilger, M., Köhler, E., Möhring, R.H., Schilling, H.: Fast Point-to-Point Shortest Path Computations with Arc-Flags. In: Demetrescu, C., Goldberg, A.V., Johnson, D.S. (eds.) Shortest Paths: Ninth DIMACS Implementation Challenge. DIMACS Book. American Mathematical Society, Providence (to appear, 2009)
15. Möhring, R.H., Schilling, H., Schütz, B., Wagner, D., Willhalm, T.: Partitioning Graphs to Speedup Dijkstra's Algorithm. ACM Journal of Experimental Algorithmics 11, 2.8 (2006)
16. Pellegrini, F.: SCOTCH: Static Mapping, Graph, Mesh and Hypergraph Partitioning and Parallel and Sequential Sparse Matrix Ordering Package (2007)
17. Delling, D., Sanders, P., Schultes, D., Wagner, D.: Highway Hierarchies Star. In: Demetrescu, C., Goldberg, A.V., Johnson, D.S. (eds.) Shortest Paths: Ninth DIMACS Implementation Challenge. DIMACS Book. American Mathematical Society, Providence (to appear, 2009) (accepted for publication)

An Application of Self-organizing Data Structures to Compression

Reza Dorrigiv, Alejandro López-Ortiz, and J. Ian Munro

Cheriton School of Computer Science, University of Waterloo, Canada
{rdorrigiv,alopez-o,imunro}@uwaterloo.ca

Abstract. List update algorithms have been widely used as subroutines in compression schemas, most notably as part of Burrows-Wheeler compression. The Burrows-Wheeler transform (BWT), which is the basis of many state-of-the-art general purpose compressors applies a compression algorithm to a permuted version of the original text. List update algorithms are a common choice for this second stage of BWT-based compression. In this paper we perform an experimental comparison of various list update algorithms both as stand alone compression mechanisms and as a second stage of the BWT-based compression. Our experiments show MTF outperforms other list update algorithms in practice after BWT. This is consistent with the intuition that BWT increases locality of reference and the predicted result from the locality of reference model of Angelopoulos et al. [1]. Lastly, we observe that due to an often neglected difference in the cost models, good list update algorithms may be far from optimal for BWT compression and construct an explicit example of this phenomena. This is a fact that had yet to be supported theoretically in the literature.

1 Introduction

It has long been observed that list update algorithms can be used in compression. In 1986, Bentley et al. [2] proposed a compression scheme that uses move-to-front as a subroutine. They proved that their compression scheme, based on move-to-front (MTF) is guaranteed to be within twice the compression ratio of the best static Huffman code. Experimentally their algorithm performs even better achieving compression ratios equal or better than Huffman's. In principle MTF can be replaced with any other online list update algorithm, which may or may not improve the compression rate. Albers and Mitzenmacher [3] studied the use of timestamp and showed theoretical and experimental evidence for its efficiency in data compression. Several online list update algorithms were compared according to their efficiency in compression by Bachrach et al. [4]. Surprisingly, their results show that some algorithms with bad competitive ratios outperform those that are optimal according to competitive analysis in terms of compression ratio.

A second application of list update is to Burrows and Wheeler compression. The Burrows-Wheeler transform (BWT) rearranges a string of symbols to one of its permutations and in doing so brings the issue of higher order entropy into

J. Vahrenhold (Ed.): SEA 2009, LNCS 5526, pp. 137–148, 2009.
© Springer-Verlag Berlin Heidelberg 2009

play. Then MTF is used to encode this transform in a way similar to the scheme proposed by Bentley et al. [2]. The resulting scheme is shown to be very effective in theory and practice and many improvements and several variants have been proposed [5,6,7,8,9,10,11,12]. The well known compression program bzip2 [13] is based on the BWT.

Our study was motivated by recent theoretical results on the impact of locality of reference assumptions for online algorithms [1]. Compression via list update hinges on an implicit assumption that the text (raw or after the BWT transform) exhibits locality of reference which can then be used advantageously by list update algorithms. In this paper we systematically study different sensible choices for the list update algorithm as well as for the basic compressor.

Our Results. We perform an experimental comparison of the latest list update algorithms for compression, both in stand alone form and as part of BWT based compression. We show that in most cases MTF is the best choice. Additionally we observe that list update algorithms optimize for a similar but different objective than a compressor and give an example of an algorithm which is a good choice for list update but not for compression, a fact that had yet to be reported in the literature.

2   Preliminaries

The List Update Problem. Consider an unsorted list of l items stored using a linked list. The input is a series of n requests to be served in an online manner. To serve a request to an item x, the algorithm should linearly search the list until it finds x at position i, for some i between 1 and l. The cost of such an access is i units. Immediately after accessing x, x can be moved to any position closer to the front of the list at no extra cost. An efficient algorithm should re-arrange the items after each access so as to minimize the overall cost of serving a sequence.

Standard List Update Algorithms. Three standard deterministic online algorithms are *move-to-front* (MTF), *transpose* (TR), and *frequency-count* (FC). MTF moves the requested item to the front of the list whereas TR exchanges the requested item with the item that immediately precedes it. FC maintains a frequency count for each item, updates this count after each access, and updates the list so that it always contains items in non-increasing order of frequency count. Sleator and Tarjan showed that MTF is 2-competitive, while TR and FC do not have constant competitive ratios [14]. Since then, several other deterministic and randomized online algorithms have been studied using competitive analysis. We only consider deterministic algorithms because randomized list update algorithms cannot be used in the compression scheme in a straightforward way. Albers introduced the algorithm *timestamp* (TS) and showed that it is 2-competitive [15]. After accessing an item a, TS inserts a in front of the first item b that appears before a in the list and was requested at most once since the last request for a. If there is no such item b, or if this is the first access to a, TS does not reorder the list.

Schulz [16] introduced an infinite (uncountable) family of list update algorithms called *sort-by-rank* (SBR). All algorithms in this family achieve the optimal competitive ratio 2 and they mediate between MTF and TS. Consider a sequence $\sigma = \sigma_1 \sigma_2 \cdots \sigma_m$ of length m. For an item a and a time $1 \le t \le m$, denote by $w_1(a, t)$ and $w_2(a, t)$ the time of the last and the second last access to a in $\sigma_1 \sigma_2 \cdots \sigma_t$, respectively. If a has not been accessed so far, set $w_1(a, t) = 0$ and if a has been accessed at most once, set $w_2(a, t) = 0$. Then we define $s_1(a, t) = t - w_1(a, t)$ and $s_2(a, t) = t - w_2(a, t)$. Note that after each access, MTF and TS reorganize their lists so that the items are in increasing order of their s_1 and s_2, respectively[1]. For a parameter $0 \le \alpha \le 1$, $SBR(\alpha)$ reorganizes its list after the tth access so that items are sorted by their α-rank function defined as $r_\alpha(a, t) = (1 - \alpha) \times s_1(a, t) + \alpha \times s_2(a, t)$.[2] More formally, upon a request for an item a in time t, $SBR(\alpha)$ inserts a just after the last item b in front of a with $r_\alpha(b, t) < r_\alpha(a, t)$. Furthermore, if there is no such item b or this is the first access to a, $SBR(\alpha)$ inserts a at the front of the list. Therefore $SBR(0)$ is equivalent to MTF and $SBR(1)$ is equivalent to TS except for the handling of the first accesses, i.e., they were equivalent if TS moves an item that has been accessed only once so far to the front of the list.

Compression Schemas. Bentley et al. [2] proposed using list update algorithms as subroutines in compression. The idea is simple enough: both the encoder and the decoder maintain a list L of all symbols in the file and agree on some online list update algorithm $\mathcal{A}$ as well as an initial arrangement for L. The encoder encodes every symbol by its current position in L and then rearranges L according to $\mathcal{A}$. It uses some variable length prefix-free binary code to transmit these integers (positions). Since the decoder knows the initial arrangement of L and the list update algorithm, it can maintain the same list as the encoder and recover all the symbols. Several variable length prefix-free binary codes can be used in this scheme, e.g., Elias encoding, δ-encoding, and ω-encoding. We refer the reader to [4] for a full description.

Burrows-Wheeler Transform. Burrows and Wheeler [5] introduced the idea of a preprocessing phase based on the BWT which is combined with a compression scheme on the transformed text. Informally, the BWT rearranges a string of symbols to one of its permutations in a reversible way so that the resulting string is "more compressible" or has more "locality of reference". The permutation is such that high order entropy is in line with locality of reference. Recall that a string has high locality of reference if when a symbol occurs in some position of the string, it is more likely to occur in a nearby position. For a detailed explanation of the BWT we refer the reader to [5,6].

---

[1] For TS, strictly speaking, this applies only to items that have been accessed at list twice.

[2] Schulz [16] denoted this by $r_t(a, \alpha)$.

3   Competitiveness of List Update Algorithms for Compression

A list update algorithm $\mathcal{A}$ incurs cost i to access the ith item of the list. However, when we use $\mathcal{A}$ as a subroutine for compression we need $\Theta(\log i)$ bits to represent that the symbol is at the ith position of the list. Other papers that have studied the use of list update algorithms in compression are silent on this issue and apparently simply assumed that competitive list update algorithms are also competitive for compression. We show via an example that this is not necessarily the case, i.e. there exist algorithms which are competitive under one model but not the other. Consider the *move-fraction (MF)* family of deterministic list update algorithms as introduced by Sleator and Tarjan [14]. Upon a request to an item in the ith position, MF(k) moves that item $\lceil i/k \rceil$-1 positions towards the front. MF(k) is known to be $2k$-competitive [14], therefore algorithm MF(2) is 4-competitive for list update. We show that under the $\Theta(\log i)$ cost model, MF(2) does not have constant competitive ratio. Let the cost of compressing for an item in the ith position be $c\lfloor \log i \rfloor + b$ for some constants c and b. For simplicity assume that we have $l = 2^p$ symbols for some integer p. Suppose that symbols are initially ordered as $a_1 a_2 \cdots a_l$ in the list. Now consider the sequence $\sigma_1 = a_l^p$. On the ith request to a_l, MF(2) incurs cost at least $c\lfloor \log \frac{2^p}{2^{i-1}} \rfloor + b = c(p - i + 1) + b$ and moves a_l to a position of index at least $\frac{2^p}{2^i}$. Therefore the cost of MF(2) on σ_1 is at least $\sum_{i=1}^{p} (c(p - i + 1) + b) = \frac{cp(p+1)}{2} + bp = \Theta(\log^2 l)$. On the other hand, MTF moves a_l to the front of the list and incurs cost $c\lfloor \log l \rfloor + b + (p - 1)b = (b + c) \log l$ on σ_1. Thus the cost of OPT on this sequence is at most $(b + c) \log l = \Theta(\log l)$. We can request the item that is now in the lth position of MF(2)'s list p times. Therefore the competitive ratio of MF(2) is at least $\frac{c \times \log l(\log l + 1)/2 + b \log l}{(b+c)\log l} = \frac{c(\log l + 1)}{2(b+c)} + \frac{b}{b+c} = \Theta(\log l)$, which is not a constant. The same holds for MF(k) for $k \geq 3$. This fact had been observed empirically by Bachrach et al. [4], who reported on the poor performance of this family for data compression purposes. It remains an open question to determine the competitive ratios of the various list update algorithms under the $c\lfloor \log i \rfloor + b$ cost of access model.

4   Experimental Results

We consider two experimental setups. The first one consists of a straightforward compression scheme similar to that of Bentley et al. [2] or Albers et al. [3]. While in practice these compression techniques are unlikely to be of use, the study of their behaviour allows us to understand their differences and advantages. The second setup consists of the realistic setting of BWT based compression. To be more precise, given a text we compute its BWT and then compare the role of various list update algorithms for compressing the transformed string.

4.1   Experimental Settings

We computed the compression ratios achieved by different list update algorithms on files in the Calgary Corpus [17] and the Canterbury Corpus [18]. These are standard benchmarks for data compression. Due to space constraints, we only present the results for the Calgary Corpus; the results for the Canterbury Corpus are similar. We considered the list update algorithms described in Section 2 as well as MTF$'$; this algorithm, on the ith access to an item a, moves a to the front of the list if i is even and does not change a's position if i is odd. We considered two implementations for frequency-count depending on the order of items with the same frequency count. In FC, an item that is less recently used precedes an item that is more recently used and has equal frequency count. FC$'$ adopts the reverse of this ordering. We performed comprehensive experiments on the compression ratios achieved by SBR(α) for different values of $0 \leq \alpha \leq 1$. These experiments showed that as α goes from 0 to 1, the behaviour of SBR(α) goes from MTF to TS. Thus we only report the results for SBR(0.5). Due to space constraints these experimental results are not included in this paper. If not explicitly mentioned otherwise, we use the standard prefix integer encoding of Elias [19] that encodes an integer i using $1 + 2\lfloor \log i \rfloor$ bits. Observe that nonetheless we propose and evaluate other alternative ways for encoding integers.

4.2   Comparing List Update Algorithms

We compare the effect of different list update algorithms on text files of the Calgary Corpus before and after BWT. Table 1 shows their performance as stand alone compression algorithms while Table 2 shows their performance as a second stage of BWT compression. From Table 1 we can see that TR and FC usually outperform MTF and TS. This is in contrast with competitive analysis in which MTF and TS are superior to TS and FC. MTF has the worst performance on all the files and TR is the best algorithm in most cases. MTF$'$ and FC$'$ always have performance close to their variants, i.e., MTF and FC, respectively. Note that the results for MTF and TS were also reported by Albers and Mitzenmacher [3], who observed that TS outperforms MTF. SBR(0.5) always mediated between the performance of MTF and TS. Thus our experimental results are not consistent with theory. This has been observed by other researches as well [4].

However, for the BWT of the files, the situation is different. Table 2 shows that in this case MTF has the best performance for most of the files. In general, MTF and TS (and thus MTF$'$ and $SBR(0.5)$) have comparable performance and always outperform FC and TR. The compression ratio they achieve after the BWT is much better than without the BWT, as one would expect given that the BWT increases the amount of locality in the string. The superiority of MTF to other algorithms is consistent with the recent result of Angelopoulos et al. proving that MTF outperforms all other online list update algorithm on sequences with high locality of reference [1]. Hence, this provides evidence that the locality of reference model proposed accurately reflects reality. We emphasize that our focus here is comparing the effect of different list update algorithms

Table 1. Compression of the Calgary Corpus without BWT

File	Size (bytes)	MTF	SBR(0.5)	TS	FC	TR	MTF$'$	FC$'$
bib	111261	95.69	89.55	89.08	**81.42**	81.64	94.16	**81.42**
book1	768771	83.82	76.64	75.67	81.34	**69.62**	81.27	81.34
book2	610856	84.35	78.36	77.55	75.74	**72.44**	82.35	75.74
news	377109	88.50	82.68	82.20	88.10	**77.87**	87.08	87.99
paper1	53161	86.79	80.96	80.35	79.48	**74.87**	85.19	79.45
paper2	82199	84.47	78.34	77.43	79.27	**71.02**	82.26	80.45
progc	39611	88.74	84.02	83.62	81.59	**77.67**	88.16	81.54
progl	71646	77.01	73.62	73.25	82.61	**69.02**	76.50	82.40
progp	49379	81.09	76.15	75.45	82.41	**71.64**	80.00	81.68
trans	93695	87.58	84.96	84.59	91.21	**83.02**	87.36	91.18

Table 2. Compression of the Calgary Corpus after BWT

File	Size (bytes)	MTF	SBR(0.5)	TS	FC	TR	MTF$'$	FC$'$
bib	111261	**30.49**	31.66	32.32	93.42	39.81	31.99	93.33
book1	768771	35.74	**34.42**	34.71	76.63	36.31	36.04	76.50
book2	610856	31.14	**31.03**	31.48	80.44	35.31	31.96	80.11
news	377109	**36.21**	37.75	38.67	85.27	44.90	38.26	85.53
paper1	53161	**34.70**	36.62	37.70	83.42	47.73	36.87	83.34
paper2	82199	**34.86**	35.35	36.04	79.00	41.28	36.17	76.46
progc	39611	**35.04**	37.32	38.54	79.03	51.09	37.54	78.91
progl	71646	**26.31**	28.52	29.43	81.23	36.18	28.33	79.77
progp	49379	**26.00**	29.08	30.22	89.11	41.13	28.57	86.08
trans	93695	**24.12**	27.64	28.71	96.08	41.52	26.76	90.22

and therefore we have not applied any post-optimizations to the compression scheme, in the presumption that these optimizations are orthogonal and hence would generally benefit all schemes equally.

We also observe that FC and FC$'$ perform badly compared to other algorithms. One explanation for this is the fact that FC considers the global rather than local environment. For example if an item is frequently accessed near the beginning and then it is not accessed at all, FC will maintain it close to the front of the list.

4.3 Alternative Techniques for Encoding of Integers

We consider other possibilities for the last step of list update based compression schemes, i.e., the prefix-free binary code for integers. As there is considerable locality of reference in the BWTs of text files intuitively a competitive list update algorithm leads to a sequence with many small integers. These algorithms assign smaller codes to small integers.

RL(1)+Elias. This algorithm combines Elias encoding with run length encoding for the value 1, i.e. when the encoded integer is 1, the following Elias-encoded integer shows the number of consecutive 1's starting from that 1. Otherwise, is the next integer encoded in Elias encoding.

RL(1)+1-2. This algorithm encodes 1 with a single bit 0, and encodes all other numbers with their binary representations prepended by 1. We need $\lceil \log_2 l \rceil$ bits for this binary representation. For most of the cases, this gives a code of length 8 for each integer greater than 1, as $64 \leq l < 128$. Also it uses run length on "1"s.

RL(1)+2-2-3: This algorithm encodes 1 and 2 with "00" and "01", respectively, and encodes all other numbers with their binary representations prepended by 1. It also uses run length on "1"s.

RL(1)+1-5-6-17: This algorithm encodes 1 by "0", 2 to 9 by "10000", "10001", ..., "10111", 10 to 17 by "110000", "110001", ..., "110111", and integers greater than 17 by their binary representation prepended by "111". Note that there are $l - 17$ such numbers, and so we can use a fixed code of length $\lceil \log_2 (l - 17) \rceil$ for their binary representations. It also uses run length on "1"s, i.e., when it encodes a "1" the following integer, encoded using the same scheme, denotes the number of consecutive ones started from that "1".

Table 3. Compression of the Calgary Corpus using RL(1)+Elias after BWT

File	Size (bytes)	MTF	SBR(0.5)	TS	FC	TR	MTF'	FC'
bib	111261	**27.87**	28.92	29.55	93.42	37.06	29.28	93.42
book1	768771	35.78	**34.50**	34.77	76.78	36.46	36.02	78.68
book2	610856	29.72	**29.56**	30.00	80.52	33.98	30.48	80.53
news	377109	**35.51**	36.82	37.71	85.33	43.96	37.37	85.50
paper1	53161	**34.60**	36.32	37.38	83.36	47.56	36.64	84.96
paper2	82199	**34.59**	35.01	35.66	79.00	41.02	35.80	78.96
progc	39611	**34.83**	36.89	38.07	79.15	50.83	37.15	82.32
progl	71646	**24.15**	26.17	27.07	81.25	33.96	26.07	84.32
progp	49379	**23.87**	26.68	27.80	89.14	38.92	26.29	91.77
trans	93695	**20.92**	24.26	25.31	95.58	38.32	23.46	102.71

Table 4. Compression of the Calgary Corpus using RL(1)+1-2 after BWT

File	Size (bytes)	MTF	SBR(0.5)	TS	FC	TR	MTF'	FC'
bib	111261	**36.36**	37.77	38.18	87.44	43.09	37.44	87.44
book1	768771	59.33	**57.89**	57.92	98.47	59.05	59.25	96.34
book2	610856	47.94	**47.89**	48.10	97.96	50.78	48.47	97.96
news	377109	**51.60**	53.52	54.16	97.87	58.28	53.09	97.87
paper1	53161	**52.15**	54.29	54.93	88.66	62.02	53.56	88.66
paper2	82199	**54.25**	54.92	55.35	99.97	59.67	55.24	99.97
progc	39611	**50.31**	53.00	53.93	85.96	61.76	52.06	99.40
progl	71646	**36.93**	40.08	41.04	99.76	47.21	38.94	99.76
progp	49379	**36.20**	39.70	40.80	99.72	48.68	37.97	99.72
trans	93695	**30.01**	34.98	35.70	90.49	45.81	31.78	99.99

Tables 4-7 show the performance of these algorithms on text files of the Calgary Corpus after BWT. According to these results, RL(1)+Elias leads to the best compression among these algorithms, then RL(1)+5-6-17, then RL(1)+2-2-3, and finally RL(1)+1-2. Comparing Table 3 to Table 2 shows that using RL(1) improves the compression factor for most list update algorithms. This can be explained by the fact that BWTs of text files have many repetitions. Each such repetition leads to a 1 in the sequence of integers. Therefore we will have many 1's and RL(1) should be effective. Also according to Tables 4-7, replacing Elias with other proposed integer encodings does not give better compression ratios.

Modified Huffman. Inspired by the fact that there are many blocks of "1"s in the integer sequence we treat them as symbols of our alphabet. Thus our alphabet is $\{1, 2, \cdots, l, 11, 111, \cdots, 1^n\}$, where 1^n means n consecutive "1"s. Then Huffman encode the elements of this alphabet. The results are shown in Table 7. Note that we should also encode the Huffman tree. This cost becomes negligible for large files, especially if one considers implicit representations of the portions of the Huffman code corresponding to 1^k. Indeed the Huffman tree has an impact

Table 5. Compression of the Calgary Corpus using Algorithm RL(1)+2-2-3 after BWT

File	Size (bytes)	MTF	SBR(0.5)	TS	FC	TR	MTF'	FC'
bib	111261	**31.74**	32.83	33.32	86.54	38.76	32.78	86.54
book1	768771	48.54	**47.29**	47.51	93.96	48.94	48.67	94.98
book2	610856	39.16	**39.05**	39.47	97.93	42.77	39.75	97.93
news	377109	**44.63**	45.94	46.66	97.89	51.72	45.98	97.89
paper1	53161	**44.31**	46.15	47.03	88.68	55.53	45.97	88.68
paper2	82199	**45.67**	46.42	47.02	88.92	52.06	46.78	88.92
progc	39611	**42.64**	44.85	45.73	83.80	55.28	44.27	86.35
progl	71646	**31.09**	33.16	33.94	86.96	41.10	32.55	86.96
progp	49379	**29.87**	32.87	33.80	97.04	43.30	31.53	99.70
trans	93695	**26.40**	29.71	30.64	88.14	41.64	27.90	92.06

Table 6. Compression of the Calgary Corpus using 1-5-6-17+RL(1) after BWT

File	Size (bytes)	MTF	SBR(0.5)	TS	FC	TR	MTF'	FC'
bib	111261	**29.54**	30.61	31.10	82.72	37.22	30.53	82.25
book1	768771	40.43	**39.41**	39.50	74.74	40.77	40.41	73.34
book2	610856	33.50	**33.49**	33.76	77.62	36.98	33.98	77.64
news	377109	**38.36**	39.68	40.44	82.37	45.62	39.69	82.68
paper1	53161	**37.80**	39.51	40.33	76.96	48.98	39.20	78.38
paper2	82199	**38.10**	38.54	39.04	77.75	43.43	38.92	77.72
progc	39611	**37.90**	39.92	40.94	75.69	51.50	39.53	84.28
progl	71646	**26.93**	29.02	29.89	80.58	35.73	28.37	83.62
progp	49379	**26.73**	29.40	30.52	85.70	40.28	28.29	86.80
trans	93695	**22.53**	26.10	27.01	90.77	38.38	24.03	96.63

Table 7. Compression of the Calgary Corpus using Modified Huffman after BWT

File	Size (bytes)	MTF	SBR(0.5)	TS	FC	TR	MTF′	FC′
bib	111261	**26.25**	27.01	27.53	65.70	33.29	27.34	65.70
book1	768771	32.54	**31.66**	31.89	56.91	33.49	32.71	56.86
book2	610856	27.70	**27.61**	27.99	59.93	31.58	28.30	59.94
news	377109	**33.44**	34.25	34.91	64.55	39.64	34.75	64.63
paper1	53161	**32.96**	34.17	35.06	59.46	42.78	34.51	59.48
paper2	82199	**32.39**	32.75	33.30	58.72	37.65	33.33	58.67
progc	39611	**33.21**	34.76	35.64	62.38	44.96	34.99	64.78
progl	71646	**23.43**	24.82	25.53	60.39	31.22	24.91	61.83
progp	49379	**23.22**	25.40	26.26	62.51	34.74	25.24	62.56
trans	93695	**20.42**	22.99	23.84	65.59	33.45	22.51	71.19

of in the order of 0.3% uniformly across the different variants for these rather modest file sizes.

According to these results, this schema outperforms all other algorithms in our study. Figure 1 reports the mean, median and variance of the comparison of other compression algorithms to the modified Huffman algorithm.

4.4 Splay Trees

List update algorithms belong to the area of self-organizing data structures. Another well known self-organizing data structure is the splay tree [20]. The splay tree is a binary search tree which applies a splay operation after each access to an item. This operation reorganizes the tree such that the most recently accessed item is moved to the root of the tree. Splay trees are believed to have good performance on sequences with high locality of reference. The working

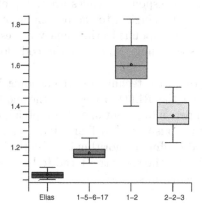

Fig. 1. Relative compression ratio versus modified Huffman. For each file, Modified Huffman equals 1.

Table 8. Compression of the Calgary Corpus using splay trees after BWT

File	Size (bytes)	Elias	RL(1)+Elias	Modified Huffman
bib	111261	37.76	35.14	31.43
book1	768771	44.91	44.94	40.40
book2	610856	38.53	37.12	33.75
news	377109	46.05	45.35	40.49
paper1	53161	43.63	43.53	38.94
paper2	82199	43.71	43.44	38.97
progc	39611	44.14	43.95	39.06
progl	71646	32.14	29.98	27.43
progp	49379	31.34	29.21	26.63
trans	93695	28.92	25.71	23.32

set theorem of [20] shows that splay trees have the working set property. The working property is based on the idea that an operation on a recently accessed item should take less time. Informally, a structure has the working set property if it performs well on sequences with high locality of reference. As stated before there is usually high locality of reference in texts (especially after applying BWT) and thus splay trees are good candidates for text compression. Jones [21] and Grinberg et al. [22] have already studies the application of splay trees to data compression, but they did not consider the BWT.

We studied the effect of using splay trees instead of list update algorithms in our compression schemas. We constructed a splay tree on the characters of the text file. Each character corresponds to a node of the tree and has a binary code that corresponds to the path from the root to its node, i.e., starting from the root, append 0 for each left traversal and 1 for each right traversal. Note that as we proceed with the compression process, the tree changes dynamically and thus the codes for characters are changing as well. Since characters can be in internal nodes, the corresponding codes are not prefix-free. To obtain a prefix-free code, we first add a single 1 to the beginning of each code. Then we consider the number that corresponds to this binary representation and encode these integers using Elias encoding. Note that the code for the root character would be 1.

We can also apply alternative techniques for encoding integers proposed in Subsection 4.3. We tested the RL(1)+Elias and the modified Huffman techniques. The compression percentages obtained by applying these schemas to the text files of the Calgary Corpus after BWT are shown in Table 8. According to these results, the modified Huffman algorithm is again the best technique for encoding integers. Furthermore, the splay trees lead to less compression compared to the good list update algorithms.

5   Conclusions

We have considered a variety of list update algorithms in the context of data compression with and without the Burrows-Wheeler transform. We observed that list update algorithms optimize for a similar but different objective than a compressor and give an example of an algorithm which is a good choice for list update but not for compression. Our experiments showed that competitive list update algorithms are not effective as compressors without BWT, while they perform well after BWT. We also considered several schemas for encoding a sequence of integers that is obtained after applying the list update algorithms. Furthermore, we experimentally tested the efficacy of splay trees in data compression and observed that they are not as effective as list update algorithms.

References

1. Angelopoulos, S., Dorrigiv, R., López-Ortiz, A.: List update with locality of reference. In: Laber, E.S., Bornstein, C., Nogueira, L.T., Faria, L. (eds.) LATIN 2008. LNCS, vol. 4957, pp. 399–410. Springer, Heidelberg (2008)
2. Bentley, J.L., Sleator, D.D., Tarjan, R.E., Wei, V.K.: A locally adaptive data compression scheme. Communications of the ACM 29, 320–330 (1986)
3. Albers, S., Mitzenmacher, M.: Average case analyses of list update algorithms, with applications to data compression. Algorithmica 21(3), 312–329 (1998)
4. Bachrach, R., El-Yaniv, R., Reinstadtler, M.: On the competitive theory and practice of online list accessing algorithms. Algorithmica 32(2), 201–245 (2002)
5. Burrows, M., Wheeler, D.J.: A block-sorting lossless data compression algorithm. Technical Report 124, DEC SRC (1994)
6. Kaplan, H., Landau, S., Verbin, E.: A simpler analysis of burrows-wheeler based compression. In: Lewenstein, M., Valiente, G. (eds.) CPM 2006. LNCS, vol. 4009, pp. 282–293. Springer, Heidelberg (2006)
7. Chapin, B.: Switching between two on-line list update algorithms for higher compression of burrows-wheeler transformed data. In: Data Compression Conference, pp. 183–192 (2000)
8. Nagy, D.A., Linder, T.: Experimental study of a binary block sorting compression scheme. In: Data Compression Conference, pp. 439–448 (2003)
9. Deorowicz, S.: Improvements to burrows-wheeler compression algorithm. Software, Practice, and Experience 30(13), 1465–1483 (2000)
10. Fenwick, P.M.: The Burrows-Wheeler Transform for block sorting text compression: principles and improvements. The Computer Journal 39(9), 731–740 (1996)
11. Balkenhol, B., Kurtz, S.: Universal data compression based on the burrows-wheeler transformation: Theory and practice. IEEE Transactions on Computers 49(10), 1043–1053 (2000)
12. Balkenhol, B., Kurtz, S., Shtarkov, Y.M.: Modifications of the burrows and wheeler data compression algorithm. In: Data Compression Conference, pp. 188–197 (1999)
13. Seward, J.: bzip2, a program and library for data compression, http://www.bzip.org/
14. Sleator, D.D., Tarjan, R.E.: Amortized efficiency of list update and paging rules. Communications of the ACM 28, 202–208 (1985)
15. Albers, S.: Improved randomized on-line algorithms for the list update problem. SIAM Journal on Computing 27(3), 682–693 (1998)
16. Schulz, F.: Two new families of list update algorithms. In: Chwa, K.-Y., H. Ibarra, O. (eds.) ISAAC 1998. LNCS, vol. 1533, pp. 99–108. Springer, Heidelberg (1998)

17. Witten, I.H., Bell, T.: The Calgary text compression corpus. Anonymous ftp from
    `ftp.cpsc.ucalgary.ca/pub/text.compression/corpus/`
    `text.compression.corpus.tar.Z`
18. Arnold, R., Bell, T.C.: A corpus for the evaluation of lossless compression algo-
    rithms. In: Data Compression Conference, pp. 201–210 (1997)
19. Elias, P.: Universal codeword sets and representations of the integers. IEEE Trans-
    actions on Information Theory 21(2), 194–203 (1975)
20. Sleator, D.D., Tarjan, R.E.: Self-adjusting binary search trees. Journal of the
    ACM 32(3), 652–686 (1985)
21. Jones, D.W.: Application of splay trees to data compression. Communications of
    the ACM 31(8), 996–1007 (1988)
22. Grinberg, D., Rajagopalan, S., Venkatesan, R., Wei, V.K.: Splay trees for data com-
    pression. In: Proceedings of the sixth annual ACM-SIAM symposium on Discrete
    algorithms (SODA 1995), pp. 522–530 (1995)

Scheduling Additional Trains on Dense Corridors*

Holger Flier[1], Thomas Graffagnino[2], and Marc Nunkesser[1]

[1] ETH Zürich, Institute of Theoretical Computer Science, Switzerland
{holger.flier,marc.nunkesser}@inf.ethz.ch
[2] SBB AG Bern, Infrastruktur/Trassenmanagement, Switzerland
thomas.graffagnino@sbb.ch

Abstract. Every train schedule entails a certain risk of delay. When adding a new train to an existing timetable, planners have to take the expected risk of delay of the trains into account. Typically, this can be a very laborious task involving detailed simulations. We propose to predict the risk of a planned train using a series of linear regression models on the basis of extensive real world delay data of trains. We show how to integrate these models into a combinatorial shortest path model to compute a set of Pareto optimal train schedules with respect to risk and travel time. We discuss the consequences of different model choices and notions of risk with respect to the algorithmic complexity of the resulting combinatorial problems. Finally, we demonstrate the quality of our models on real world data of Swiss Federal Railways.

1 Introduction

The demand for passenger train transportation has been increasing steadily in Switzerland since the introduction of Rail 2000 [12]. As a consequence, Swiss Federal Railways (SBB) has to operate more trains. It seems difficult if not impossible to expand track resources at the same rate as demand increases. Therefore, railway traffic is becoming denser, making both resource scheduling and delay management more difficult and of major importance.

In this paper, we address the recurring problem of adding a train path, i.e., a schedule for a single train in terms of track allocation in space and time, to a given dense timetable on a corridor, i.e., an important subnetwork in form of a path between two major stations. In particular, we are interested in finding robust train paths in the sense that the additional train has a low risk of delay upon arrival at its final station. For related work, although without explicit coverage of robustness, see, e.g., [3,4,7]. A general notion of robustness is proposed in [10]. Currently, planners use a mixture of domain knowledge and past experience to come up with potential solutions which then undergo detailed simulations to select the most appropriate solution. We present a model that

---

* This work was partially supported by the Future and Emerging Technologies Unit of EC (IST priority - 6th FP), under contract no. FP6-021235-2 (project ARRIVAL).

J. Vahrenhold (Ed.): SEA 2009, LNCS 5526, pp. 149–160, 2009.

supports railway planners by computing a set of recommended train paths for a given train request.

A novelty of our approach is that we use extensive historic delay data to compute such recommendations. The underlying data have been recorded by SBB during operations. We combine risk predictions with a combinatorial model that can answer the planners' queries very quickly. As there is a trade-off between risk and travel time of a train path, not only a single solution is computed, but a set of Pareto optimal solutions with respect to travel time and expected delay. Thus, as an advantage over simulating just a few scenarios, the planners get a range of different, efficient solutions. Another advantage is that most of the necessary data are available from the database. The data implicitly contain a wealth of information, e.g., dependencies between trains, resource bottlenecks, or dispatching decisions. We will show how to profit from these information.

The paper is structured as follows. In Section 2 we present the problem more formally. Our solution approach consists of two main steps. In the first step, as described in Section 3, we extract predictors from the historic delay data in order to compute a series of linear regression models for risk prediction. In the second step, a set of Pareto optimal train paths is computed for a given request. We give the algorithmic details and complexity results of the second step in Section 4. In Section 5 we present experiments that show the quality of our approach.

2   Problem Description

A typical requests that planners have to deal with is, e.g., "add one train in the morning rush hour between Bern and Zurich". The planner's task is to add a train path that satisfies the client's request, is feasible with respect to operational safety constraints, and has a low risk of delay, to a timetable that has been in operation over a period of time.

More formally, a request for an additional train Θ specifies the type $\tau(\Theta)$ (e.g. local, regional, long distance), the corridor $\langle S_1, S_2, \ldots, S_\ell \rangle$, and the dates on which the train should run (e.g. weekdays, weekends). Further, the request comprises earliest and latest departure times $[\overline{d}, \underline{d}]$ and arrival times $[\overline{a}, \underline{a}]$ at the first station S_1 and the last station S_ℓ, respectively, as well as intermediate stops, if any, at stations along the corridor.

A train path π is characterized by arrival and departure times t_i^a and t_i^d, respectively, at stations S_i, $i \in \{1, \ldots, \ell\}$, where the train stops, and by pass-through times t_i^p at stations where it does not, e.g., $\pi = \left(t_1^d, t_2^p, t_3^p, t_4^a, t_4^d, t_5^p, t_6^a\right)$.

3   Regression Models

In this section, we first describe how to compute a series of linear regression models from historic delay data. These models allow us to "predict" delays of *historic* train paths during the period in which the timetable has been operational. Thus, the predicted delays are not in the future but in the past! The purpose of our prediction is not to be able to predict delays of current trains, but to be able to evaluate how an additional train would have been delayed if it had run on a specific planned train path on a day in the timetable period.

Second, we define the *risk* of a train path as an aggregated value of the delay predictions for all days of the recorded period. This definition allows to associate a risk with every feasible train path. An appropriate choice of the prediction model will allow us to search efficiently for optimal train paths by a special shortest path computation in a time expanded graph. For the statistical terminology used in this section, see any textbook on linear models, for example [9].

3.1 Predictors and Linear Regression Models

The first step towards prediction of delays and thus the final goal of a conclusive risk measure is to identify relevant predictors that can be extracted from the recorded delay data. In cooperation with planners from SBB we identified the following potential "causes" for the delay of a train Θ on train path π upon planned arrival at station S_i at time t_i^a on day d of the recorded period:

previous delay propagation of delay $\delta_{i-1}(\Theta, d, \pi)$ of train Θ at the previous station S_{i-1} on day d

type of train $\tau(\Theta)$, a set of indicator variables for each possible train type

train density number of actual train arrivals or departures at S_i in an interval around the planned arrival time t_i^a, denoted by $\text{window}_q(t_i^a, d)$, $q \in (I \times J)$ for a set of time intervals (windows) I and a set of cases J of trains arriving/departing in the same/opposite direction as train Θ

timetable measures time $\Delta_j^{\text{prev}}(t_i^a, d)$ to the j-th previous/next planned train having the same direction as train Θ; slack time $\text{slack}(\Theta, t_{i-1}^d, t_i^a)$ with respect to the minimum driving time of train Θ between S_{i-1} and S_i

delays of neighboring trains delays of the trains that are scheduled directly before/after the planned arrival time t_i^a, e.g., $\delta_4^{\text{prev}}(t_i^a, d)$ is the delay of the fourth train before t_i^a on day d according to plan

properties of the tracks $\text{track-loss}(t_i^a, d)$, the average net change in delay of trains between S_i and S_ℓ during one hour around t_i^a on day d

Our goal is to use the most relevant of the above predictors in the linear regression models. We emphasize that we are not mainly interested in the exact type of dependence of predictors and dependent variable but rather in a model that predicts well and that blends well with our combinatorial search for a low risk path. To get meaningful models with well-balanced bias and variance we select a subset of good predictors that lead to models with adequate Akaike information criterion (AIC) [2]. The AIC is an established tool for model selection that tends to avoid overfitting problems. To find these models we used the greedy stepAIC algorithm of Venables and Ripley [13] implemented in R [11]. We stopped this algorithm after 20 steps when usually no significant further improvement in terms of AIC was made. We note that due to the large amount of available data, overfitting is not very likely to occur in our case even if we include the full set of predictors.

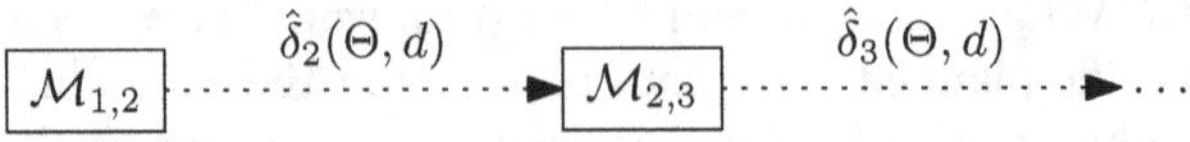

$$\boxed{\mathcal{M}_{1,2}} \cdots\cdots\cdots\cdots\cdots\cdots\xrightarrow{\hat{\delta}_2(\Theta,d)} \boxed{\mathcal{M}_{2,3}} \cdots\cdots\cdots\cdots\cdots\cdots\xrightarrow{\hat{\delta}_3(\Theta,d)} \cdots$$

Fig. 1. Sequence of between-station models where the delay predicted by the previous model is a predictor for the next

Given the stations $\{S_1, \ldots, S_\ell\}$ along the corridor, we set up a linear regression model $\mathcal{M}_{i-1,i}$ for each pair of consecutive stations (S_{i-1}, S_i), $i \in \{2, \ldots, \ell\}$, to predict the delay $\delta_i(\Theta, d, \pi)$ upon arrival at station S_i. We call these models the *between-stations* models. We also set up an *in-station model* $\mathcal{M}_{i,i}$ for each intermediate station S_i in which the train is requested to stop, by analogous definitions of predictors, but omit their description for brevity. Model $\mathcal{M}_{i-1,i}$ uses the set of predictors sketched above, in particular the previous delay as predicted by model $\mathcal{M}_{i-2,i-1}$ (or $\mathcal{M}_{i-1,i-1}$). This means that the prediction of the last model is used as predictor for the next model in the sequence, as illustrated in Figure 1.

The models $\mathcal{M}_{i-1,i}$ are basically of the form

$$\delta_i(\Theta, d, \pi) = \alpha + \sum_k \beta_k \text{predictor}_k + \varepsilon_{i,\Theta,d}$$
$$= \text{model}(\Theta, t_i^a, t_{i-1}^d, d, \delta_{i-1}) + \varepsilon_{i,\Theta,d} \tag{1}$$

where predictor_k denotes the predictors described above, and under the usual assumptions for linear regression models [8].

Note the dependency of the fitted value for $\hat{\delta}_i(\Theta, d, \pi)$ on Θ, t_i^a, t_{i-1}^d, d, and $\delta_{i-1}(\Theta, d, \pi)$ as indicated by the term $\text{model}(\ldots)$.

3.2   Series of Regression Models

In order to compute a risk measure for a planned train path $\pi = \left(t_1^d, \ldots, t_\ell^a\right)$ we use the sequence of regression models $(\mathcal{M}_{1,2}, \ldots, \mathcal{M}_{\ell-1,\ell})$ to predict the delay on each day d of the recorded period of time:

$$\hat{\delta}_i(\Theta, d, \pi) = \begin{cases} \text{model}(\Theta, t_i^a, t_{i-1}^d, d, \hat{\delta}_{i-1}) & \forall i > 1 \\ \delta_0(\Theta, d, \pi) \end{cases} \tag{2}$$

where $\delta_0(\Theta, d, \pi)$ is an estimation of the start delay of the train (for example the average delay of trains in that hour of the day). As a risk measure we propose an aggregated value of these values:

Definition 1 (risk). *For a given train Θ and a train path π on a corridor $\langle S_1, S_2, \ldots, S_\ell \rangle$ we define its risk with respect to a recorded period D and a given prediction model as $\text{risk}(\pi) = \frac{1}{|D|} \sum_{d \in D} \hat{\delta}_\ell(\Theta, d, \pi)$, where the $\hat{\delta}_\ell(\Theta, d, \pi)$ values are obtained via regression models as in (2).*

There are different types of possible regression models. In particular, if we restrict the model above to a subset of the predictors, we can limit its dependency on the data. A very basic model depends only on $\hat{\delta}_{i-1}$, t_i^a, Θ and d. Such a model could look as follows, using dummy-variable regression for the categorical predictor $\tau(\Theta)$:

$$
\begin{aligned}
\textbf{basic:} \quad \hat{\delta}_i(\Theta, d, \pi) &= \text{model}(t_i^a, \Theta, d, \hat{\delta}_{i-1}) \\
&= \alpha + \beta_1 \hat{\delta}_{i-1}(\Theta, d, \pi) + \beta_2 \tau(\Theta) + \sum_q \beta_{3,q} \text{window}_q(t_i^a, d) \\
&\quad + \sum_j \left(\beta_{4,j} \Delta_j^{\text{prev}}(t_i^a, d) + \beta_{5,j} \delta_j^{\text{prev}}(t_i^a, d) \right) \\
&\quad + \sum_{j'} \left(\beta_{6,j'} \Delta_{j'}^{\text{next}}(t_i^a, d) + \beta_{7,j'} \delta_{j'}^{\text{next}}(t_i^a, d) \right) \\
&= \beta_1 \hat{\delta}_{i-1} + b(t_i^a, \Theta, d)
\end{aligned} \tag{3}
$$

Here $b()$ is a value that depends only on the indicated terms. More advanced models depend also on t_{i-1}^d, use power transformed predictors, or involve interaction terms not containing $\hat{\delta}_{i-1}$. Interaction terms are basically products of predictors, see [9] for more details. Such models can for example take into consideration the interaction between track loss and slack. This could model the potential situation that trains with high slack between two stations are not affected by high track losses of other trains, whereas trains with low slack are.

$$
\begin{aligned}
\textbf{advanced:} \quad \hat{\delta}_i(\Theta, d, \pi) &= \text{model}(\Theta, t_i^a, t_{i-1}^d, d, \hat{\delta}_{i-1}) \\
&= \text{basic} + \beta_8 \text{track-loss} + \beta_9 \text{slack}(\Theta, t_{i-1}^d, t_i^a) \\
&\quad + \beta_{10} \text{track-loss} : \text{slack}(\Theta, t_{i-1}^d, t_i^a) + \ldots \\
&= \beta_1 \hat{\delta}_{i-1} + b(t_i^a, t_{i-1}^d, \Theta, d)
\end{aligned} \tag{4}
$$

If one wants to model that different types of train can catch up differently on delays one would also have to include interaction terms involving $\hat{\delta}_{i-1}$. Another example would be the idea that track loss and previous delay interact, i.e., in situations with high track loss a high previous delay will lead to a high delay at the current station, whereas with low track loss it will have a much lower effect.

$$
\begin{aligned}
\textbf{all interactions:} \quad \hat{\delta}_i(\Theta, d, \pi) &= \text{advanced} + \ldots + \hat{\delta}_{i-1} : \tau(\Theta) \\
&= a(t_i^a, t_{i-1}^d, \Theta, d)\hat{\delta}_{i-1} + b(t_i^a, t_{i-1}^d, \Theta, d)
\end{aligned} \tag{5}
$$

For constant $(t_i^a, t_{i-1}^d, \Theta, d)$, the models above all boil down to a simple linear function in $\hat{\delta}_{i-1}$. One can easily think of further refinements leading to "arbitrary" functions in $\hat{\delta}_{i-1}$, see [8]. Such models would possibly allow for a better quality of prediction, but already models of type "all interactions" can make the combinatorial model discussed in the next section NP-hard.

4   Shortest Path Algorithm

The search for train paths with low risk, as defined in the last section, leads to a shortest path problem on an appropriately defined time expanded graph. In the following, we describe how this graph is constructed, discuss the algorithmic complexity of possible shortest path models, and describe the algorithm used in our experiments. All proofs are given in the corresponding technical report [8].

4.1   Time Expanded Graph Model

We want to construct a graph such that every path in the graph corresponds to a feasible planned train path w.r.t. the most important operational and safety constraints. For our purposes, these are minimum/maximum driving times, the number of available parallel (bidirectional) tracks between stations, and headway times, i.e., the security requirement that a train can follow another one on the same track only after a certain time span.

Given a train request r, a layered time expanded graph $G_r = (V_1 \uplus V_2 \uplus \ldots \uplus V_\ell, E)$ is constructed as follows. Each node $u_i^t \in V_i$ represents a station S_i at a certain point in time t. The number of nodes in each V_i depends on r and on the chosen granularity, e.g., 10 nodes per minute. Every edge $(u_i^t, u_{i+1}^{t'}) \in E$ represents a driving activity between two stations or a dwelling activity within a station. For simplicity, we denote u_i^t simply by u_i and the edge $(u_i^t, u_{i+1}^{t'})$ by $e_{i,i+1}$. Thus, every u_1-u_ℓ-path in G_r with $u_1 \in V_1, u_\ell \in V_\ell$, corresponds to a train path π.

To model realistic driving times, we distinguish between three types of nodes representing the state of the train, namely *arrival* (arr), *departure* (dep), and *pass-through* (pass) nodes. Based on these types, we extract minimum and maximum driving times along the corridor from the historic delay data and include only those edges in G_r that respect them.

Track capacities and headway constraints are modeled by omitting edges of G_r which would cause a train path to be infeasible w.r.t. the current timetable. Hence, for every potential edge we need to decide if it would be possible to schedule an additional train on the track segment and at the time specified by that edge. For our purposes, the time during which a track is blocked by a train can be modeled by a trapezoid, as shown in Figure 2(a). The problem of deciding whether an edge e of G_r is feasible reduces to the chromatic number problem in trapezoid graphs, which can be solved in time $O(n \log n)$, see [5,6]. Assuming that the existing timetable is feasible w.r.t. headway constraints, it even suffices to compute the size of the maximum clique containing the vertex corresponding to the trapezoid of e, see Figure 2(b) and (c).

4.2   Model Choice and Algorithmic Complexity

In this section we discuss the complexity of several shortest path problems arising from different regression models. To compute Pareto optimal paths in G_r, we need to assign *costs* to paths, reflecting their risk according to Definition 1. First,

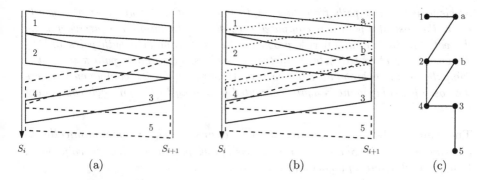

Fig. 2. (a) Trapezoidal representation of a schedule between two consecutive stations S_i and S_{i+1}. Each trapezoid represents the resource utilization (headway constraints) of a single train in time (vertical axis) and space (horizontal axis). Here, two tracks suffice, as trains 1, 2, and 3 can be scheduled on one track and trains 4 and 5 on the other without conflict. (b) Requests for additional trains a and b. (c) Corresponding trapezoid graph for (b). Trains 1 to 5 and a could still be scheduled on two parallel tracks, since the size of its maximum clique in the trapezoid graph is two, whereas to add train b, three parallel tracks would be necessary.

we consider the cost of a path for a single day $d \in D$ only. In this case, we let the risk equal the predicted delay $\hat{\delta}_\ell(\Theta, d, \pi)$. The structure of Equations 3, 4, and 5 leads to a cost structure, in which on each edge $(u_i^t, u_{i+1}^{t'})$ the accumulated delay $\hat{\delta}_i(\Theta, d, \pi)$ at u_i^t is multiplied with a constant $a(e)$ and then a constant $b(e)$ is added to this value to yield $\hat{\delta}_{i+1}(\Theta, d, \pi)$. Hence, to define the cost structure for the minimum risk computation for a single day, one can annotate the edges with these pairs (a, b). More formally, mirroring Equation 2, the cost of a path $\pi = \{u_1, u_2, \ldots, u_\ell\}$ in G_r can be recursively defined as follows:

$$\text{cost}(u_1, u_2, \ldots, u_i) = \begin{cases} a(u_{i-1}, u_i)\text{cost}(u_1, u_2, \ldots, u_{i-1}) + b(u_{i-1}, u_i) & \text{for } i > 1 \\ \delta_0(\Theta, d, \pi) & \text{for } i = 1 \end{cases}$$

(6)

which yields

$$\text{cost}(\pi) = \delta_0(\Theta, d, \pi)a(e_{1,2})a(e_{2,3}) \cdots a(e_{\ell-1,\ell}) + b(e_{1,2})a(e_{2,3}) \cdots a(e_{\ell-1,\ell})$$
$$+ \ldots + b(e_{\ell-2,\ell-1})a(e_{\ell-1,\ell}) + b(e_{\ell-1,\ell}) \quad (7)$$

The risk computation for the whole period D can be carried out by doing the above computation for each day $d \in D$ resulting in $|D|$ delay predictions, which can be read as a vector $(\hat{\delta}_\ell(\Theta, 1, \pi), \ldots, \hat{\delta}_\ell(\Theta, |D|, \pi))^T$. According to Definition 1, the risk is the average over the entries of this vector. It follows that the full shortest path problem is a problem over vectors of dimension $|D|$, which we formalize in the following definition.

Definition 2. *Given a layered time-expanded graph $G_r = (V_1 \uplus V_2 \uplus \ldots \uplus V_\ell, E)$ with edges $e = (u_i, u_{i+1})$ labeled by $(a(e), b(e))$. The one day minimum risk* *problem asks for a path from layer 1 to layer ℓ of minimum cost at layer ℓ, where the cost of a path π is computed according to Equation 7 recursively along the path.* In the minimum risk *problem edges are annotated with pairs of $|D|$-dimensional vectors $(\boldsymbol{a}(e), \boldsymbol{b}(e))$ instead of scalars. For a given path π its cost is computed as the average over the costs $c_i, 1 \leq i \leq |D|$ for the components, where the cost of a component is again computed according to (7) for each component separately.*

Theorem 1. *As long as the prediction functions model() are monotonically increasing in $\hat{\delta}_{i-1}$ the one day minimum risk problem can be solved in polynomial time by a label setting algorithm.*

This theorem characterizes in a sense "well-behaved" models. If the models predictions are not monotonically increasing in $\hat{\delta}_{i-1}$ a model might predict that a train arrives earlier at station i for larger delays at station $i - 1$. Note that G_r is acyclic.

For an efficient algorithm for the minimum risk problem we need more than just the efficient computation of the one day problem.

Theorem 2. *If all components of the cost vectors $\boldsymbol{a}(e_{i,i+1})$ are equal to a single value $a_{i,i+1}$ for each layer $1 \leq i < \ell - 1$ of G_r, i.e., $\boldsymbol{a}(e_{i,i+1}) = a_{i,i+1}\boldsymbol{1}$, then the minimum risk problem can be solved by a label setting algorithm in polynomial time.*

Fortunately, this condition is met by the "basic" models, by the "advanced" models and even by models that include interactions of $\hat{\delta}_{i-1}$ and predictors that do not depend on d like $\tau(\Theta)$ or slack$(\Theta, t_{i-1}^d, t_i^a)$. The above theorem is complemented with an NP-hardness proof for models with varying $\boldsymbol{a}$.

Theorem 3. *The general minimum risk problem (without the condition of Theorem 2 on the $\boldsymbol{a}$ vectors) is NP-hard.*

This concerns models of type "all interactions" that include for example interaction terms of $\hat{\delta}_{i-1}$ and some window variables or any other predictor that depends on d.

As far as the aggregation function in the risk computation is concerned, SBB planners prefer to work with the more robust median. Therefore, one could also define

$$\widetilde{\text{risk}}(\pi) = \text{median}\{\hat{\delta}_\ell(\Theta, d, \pi) \mid d \in D\} \tag{8}$$

as the median of the delay predictions for the last station. This choice, however, leads to an NP-hard shortest path problem, as the following theorem shows.

Theorem 4. *For the median as an aggregation function in the risk computation the classical shortest path problem with respect to this cost measure $\widetilde{\text{risk}}(\pi)$ is NP-hard already for additive vector valued edge costs and therefore also for all variants discussed here.*

As every u_1-u_ℓ-path is only a suggestion for the planner, who may have to take further feasibility requirements into account, we would like to provide a set of k

"best" solutions. Calculating the k-shortest paths, however, would lead to a set of solutions that are very similar to each other. Instead, we propose to compute a Pareto frontier with respect to the trade-off between risk and travel time, which is a natural choice in this context.

Lemma 1. *The size of the Pareto frontier is proportional to the difference of the minimal and maximal travel time of u_1-u_ℓ-paths in G_r.*

Note that although the size of the Pareto frontier is pseudo-polynomial in the size of the input, the range of possible travel times of u_1-u_ℓ-paths is limited in practice.

4.3   Algorithm

We sketch the algorithm to find a set of Pareto optimal u_1-u_ℓ-paths in G_r. Since G_r is acyclic, it suffices to consider each edge once in the order given by any topological sorting of the nodes and to apply a reaching algorithm [1]. Note that a topological sorting of the nodes is readily available by the order of the stations along the corridor, as each node is associated with one station.

First, G_r is created as defined above. For performance reasons, all nodes and edges that are not on a u_1-u_ℓ-path can be removed in a preprocessing step. Associated with every node u_i is a Pareto frontier $F(u_i)$ of paths from S_1 to S_i with associated labels, i.e., the accumulated risk up to S_i. For each edge (u_i, u_j), the algorithm checks – by evaluating the paths in $F(u_i)$ with model $\mathcal{M}_{i,j}$ – whether labels in $F(u_i)$ can be extended to u_j, such that they dominate labels in $F(u_j)$. Whenever this is the case, a new label is inserted into $F(u_j)$.

Once all edges have been considered, the algorithm constructs the final Pareto frontier F from the labels of all $F(u_\ell)$ at the last station S_ℓ, eliminating all dominated labels. Thus, F contains only Pareto optimal u_1-u_ℓ-paths.

Note that even though the estimators provided by the linear regression models are unbiased, this unbiasedness is lost in the search for minimum risk paths. We explain this effect in [8] and also justify, why it is negligible in our case.

5   Experiments

To demonstrate the quality of the models, we created between-station models for the Zofingen-Lucerne corridor in Switzerland, as listed in Table 1. The residual error S_E is less than 30 seconds for the majority of the models, and not more than 50 seconds for any model. Apart from some outliers, which one would expect, the residuals are very moderate. Against the background of complicated dependencies between trains in real world operations, the results are very encouraging.

To get a better impression on how the residuals are typically distributed, see Figure 3. The between-stations model for Wauwil-Sursee in Figure 3(a) has a very good fit, which also mirrors the fact that it does not seem to be a "critical" station. On the other hand the in-station model for Olten in Figure 3(b) has a

Table 1. Quality of between-stations models on the Zofingen-Lucerne corridor. The residual standard error is denoted by S_E, the degrees of freedom by DoF, the i-th quartile by iQ. Residuals and S_E are given in seconds.

| from | to | S_E | DoF | residuals | | | | | Multiple r^2 | Adjusted r^2 |
				min	1Q	2Q	3Q	max		
ZF	BRIT	9.2	11903	-27.8	-5.1	-1.2	3.7	126.0	0.9931	0.9931
BRIT	DAG	12.6	11903	-61.5	-4.8	-1.3	3.5	295.9	0.9874	0.9874
DAG	NEB	8.2	11902	-57.5	-3.5	-0.8	2.3	219.2	0.9948	0.9948
NEB	WAU	6.2	11902	-33.8	-3.4	-0.5	2.7	177.2	0.9971	0.9971
WAU	SS	18.3	11898	-105.1	-8.5	-2.3	5.4	370.5	0.9778	0.9778
SS	SEM	27.8	14273	-108.0	-14.6	-2.3	11.4	708.9	0.9403	0.9402
SEM	RBG	21.1	14274	-141.3	-7.9	-2.2	3.5	1032.0	0.9686	0.9685
RBG	HUEB	25.2	14274	-102.1	-11.4	-0.8	8.3	382.9	0.9582	0.9581
HUEB	EBR	16.3	19964	-58.5	-7.7	-1.7	4.6	663.7	0.9820	0.9820
EBR	GTS	49.0	19966	-223.1	-27.4	-5.9	18.9	587.5	0.8342	0.8340
GTS	LZ	41.2	37761	-202.4	-22.5	-7.3	14.0	760.5	0.8769	0.8768

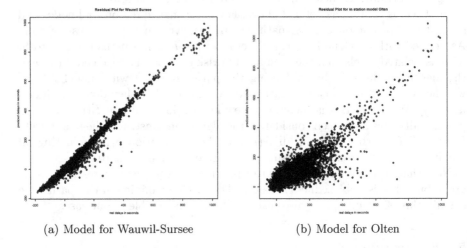

(a) Model for Wauwil-Sursee   (b) Model for Olten

Fig. 3. Residual plots depicting real delay (x-axis) vs. predicted delay (y-axis)

less good fit, which might come from the more complicated structure of delays in Olten. In both plots outlying points are rather below than above the diagonal, which is exactly what one would expect from a delay prediction model: some delays are simply unpredictable.

The two residual plots also help to see to what extent the standard assumptions of linear regression modeling are satisfied. From both plots one can see that the linearity assumption $\mathbb{E}(\varepsilon_{i,\Theta,d}) = 0$ seems to hold. On the other hand, the constant variance assumption does not seem to hold, the residuals look heteroscedastic. As this does not influence the unbiasedness and consistency of the used least squares estimators but rather the efficiency, this does not invalidate

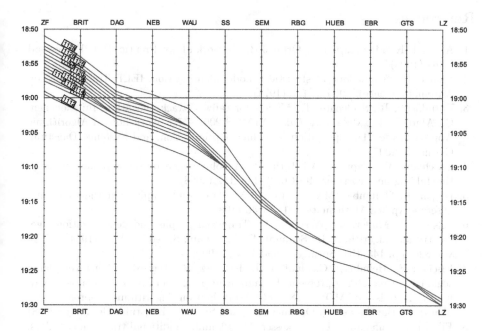

Fig. 4. Pareto optimal train paths for a (hypothetical) request on the Zofingen-Lucerne corridor during the evening hours, not showing the existing timetable

our approach: Given the very large amount of data that the models are estimated from, statistical efficiency is not our primary concern.

To demonstrate the possible quality of predicting a whole train path, we performed a cross validation on an extra train that drove on only 9 days, which were removed from the data before the models were learned, see [8] for details.

Continuing our example of the Zofingen-Lucerne corridor, we briefly present a resulting Pareto optimal set of train paths for a hypothetical user requests in Figure 4. The request was to add a fast train from Zofingen to Lucerne with earliest departure time 17:00, and latest arrival time 20:00, no intermediate stops, maximum driving time of 150% (w.r.t. the minimum driving time). Interestingly enough, there is no path in the set departing before 18:50, and hence, all such paths are dominated by the depicted solutions. Although the time windows and maximum driving time chosen for this example are very large, the solutions give a clear indication of where the train should be added.

Acknowledgements

We are grateful to Marloes Maathuis and Werner Stahel for their valuable advice regarding the statistical models, and would like to thank the anonymous referees for their helpful comments.

References

1. Ahuja, R.K., Magnanti, T.L., Orlin, J.B.: Network Flows. Prentice Hall, Englewood Cliffs (1993)
2. Akaike, H.: A new look at statistical model identification. IEEE Transactions on Automatic Control 19, 716–723 (1974)
3. Borndörfer, R., Schlechte, T.: Models for railway track allocation. In: Liebchen, C., Ahuja, R.K., Mesa, J.A. (eds.) ATMOS 2007 – 7th Workshop on Algorithmic Approaches for Transportation Modeling, Optimization, and Systems, Dagstuhl, Germany (2007)
4. Cacchiani, V., Caprara, A., Toth, P.: A column generation approach to train timetabling on a corridor. 4OR 6(2), 125–142 (2008)
5. Dagan, I., Golumbic, M.C., Pinter, R.Y.: Trapezoid graphs and their coloring. Discrete Apllied Mathematics 21, 35–46 (1988)
6. Felsner, S., Müller, R., Wernisch, L.: Trapezoid graphs and generalizations, geometry and algorithms. In: Schmidt, E.M., Skyum, S. (eds.) SWAT 1994. LNCS, vol. 824, pp. 143–154. Springer, Heidelberg (1994)
7. Fischer, F., Helmberg, C., Janßen, J., Krostitz, B.: Towards solving very large scale train timetabling problems by lagrangian relaxation. In: Fischetti, M., Widmayer, P. (eds.) ATMOS 2008 – 8th Workshop on Algorithmic Approaches for Transportation Modeling, Optimization, and Systems, Dagstuhl, Germany (2008)
8. Flier, H., Graffagnino, T., Nunkesser, M.: Planning additional trains on corridors. Technical Report 164, ARRIVAL Project (2008), http://arrival.cti.gr/
9. Fox, J.: Applied Regression Analysis and Generalized Linear Models, 2nd edn. SAGE, Thousand Oaks (2008)
10. Liebchen, C., Lübbecke, M., Möhring, R.H., Stiller, S.: Deliverable 1.2: New theoretical notion of the prices of robustness and recoverability. ARRIVAL Project (2007), http://arrival.cti.gr/
11. R Development Core Team. R: A Language and Environment for Statistical Computing. R Foundation for Statistical Computing, Vienna, Austria (2008), http://www.R-project.org/
12. SBB. Rail 2000 - a public transport network for the third millenium (2000), http://www.sbb.ch/rail2000/
13. Venables, W.N., Ripley, B.D.: Modern Applied Statistics with S. Statistics and Computing Series. Springer, Heidelberg (2002)

Broadword Computing and Fibonacci Code Speed Up Compressed Suffix Arrays

Simon Gog

Insitute of Theoretical Computer Science
Ulm University

Abstract. The suffix array of a string s of length n over the alphabet Σ is the permutation that gives us the lexicographic order of all suffixes of s. This popular index can be used to solve many problems in sequence analysis. In practice, one limitation of this data structure is its size of $n \log n$ bits, while the size of the text is $n \log |\Sigma|$ bits. For this reason compressed suffix arrays (CSAs) were introduced. The size of these CSAs is asymptotically less than or equal to the text size if the text is compressible, while maintaining $O(\log^\epsilon n)$ access time to the elements $(0 < \epsilon \le 1)$. The goal of a good CSA implementation is to provide fast access time to the elements while using minimal space for the CSA. Both access time and space depend on the choice of a self-delimiting code for compression. We show that the Fibonacci code is superior to the Elias δ code for strings that are low compressible. Our second contribution are two new broadword methods that support the decoding of Fibonacci encoded numbers on 64 bit architectures. Furthermore, our experiments show that the use of known broadword methods speed up the decoding of Elias δ code for strings that are high compressible, like XML. Finally, we provide a new efficient C++ library for succinct data structures which includes a generic CSA for further experiments.

1   Introduction

Text indexes like the suffix array [10] or the enhanced suffix array [1] are powerful tools to efficiently answer a large range of different queries. However, the size of these data structures is a real problem. While the text itself occupies n bytes (over the ASCII alphabet) the suffix array takes $n \log n$ bits, which is equivalent to $4n$ bytes for text lengths less than four gigabyte. Besides the drawback of the large space consumption to store the index, the case in which the whole index does not fit in main memory is most severe, because query times slow down due to the disk access. For this reasons, much theoretical work has been done at the beginning of this decade, resulting in several *compressed indexes* (see [4,5,9,14]). Compressed means that the size of the index depends on the entropy of the indexed text while preserving the efficient (constant or logarithmic time) access to the elements. Furthermore, most of the indexes are *self-indexes* meaning that the text can be reconstructed from the index. Navarro and Mäkinen [12] provide a good overview over the whole topic.

J. Vahrenhold (Ed.): SEA 2009, LNCS 5526, pp. 161–172, 2009.

In this paper, we present and apply *broadword computations* to speed up the access time to the self-index of Sadakane[14]: The compressed suffix array (CSA). Here, the term *broadword computation* means parallel programming in a register/word of size greater than or equal to 64 bit. This is also known as *SIMD*[1] Within A Register (*SWAR*). The development and use of the new broadword computation methods was inspired by a recent work of Vigna [15] in which SWAR was successfully applied to accelerate rank and select data structures in practice.

In addition to the theoretical results to CSAs, there are efforts to provide efficient implementations. For example, the *Pizza&Chilli*[2] site of Ferragina and Navarro [3] offers a *C* interface, existing solutions, and a set of test cases. This gives us the opportunity to compare our solution to the original implementation of Sadakane.

2   Preliminaries

2.1   Notations and Definitions

A *string* $t = t[0, n-1] = t_0 t_1 \ldots t_{n-1}$ is a sequence of n characters over an *ordered alphabet* Σ (of size σ). We denote the empty string by ϵ. Each string of length n has n *suffixes* $s_i = t[i, n-1]$ $(0 \le i < n)$. We define the *lexicographic order* "$<$" on strings as follows: ϵ is smaller than all other strings. Now $t < t'$ if $t_0 < t_0'$ or $t_0 = t_0'$ and $s_1 < s_1'$.

A *suffix array* SA of a text t is the lexicographically sorted array of all the suffixes of t. The uncompressed version occupies $O(n \log n)$ bits of space. Sadakane [14] presented a CSA that occupies $O(nH_0(T) + n \log \log \sigma)$ bits of space while providing $O(\log^\epsilon n)$ random access time to the entries for $0 < \epsilon \le 1$. Compression is achieved by using the Ψ-function to store the information about the suffix array. Ψ is defined as follows:

$$\Psi(i) = SA^{-1}[(SA[i] + 1) \mod n]$$

where SA^{-1} is the inverse suffix array. That is, $\Psi[i]$ equals the position of $s_{SA[i]+1}$ (the suffix following suffix $s_{SA[i]}$ in the text). The key observation for compression is that Ψ consists of at most σ piecewise increasing sequences [14].

In theory, the access time on *CSA* depends on two data structures that solve the following queries in constant time. Given a bit vector b, the functions $rank(b, i)$ tells us the number of ones in the prefix $b[0, i-1]$ of b and $select(b, j)$ tells us the position where the jth one bit occurs in b. There exist constant time solutions for both query types (see [6,11,13]). These solutions require only $o(n)$ bits extra space on top of the n bits of the bit vector. We call the corresponding data structures `rank_support` and `select_support`. Finally, we will define some notations concerning a 64 bit word w. The least significant bit of w is indexed with 0. A *bit pair* bp_i consists of two consecutive bits $w[i]w[i+1]$. We call it *even* (*odd*) if i is even (odd).

---

[1] Single Instruction, Multiple Data.
[2] http://pizzachili.di.unipi.it/

2.2   Self-delimiting Codes

Definition 1. *Let x be a positive integer. A code $c(x) \in \{0,1\}^*$ is a self-delimiting code if the following conditions hold:*

(a) $c(x)$ is not prefix of $c(y)$ for any integer $y > 0$ and $y \neq x$;

(b) $|c(x)| \leq \alpha \log x + g(x)$, where $g(x) = o(\log x)$, and α is a constant;

Examples of random access self-delimiting codes are *Elias γ-code* and *δ-code* [2](δ-code for short) that are defined as follow (see [16] for more examples):

$$c_\gamma(x) = \underbrace{0\cdots 0}_{|b(x)|-1} 1\tilde{b}(x) \qquad \text{and} \qquad c_\delta(x) = \underbrace{0\cdots 0}_{|\tilde{b}(b(x))|} 1\tilde{b}(|b(x)|)\tilde{b}(x)$$

where $b(x)$ is the binary representation of x and $\tilde{b}(x)$ is $b(x)$ minus the most significant bit of $b(x)$. Thus, the size to encode an integer x in Elias δ-code is about $\log x + 2 \log \log x$. Sadakane used this code in his implementation. We will now present two additional random access self-delimiting codes for which we will present efficient calculation of decoding information in section 3.

The *Fibonacci code c_Φ* (Φ-code for short) is derived from the Fibonacci sequence defined (for this purpose) as $F_0 = 1$, $F_1 = 2$ and for $i \geq 2$ as $F_i = F_{i-1} + F_{i-2}$. *Zeckendorf's theorem* [17] states that we can represent every positive integer x with the Fibonacci code $c_\Phi(x) = \{0,1\}^k 1$ which meets following conditions: (a) $F_k > x$ and $c_\Phi(x)[k-1] = 1$, (b) $x = \sum_{i=0}^{k-1} c_\Phi(x)[i] \cdot F_i$, and (c) $c_\Phi(x)[i-1] \cdot c_\Phi(x)[i] = 0$ for $0 < i < k$. Note that this code is a prefix code, since only the last bit pair bp_{k-1} equals '11'. The length of $c_\Phi(x)$ follows from condition (a) and $F_k \approx \Phi^{k+2}/\sqrt{5}$, where $\Phi = (1 + \sqrt{5})/2$ is the golden ratio. We need about $1 + \log_\Phi x \approx 1 + 1.44 \log x$ bits. This is equal to or less than the length of c_δ for values between 2 and $F_{18} = 6765$. See Fig. 1 for some examples. The *ternary code c_Δ* (Δ-code for short) is based on the ternary system. The integer x is considered as a number in the ternary system that consists of the three symbols $\{00, 01, 10\}$. The terminating bit pair is '11' and the code length of $c_\Delta(x)$ is therefore $2 + \log_3 4 \log x \approx 2 + 1.26 \log x$ bits. This is asymptotically even better than the other codes but in our applications the values are so small that this code will consume too much space.

x	$c_\Phi(x)$	$c_\Delta(x)$	$c_\delta(x)$	x	$c_\Phi(x)$	$c_\Delta(x)$	$c_\delta(x)$
1	11	10 11	1	6	10011	00 01 11	01 1 10
2	011	01 11	01 0 0	7	01011	10 01 11	01 1 11
3	0011	00 10 11	01 0 1	8	000011	01 01 11	001 00 000
4	1011	10 10 11	01 1 00	9	100011	00 00 10 11	001 00 001
5	00011	01 10 11	01 1 01	10	010011	00 00 01 11	001 00 010

Fig. 1. Example of different self-delimiting codes with encoded values

2.3   Sadakane's Compressed Suffix Array Revisited

The most important data structure for the practical implementation of the *CSA* is the compressed form of the Ψ-function supporting constant time random access to its elements. As this data structure can be used in general for integer arrays, we will refer to it as enc_vector throughout the paper. Sadakane [14] proposed the following construction of the enc_vector (see also Fig. 2). We first create an array d_Ψ which contains the differences of two consecutive Ψ values. These differences are then encoded with a self delimiting code and written in a bit vector z. Not all differences are encoded, as we want to provide constant time access. We sample $\Psi[i]$ if $d_\Psi[i]$ is not positive or if the encoded bits between two samples is greater than a parameter $s \in O(\log n)$. The bit vector is_sample indicates sampled Ψ values, the array sample stores the sample values, and sample_pointer stores pointers to the next compressed value in z. In addition to that, is_sample is augmented by rank_support and select_support. Constant time random access to an encoded value $\Psi[i]$ is now easy to realize: Test whether the ith value is sampled. If so, return sample$[rank(i)]$. Otherwise, we retrieve the greatest index $j < i$ which is sampled by $j = select(rank(i))$. The result is the sum of sample$[rank(i)]$ plus the sum of $i - j$ decoded values starting at position sample_pointer$[rank(i)]$ in z. As there are at most $O(\log n)$ bits to decode for calculating the (prefix) sum of the $i - j$ values, one could use a lookup table of size $2^{O(\log n)}$ to perform this task. In practice, not one lookup table is used since a good compression rate of the enc_vector forces us to use large constants inside the $O(\log n)$ term. Hence, the lookup table would not fit into main memory and the table access would be really slow. Therefore Sadakane proposed to use a constant number of small lookup tables of size $o(n)$ and decode with $O(1)$ accesses to this small tables. Furthermore, the use of the is_sample array with the rank/select data structures is not competitive in the access time/space tradeoff to a simpler solution: Every s_Ψ-th value of Ψ is sampled and the few $(\leq \sigma)$ negative values of d_Ψ are considered modulo n. We call this simplified data structure enc_vector_prac.

So, the size of enc_vector_prac mainly depends on the choice of the self-delimiting code and the access speed depends on the decoding speed for this code.

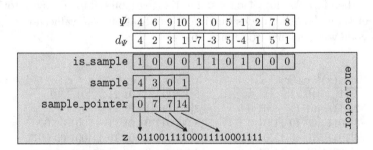

Fig. 2. Example for the enc_vector data structure. Φ-code is used to represent the d_Ψ-values $2, 3, 1, 5, 1, 5, 1$ of the not sampled ψ-values.

3   Guiding Information to Decode Self-delimiting Codes

Let us now introduce our SWAR methods to calculate *guiding information* to decode prefix sums of self-delimiting codes. Guiding information means information that either reduces the size for lookup tables supporting the decoding or the time spent on the decoding. We developed two new methods supporting the Φ-code and show how to use known methods to support Δ- and δ-code. In the following we will consider these four SWAR methods:

- $\texttt{b1Cnt(x)}$ Count the number of ones in x (a.k.a. sideway addition).
- $\texttt{i1BP(x,i)}$ Compute the position of the ith one in x.
- $\texttt{b11Cnt(x)}$ Count the number of Φ-encoded numbers in x.
- $\texttt{i11BP(x,i)}$ Get the end position of the ith Φ-encoded number in x.

Knuth presented a solution for $\texttt{b1Cnt}$ in [7] (see algorithm 1). After each of the first three steps, x_i will contain $64/2^i$ sums of 2^i consecutive bits. In the last step the 8 sums of the 8 bytes are multiplied with the constant $0x0101010101010101$. The result is an integer y containing the prefix sum of the j least significant bytes of x_3 in the jth least significant byte. This means the most significant byte of y equals $\texttt{b1Cnt}$ and the prefix sums could be used to calculate $\texttt{i1BP}$ (see [15] for details). Both methods $\texttt{b1Cnt}$ and $\texttt{i1BP}$ take $O(\log \log d)$ steps, where d is the length of the word in bits.

Algorithm 1. $\texttt{b1Cnt(x)}$: Counting the number of ones in x

```
1   x₁ = x-( (x≫1)&0x5555555555555555)
2   x₂ = (x₁&0x3333333333333333)+((x₁≫2)&0x3333333333333333)
3   x₃ = (x₂+(x₂≫4))&0x0F0F0F0F0F0F0F0F ;
4   return (x₃*0x0101010101010101)≫56
```

Counting Fibonacci encoded numbers. The following procedure counts the number of Φ-encoded numbers in a word x, which is equivalent to counting the number of non overlapping '11' bit pairs in x. We say that a '11' bit pair starts at position i in x if one of the following conditions holds: (1) $i = 0$, (2) there is a zero at position $i - 1$, or (3) a '11' pair starts at position $i - 2$. Consequently, a '11' ends at position i if it starts at position $i - 1$.

In the first step of our algorithm we generate three words $\texttt{ex11}$, $\texttt{ex01}$, and $\texttt{ex10}$ (see Fig. 3). Each of these words extracts information of even bit pairs $bp_j(x)$ in x. Bit pair $bp_j(\texttt{ex11})$ in $\texttt{ex11}$ is set to '01' if $bp_j(x)$ equals '11' and is set to '00' otherwise. Note that the number of ones in $\texttt{ex11}$ is a lower bound on the encoded numbers as each 1 in $\texttt{ex11}$ indicates that a '11' bit pair starts or ends at this position. The remaining task is to add the number of '11' pairs that end at odd positions and are not already captured by $\texttt{ex11}$ (e.g. see the bit pair ending at position 8 in Fig. 3). We use $\texttt{ex01}$ (which contains either '01' if $bp_j(x)$ equals '01' or '00' otherwise) and $\texttt{ex10}$ (which contains either '10' if $bp_j(x)$ equals '10' or '00' otherwise) to calculate this number. The missing '11'

Algorithm 2. b11Cnt(x): Counting the number of Φ-encoded numbers in x

---

```
1   ex11 = (x&(x≫1))&0x5555555555555555 ;
2   ex01 = (x⊕(x≫1))&x&0x5555555555555555 ;
3   ex10 = (x⊕(x≪1))&x&0xAAAAAAAAAAAAAAAA;
4   x₁ = (((ex11|(ex11≪1))+(ex10≪1))&ex01);
5   x₂ = ex11 | x₁;
6   x₃ = (x₂&0x3333333333333333)+((x₂≫2)&0x3333333333333333);
7   x₄ = (x₃+(x₃≫4))&0x0F0F0F0F0F0F0F0F;
8   return (0x0101010101010101*x₄)≫56;
```

---

pairs have the following structure: A bit pair $bp_j(x)$ equals '10', followed by $k \geq 0$ bit pairs each equals '11', and finally a bit pair $bp_{j+k+1}(x)$ set to '01'. If k equals zero it is easy to get all missing occurrences: $y = ex01\&(ex10 \ll 1)$ will have a one at each end of a missing '11' bit pair. If $k > 0$ we use addition to propagate the one of $bp_j(ex10)$ over the k occurences of '11's in x. See line 4 in algorithm 2 for details. Finally, we use sideway addition (line 6-8) to count the ones in the union of $ex11$ and y. As we only use a constant number of operations on top of the sideway addition, b11Cnt also takes $O(\log \log d)$ steps.

Locating Fibonacci encoded numbers. The next problem is to locate the end of the ith Φ-encoded number in a word x. As we have already a procedure to calculate i1BP, we will reduce our problem to this case. So, we will show how to calculate a word w having ones at those positions where Fibonacci numbers end in x. Once again we first compute ex11, ex10, and ex01. Now consider ex11 (see Fig. 3 for an example). ex11 corresponds to w in the following case: (a) $bp_j(x)$ equals '11' and there is a consecutive odd number of ones to the right of $bp_j(x)$, like $bp_1(x)$ and $bp_3(x)$ in Figure 3. However, if (b) $bp_j(x)$ equals '11' and there is an even number of consecutive ones to the right of $bp_j(x)$, ex11 does not match with w, as in this case ex11 marks one position to the right of the actual end of the Φ-encoded number. In addition, (c) if $bp_j(x)$ equals '01' and is preceded

	bp9		bp8		bp7		bp6		bp5		bp4		bp3		bp2		bp1		bp0	
x	1	0	1	1	0	1	1	0	1	0	0	1	1	1	1	0	1	1	1	0
ex11	0	0	0	1	0	0	0	0	0	0	0	0	0	1	0	0	0	1	0	0
ex01	0	0	0	0	0	1	0	0	0	0	0	1	0	0	0	0	0	0	1	0
ex10	1	0	0	0	0	0	1	0	1	0	0	0	0	0	1	0	0	0	0	0
x_1	0	0	0	0	0	1	0	0	0	0	0	1	0	0	0	0	0	0	0	0
w	0	0	1	0	0	1	0	0	0	0	0	1	0	1	0	0	0	1	0	0

Fig. 3. Example for intermediate results in b11Cnt (x_2) and i11BP (w)

Algorithm 3. i11BP(x): Get the end position of the ith Φ-encoded number in x

---

```
1   ex11 =  (x&(x≫1))&0x5555555555555555 ;
2   ex01 =  (x⊕(x≫1))&x&0x5555555555555555 ;
3   ex10 =  (x⊕(x≪1))&x&0xAAAAAAAAAAAAAAAA;
4      m =  ((ex11|(ex11≪1))+(ex10≪1))  &  (ex01  |  ex11);
5      w =  ex11 + m;
6   return i1BP(w, i);
```

---

by an odd number of consecutive ones, $bp_j(w)$ has to be '01', while $bp_j($ex11$)$ equals '00'.

We calculate the correction word m with the following properties to handle the cases where ex11 does not correspond to w (cases (b) and (c), see line 4 in algorithm 3). A bit pair $bp_j(m)$ is set to '01' if either $bp_j(x)$ equals '11' and is preceded by an even number of consecutive ones or $bp_j(x)$ equals '01' and is preceded by an odd number of consecutive ones. Otherwise, $bp_j(m)$ equals '00'. Now, adding m to $ex11$ results in the desired word w. Note again that we only add a constant number of operations to the known SWAR operation b1BP. So b11BP takes again $O(\log \log d)$ steps. We also extended both methods to take a carry bit of another word. We refer to our implementation documentation for details.

Guiding information. While b11Cnt and i11BP support Φ-code, b1Cnt and i1BP support the Δ-code as the end markers of the Δ-code are always even bit pairs. The virtue of these procedures is that one gains information about large areas (up to 128 bits in recent CPUs) of the encoded data instead of only small pieces (about 16 bits in practice) with lookup tables. Suppose we work with lookup tables that can decode ℓ bits to calculate the prefix sum of up to k Φ-encoded numbers. This would take us $k \cdot 2^\ell$ words for k lookup tables of size 2^ℓ. One additional call of i11BP(x, k) and masking of the relevant bits reduces the space for lookup tables to 2^ℓ. Similarly, one can decide with one call of b11Cnt whether one can apply the lookup table or has to decode the region without the lookup table since there are more than k numbers encoded in ℓ bits.

We also use the known SWAR method r1BP (rightmost position where a bit is set to one) [7] to speed up the decoding of δ-code used in CSAs for high compressible strings. For those CSAs, the bit vector z of enc_vector_prac consists of many ones. So we could decode runs of ones by applying r1BP on the inverted bit vector z.

4   Implementation and Experimental Results

Implementation. We have implemented a C++ template library for succinct data structures (called *sdsl*[3]) following the guidelines in [8]. It includes basic

---

[3] The source code as well as the documentation is available under http://www.uni-ulm.de/in/theo/research/sdsl. The source code is licensed under the GNU Public License.

data structures like a bit vector, a vector for integers (of size $\ell \leq 64$, i.e. not necessarily byte-aligned), and different rank and select data structures supporting bit vectors. In addition, we provide a class for efficient broadword operations (see section 3) and several classes for self-delimiting codes (see section 2.2). More complex classes like the enc_vector and enc_vector_prac are composed of these basic classes. More precisely, all complex classes can be parametrized with basic classes, e.g. one can specify the type of the rank data structure and the sample density of enc_vector. We implemented two generic classes for the CSA in the *sdsl*. The first one, csa_sada_theo is a one-to-one implementation of the (hierarchical) data structure described in [14] with our data structures. The second one is csa_sada_prac which follows the advices in [12] to implement the CSA of Sadakane in practice (i.e. no hierarchical data structure). Both implementations, as well as most other data structures, are designed to be immutable STL (Standard Template Library) conform containers.

Experimental setup. We executed our experiments on a Linux-based system using gcc 4.2.1, using compilation options -O9 -ffast-math for all programs. The machine was equipped with a Dual-Core 64-bit Opteron processor running at 1000 MHz with 1MB L1 cache. As most of our measured operations only take a few microseconds, we repeated the operations millions of times and calulated the average value. The system function getrusage() was used to get the user time.

We used the text collection from the *Pizza&Chili* site of size 50 and 200 MB as test cases. The corpus cover a representative set of different application areas, e.g. bioinformatics and XML-processing. Two additional inputs (random_k128 and random_k26) are uniformly distributed texts over alphabets of size 26 and 128 (see Table 1 for more information about the corpus). We compared our implementation with the original implementation of Sadakane (optimized C code) which is also available from the *Pizza&Chili* site.

Experiments. In the experiments, we consider CSAs as a replacement for uncompressed suffix arrays (i.e. as black boxes). Therefore, we measure the average access

Table 1. Statistics for inputs and resulting CSAs. We used the Partial-Match-based compressor PPMDi to get an idea of the compressibility of the inputs.

Test case	σ	Compression by ppmdi -l 9	Encoded ones in z	Encoded values < 32 in z	Access speed sada_64δ/sada_orig
random_k128.50MB	128	0.894	< 0.01	≈ 0.21	-
random_k26.50MB	26	0.698	≈ 0.03	≈ 0.70	1.08
proteins.50MB	27	0.421	≈ 0.36	≈ 0.82	1.40
pitches.50MB	133	0.305	≈ 0.37	≈ 0.83	1.55
dna.50MB	16	0.243	≈ 0.57	≈ 0.99	1.65
english.50MB	239	0.242	≈ 0.68	≈ 0.94	1.68
sources.50MB	230	0.167	≈ 0.76	≈ 0.94	1.90
dblp.xml.50MB	97	0.092	≈ 0.85	≈ 0.97	2.25

time to one suffix array element. The size of a CSA depends on the choice of a self-delimiting code and on two parameters s_Ψ and s_A. As already mentioned in section 2.3, we sample every s_Ψth value in enc_vector_prac. The CSA itself stores every s_Ath value of the uncompressed suffix array in the array sa_samples. One element access in the CSA now requires one access of sa_samples plus $i \mod s_A$ accesses of enc_vector_prac. To show different time/space tradeoffs, we parametrized the CSAs with $s_\psi = 128$ and $s_A = \{2, 4, 6, 8, 12, 16, 24, 32, 48, 64, 96\}$. We have fixed s_Ψ as further increasing does not affect the compression of the CSAs substantially and the access time to enc_vector_prac increases linear with the parameter.

Results. Figure 4 contains the experimental results for the following three CSAs:

– Sadakane's original implementation which uses δ-code (sada_orig).
– csa_sada_prac parametrized with δ-code (sada_64δ).
– csa_sada_prac parametrized with Φ-code (sada_64Φ).

Each diagram shows the time/space for the CSAs for texts of size 50 MB (the 200 MB test cases show the same result). Note that the abscissa corresponds to the space usage of the CSA in ratio to the text size. An uncompressed suffix array (for a string of size $\leq$ 4GB and $\sigma \leq 256$) occupies four times the text size. The diagrams (a) to (h) are sorted in increasing order of their compressibility. First, we consider only the graphs of sada_64δ and sada_64Φ relative to each other. We observe that the graph of sada_64Φ moves from plot (a) to (h) from left to right. I.e. Φ-code uses less space than δ-code for inputs that are not good to compress and more space for high-compressible inputs. As mentioned in section 2.2, Φ-code uses less space for encoding values than δ-code if the values are between two and 6765. Since over 99.8% of the encoded values in each test case are less than 6766, sada_64Φ uses less space if there are few ones in z (see Table 1 for the quantities). Since the decoding speed of both implementations is approximately equal for the same choice of (s_A, s_Ψ), it follows that sada_64Φ results in a better time/space tradeoff for the first three cases.

Second, we compare the graphs of sada_orig and sada_64δ. Except for plot (a), the sizes of the CSAs are approximately equal for the same choice of (s_A, s_Ψ). The reason for the difference in (a) is that sada_orig stores every value $d_\Psi[i]$ in z while we do not store $d_\Psi[i]$ if $\Psi[i]$ is sampled. Whereas most additional stored values in cases (b)-(h) are really small (e.g. at least 70% are less than 32, see Table 1) most values in case (a) are big. For this reason, we only compare the time/space tradeoff for cases (b)-(h). Table 1 contains the minimal ratio of access speeds for all parameter pairs (s_A, s_Ψ) for the two implementations. The ratio increases from 1.08 for random text to 2.25 for the XML file. This result was expected, as with every test case the probability of long runs of ones in z increases and with this probability that the broadword method in sada_64δ is used successfully.

Finally, we have measured the effects of adding the broadword methods in sada_64Φ and get results as we have seen for δ-code. Broadword methods decrease the speed (about 10%) in the first cases ((a)-(b)) and increase the speed

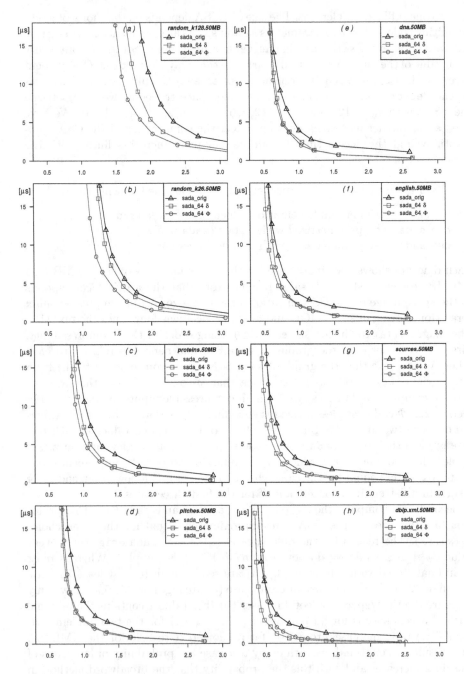

Fig. 4. Comparison of three CSA implementations. The x-axis corresponds to the ratio of space usage to text size and the y-axis is the average access time. The diagrams (a) to (h) are sorted in increasing order of text compressibility.

ratio for the last cases by factor of up to three. The decrease of speed in case (a) and (b) is due to the fact that the gained information could not be used to directly decode values and therefore we have only additional calculation overhead. In the latter cases, we can use the gained information to directly decode runs of ones or other patterns without using lookup tables. This explains the significant speed-up.

5 Conclusion

We presented two new broadword methods that can be applied to generate guiding information for the decoding of Fibonacci encoded numbers. We used those new and known broadword methods in our implementation of Sadakane's CSA and showed that the methods accelerate the access time for good compressible inputs. We also showed that the use of Fibonacci code for CSAs for text of low compressibility improves the time/space tradeoff. We think broadword methods for itself are interesting and gain more and more importance as modern CPUs support them. We have one problem left which also could accelerate decoding Φ- or Δ-code in practice if we find an efficient broadword method for it: Calculate the length of the minimal gap between two ones in a 64 bit word. The result can be used to determine the length of the greatest encoded value in the 64 bit word. Thus we could decide if there are some values that could not be decoded with a lookup table of fixed length.

References

1. Abouelhoda, M.I., Kurtz, S., Ohlebusch, E.: Replacing suffix trees with enhanced suffix arrays. J. Discrete Algorithms 2(1), 53–83 (2004)
2. Elias, P.: Universal code word sets and representations of the integers. IEEE Transactions on Information Theory 21(2), 194–203 (1975)
3. Ferragina, P., González, R., Navarro, G., Venturini, R.: Compressed text indexes: From theory to practice! arXiv:0712.3360v1 [cs.DS] (2007)
4. Ferragina, P., Manzini, G.: Opportunistic data structures with applications. In: FOCS, pp. 390–398 (2000)
5. Grossi, R., Vitter, J.S.: Compressed suffix arrays and suffix trees with applications to text indexing and string matching. SIAM J. Comput. 35(2), 378–407 (2005)
6. Jacobson, G.: Space-efficient static trees and graphs. In: FOCS, pp. 549–554. IEEE, Los Alamitos (1989)
7. Knuth, D.E.: The Art of Computer Programming. Pre-Fascicle 1a. A Draft of Section 7.1.3: Bitwise Tricks and Techniques (2008)
8. Lakos, J.: Large-Scale C++ Software Design. Eddison-Wesley (1996)
9. Mäkinen, V., Navarro, G.: Compressed compact suffix arrays. In: Sahinalp, S.C., Muthukrishnan, S.M., Dogrusoz, U. (eds.) CPM 2004. LNCS, vol. 3109, pp. 420–433. Springer, Heidelberg (2004)
10. Manber, U., Myers, E.W.: Suffix arrays: A new method for on-line string searches. SIAM J. Comput. 22(5), 935–948 (1993)
11. Ian Munro, J.: Tables. In: Chandru, V., Vinay, V. (eds.) FSTTCS 1996. LNCS, vol. 1180, pp. 37–42. Springer, Heidelberg (1996)

12. Navarro, G., Mäkinen, V.: Compressed full-text indexes. ACM Comput. Surv. 39(1) (2007)
13. Rahman, N., Raman, R.: Rank and select operations on binary strings. In: Encyclopedia of Algorithms. Springer, Heidelberg (2008)
14. Sadakane, K.: New text indexing functionalities of the compressed suffix arrays. J. Algorithms 48(2), 294–313 (2003)
15. Vigna, S.: Broadword implementation of rank/select queries. In: McGeoch, C.C. (ed.) WEA 2008. LNCS, vol. 5038, pp. 154–168. Springer, Heidelberg (2008)
16. Witten, I.H., Moffat, A., Bell, T.C.: Managing Gigabytes, 2nd edn. Morgan Kaufmann Publishers, San Francisco (1999)
17. Zeckendorf, E.: Représentation des nombres naturels par une somme de nombres de Fibonacci ou de nombres de Lucas. Bull. Soc. R. Sci. Liège 41, 179–182 (1972)

Speed-Up Techniques for the Selfish Step Algorithm in Network Congestion Games[*]

Matthias Kirschner, Philipp Schengbier, and Tobias Tscheuschner

Universität Paderborn, Fürstenallee 11, 33102 Paderborn
{maki82,erkan,chessy}@upb.de

Abstract. Recently, many speed-up techniques were developed for the computation of shortest paths in networks with rather static edge latencies. Very little is known about dealing with problems which rely on the computation of shortest paths in highly dynamic networks. However, with an increasing amount of traffic, static models of networks rather sparsely reflect realistic scenarios. In the framework of network congestion games, the edge latencies depend on the number of users traveling on the edges. We develop speed-up techniques for the selfish step algorithm to efficiently compute (pure) Nash equilibria in network congestion games. Our approaches

1. periodically compute estimations for lengths of shortest paths during the advance of the selfish step algorithm with the purpose to use A^* for many path computations, and
2. completely save many path computations or substitute them by more efficient tests.

In comparison to an implementation of the selfish-step algorithm using Dijkstra's algorithm we improve the total running time by a factor of 4 up to 9 on highway networks and grids.

Keywords: speed-up techniques, selfish step algorithm, Nash equilibria, game theory, network congestion games, shortest paths.

1 Introduction

For systems, in which there is no central control guiding the users, it is a natural approach to assume that the users optimize their own objectives. If the systems are very large-scaled, it is also not devious to assume that they do not coordinate themselves or even cooperate. If, on the other hand, their actions have influence on the prices of shared resources, then questions like the following arise:

- How will the system develop from a given state?
- Which resources will suffer a high price for given user demands?
- At what price will the average user reach a predetermined goal?
- How will the system behave if we add or remove some resources?

---

[*] This research has (partially) been supported by the DFG Sonderforschungsbereich 614 "Selbstoptimierende Systeme des Maschinenbaus."

J. Vahrenhold (Ed.): SEA 2009, LNCS 5526, pp. 173–184, 2009.

A suitable framework for modeling scenarios of non-cooperative users which share resources are congestion games which were introduced by Rosenthal in [9]. In a congestion game players choose among given subsets of a set of common resources, i.e. they choose strategies. Hereby, the resources have associated delays (or prizes), which depend on the number of players selecting a strategy which contains the particular resource. The aim of the players is to minimize the delay of the resources contained in their strategies.

For this framework one is interested in so-called (pure) Nash equilibria. Nash equilibria are states, in which no player has an incentive to unilaterally deviate from her strategy. These states can represent some kind of forecast for the behavior of the system and knowledge about them may help the users and the designers of the system to make better decisions. Rosenthal showed in [9] that every congestion game possesses a pure Nash equilibrium. In his proof he uses a potential function which provides for every state an upper bound on the number of consecutive improving strategy changes (so called selfish steps) of the players. Moreover, the proof shows that the simple algorithm of consecutively performing selfish steps on an arbitrary initial solution computes a Nash equilibrium for every congestion game.

Network congestion games are a well-established game theoretic model for the analysis of traffic scenarios. In network congestion games the strategies of the players correspond to paths through an underlying network. If one performs the consecutive selfish steps of the players on the network with the purpose to compute a Nash equilibrium, then this may lead to a highly dynamic behavior of the edge latencies, since every selfish step changes the latencies on the old and on the new path of the player who changes her strategy.

2   Related Work

Fabrikant et al. [5] showed that a Nash Equilibrium can be computed in polynomial time for symmetric network congestion games, i.e. network congestion games in which all players choose their strategies among the same set of strategies. For asymmetric network congestion games they showed that the problem of finding a pure Nash equilibrium is complete for the complexity class PLS and as a byproduct, their reduction shows that there are states which are exponentially many selfish steps away from any Nash equilibrium. An easier proof can be found in Ackermann et al. [1].

In [8] Panagopoulou et al. studied the case of weighted players in the symmetric congestion game with edge latencies equally to the load. For this model they show experimentally that the selfish step algorithm requires a significantly lower number of selfish steps if the initial solution that is given to the selfish step algorithm is constructed in a *shortest-path-allocation* way compared to a randomly chosen initial solution.

Since the strategies of the players correspond to paths and the players aim to minimize their costs, speed-up techniques for shortest path algorithms relate to this work. In the recent past, many techniques were engineered for the case of static edge latencies (for an overview see [11]): Gutman [6] introduces the *reach*-based search in which during the search of a shortest path nodes can be pruned based on lower bounds on shortest path distances which are too far away from a possible shortest path — they are out of reach. Sanders and Schultes ([12], [13]) introduce highway hierarchies which gradually coarsen the network by identifying paths which are replaced by shortcut-edges in higher levels of the hierarchy. Using these precomputed hierarchies they significantly decrease the computation times for a shortest path query. Basing on highway hierarchies Bast et al. ([2], [3]) develop a technique called *transit node routing*. In that they reduce the computation of a shortest path down to a small number of lookups in a precomputed table of shortest paths between the origin node and the transit nodes, between the transit nodes themselves, and finally between the transit nodes and the destination node.

For the case that at most a "moderate" number of edges change their weights or their latency function Schultes et al. [14] and Delling et al. [4] developed speed-up techniques. But, to our knowledge there are no speed-up techniques known for shortest path computations on highly dynamic networks.

3   Model

A network congestion game is defined as follows:

Definition 1 (Network congestion game)
A network congestion game Γ is a tuple $\Gamma = ([K], G, \{c_e\}, \{(s_i, t_i)\})$, where

- *$[K] := \{1, \ldots, K\}$ is a finite set of players,*
- *$G = (V, E)$ is graph with a finite set V of nodes and a set $E \subseteq V \times V$ of edges.*
- *$\{c_e\}$ is a set of $|E|$ cost functions. Each cost function is associated with an edge $e \in E$. For $e \in E$, $c_e : \mathbb{N} \mapsto \mathbb{R}_+$ and c_e is monotonically increasing.*
- *$\{(s_i, t_i)\} \subseteq V \times V$ is a multiset of K origin-destination pairs. Each player is associated with an origin-destination pair.*

The *strategy set* Σ_i of player $i \in [K]$ is implicitly defined by her origin-destination pair (s_i, t_i):

$$\Sigma_i = \{p \mid p \text{ is a simple path from } s_i \text{ to } t_i\} \subseteq \text{Pow}(E).$$

The set of all origin nodes of the game is denoted by S, the set of all destination nodes is denoted by T.

A tuple $\sigma = (\sigma_1, \ldots, \sigma_K) \in \Sigma = \times_{k \in [K]} \Sigma_k$ is called a strategy profile. A strategy profile is a possible outcome of the game, in which each player plays one of her strategies. For a strategy profile $\sigma = (\sigma_1, \sigma_i, \ldots, \sigma_K)$, σ_{-i} denotes the tuple $(\sigma_1, \ldots, \sigma_{i-1}, \sigma_{i+1}, \ldots, \sigma_K)$. Using this notation, we write σ as (σ_i, σ_{-i}).

The *congestion* $f_\sigma(e)$ of an edge $e \in E$ is the number of players using edge e in σ in profile σ, i.e. $f_\sigma(e) = |\{i \in [K] \mid e \in \sigma_i\}|$. Player i's *private costs* are defined by $\text{PC}_i(\sigma) = \sum_{e \in \sigma_i} c_e(f_\sigma(e))$. Each player tries to minimize her private costs.

If the current profile is $\sigma = (\sigma_i, \sigma_{-i})$ and there is a strategy $\sigma_i' \in \Sigma_i$ such that $PC_i(\sigma_i', \sigma_{-i}) < PC_i(\sigma)$, i has an incentive to switch to strategy σ_i'. A change of i's strategy is called a *selfish* step, since it reduces i's costs but may increase the other players' costs. For player i a strategy σ_i^* such that $\sigma_i^* \in$ argmin$_{\sigma_i \in \Sigma_i} PC_i(\sigma, \sigma_{-i})$ is a *best response* to σ_{-i}. A profile σ in which each player plays a best response is called a (pure) *Nash equilibrium* (in the literature there are also *mixed* Nash Equilibria, but in this paper we do solely consider pure Nash Equilibria).

As mentioned in the introduction, a Nash equilibrium always exists in network congestion games and can be computed by iteratively performing selfish steps until no player has an incentive to unilaterally deviate from her strategy.

4   Data Structures and Algorithms

In 4.1 we describe a basic implementation of the Selfish Step Algorithm and point out where there is potential for improvement. In the remaining subsections we present our speed-up techniques: In section 4.2 we present several heuristic that allow us to use the A^* search algorithm rather than Dijkstra's algorithm to compute improving strategy changes for the players. While this approach speeds up the computation of the paths, we introduce in section 4.3 efficient tests which replace a huge portion of path computations. Lastly, we show in section 4.4 how to avoid many path computations and many tests by arranging players in a hierarchical data structure.

4.1   Basic Selfish Step Algorithm

First of all, an initial feasible solution is computed. An obvious way to do this is the *shortest path allocation*: The players are added successively into the game, using an arbitrarily chosen permutation of the player set (in our implementation we used a randomly chosen permutation). To insert player i, her shortest path is computed, which depends on the congestion caused by all players inserted before i is inserted. For the most part, this allocation corresponds to the allocation described in [8], but since we consider the unweighted case, there is no need to sort the players by weights.

After the allocation, the selfish step loop begins: For each player i, a shortest path is computed using a shortest path algorithm. In our implementation, we used Dijkstra's algorithm (unidirectional). If the costs of the computed path p is strictly lower than the current private costs of i, then i performs a selfish step. Otherwise, i's current path is an optimal path. In this case the computation of the shortest path *verifies* i's current strategy as a best response.

Whenever player i performs a selfish step, the congestion on some edges of the underlying graph changes. This may affect the optimality of the path of any player $j \neq i$. Therefore, in order to guarantee convergence in a Nash equilibrium, the algorithm terminates if and only if during a complete iteration through the entire player set no player with suboptimal path was found. A simple improvement incorporated in our basic implementation of the selfish step algorithm is to treat players who are playing the same strategy as a group, which saves shortest path computations if a path used by multiple players is optimal.

It is obvious that the running time of the basic selfish step algorithm is almost exclusively dominated by computations of shortest paths. Moreover, we observed in our tests of the algorithm that most shortest path computations verify the players' strategies and only few improvements were found. Based on these observations, we pursued the following two approaches to improve the performance of the selfish step algorithm:

1. Improve the performance of computing the players strategies (section 4.2).
2. Reduce the number of path computations (sections 4.3 and 4.4).

4.2 Heuristic Shortest Path Computations

To speed up shortest path computations, we replaced Dijkstra's Algorithm by the A^* search algorithm [7]. The A^* search algorithm computes a shortest path between a pair $(s,t) \subseteq V \times V$. It expects as additional input a heuristic function $h : V \rightarrow \mathbb{R}_{\geq 0}$ which estimates the travel costs from a given node $v \in V$ to the destination node t. The function h is used by the A^* search algorithm to guide the search in a particular direction. It can be shown that the A^* search algorithm computes shortest paths if h does not overestimate the actual distance from v to t. A heuristic h satisfying this property is called *admissible*. We sometimes use heuristics for the computation of selfish steps that are not admissible, but if we do so, then we compare the computed path with the current strategy of the considered player and perform a strategy change only if it is improving.

Since in asymmetric network congestion games we deal with many destination nodes, we are interested in a binary rather than a unary heuristic function, that is, we let h map from $V \times T$ to $\mathbb{R}_{\geq 0}$. We present two such heuristics that are dedicated to the selfish step algorithm.

Online heuristic. The idea behind the online heuristic is to learn estimations of the travel costs during the execution of the algorithm. We initialize our estimates by setting $h_{\mathrm{On}}(u, v) = 0$ for all $u, v \in V$. After computing a shortest path $\sigma_i = \{e_1, ..., e_t\}$, $e_i = (u_i, v_i)$, from s_i to t_i for player i, we compute the costs of all subpaths of σ_i and update the corresponding values, that is, we set $h_{\mathrm{On}}(u_j, v_k) = \sum_{\nu=j}^{k} c_{e_\nu}(f_\sigma(e_\nu))$ for all $1 \leq j \leq k \leq t$. The costs for this update strategy are quadratic in the length of the path. During the advance of the selfish step algorithm, the online heuristic is not necessarily admissible since every selfish step may influence the length of a shortest path of any other player. Therefore, it is essential to reoptimize the computed profile with an exact shortest path algorithm. However, it can easily be shown that the online heuristic is admissible during the shortest path allocation. This is due to the fact that during this phase of the algorithm the latencies of all edges are monotonically increasing.

Shortest Path heuristic. The shortest path heuristic in profile σ is defined by $h_{\mathrm{SP}}(u, v) = \delta_\sigma(u, v)$. Here, $\delta_\sigma(u, v)$ denotes the cost of a shortest path between

$u \in V$ and $v \in V$ in a network with edge latencies of $w_\sigma(e) = c_e(f_\sigma(e))$. Since $w_\sigma(e)$ is a lower bound for the cost of using edge e for every player i in profile σ, $h_{\mathrm{SP}}(u, v)$ is an admissible heuristic in profile σ. Moreover, if $c_e(f_\sigma(e) + 1)$ is not much larger than $c_e(f_\sigma(e))$ for every edge e, the h_{SP} provides good estimations. Note that the above assumption is realistic in the context of congestion on roads, since the presence of a single car has only small influence on the total congestion.

A drawback of the shortest path heuristic is its computational costs. Since we need all values $\delta_\sigma(u, t)$ for $u \in V$ and $t \in T$, we have to solve at least $|T|$ shortest path problems. In most cases, it is too expensive to recompute the estimations after each selfish step. Therefore, we allow the computation of the shortest path heuristic only for few strategy profiles that evolve during the algorithms execution. In other profiles, we simply reuse the heuristic of a preceding strategy profile. In general, the heuristic h_{SP} computed in a strategy profile σ is also no more admissible in a profile σ' evolving from σ by performing a selfish step, but can still provide good estimations if the latencies on few edges have changed. We recompute the shortest path heuristic after each iteration through the player set and therefore we can guarantee that a Nash equilibrium is reached if after an entire iteration through the player set no selfish step was performed.

4.3   Path Filter

A path filter is a function that maps a pair (σ_i, σ_{-i}) to a boolean value. It is supposed to return *true* if it decides that σ_i is a best response to σ_{-i}, *false* otherwise. But we allow a path filter to be heuristic, that is, its decision may have false positives and false negatives. A path filter is used to reduce the number of path computations: Only if a filter maps the player i's path σ_i to *false*, we allow the selfish step algorithm to initiate a computation of an improving path for player i. Note that a false negative output of a path filter does not affect the correctness of the selfish step algorithm, since it only allows the invocation of a (non-heuristic) shortest path computation for a path which is already a shortest path. In the following paragraphs, we present three different path filters: The cost filter, the detour filter and the edge filter.

Cost Filter. In 4.2 we argued that $\delta_\sigma(u, v)$ is a lower bound for the costs of an u-v-path for all players i and took advantage of this fact by using $\delta_\sigma(u, v)$ as a heuristic for the A^* search. But if in profile $\sigma = (\sigma_i, \sigma_{-i})$ we have $\mathrm{PC}_i(\sigma) = \delta_\sigma(s_i, t_i)$, then this proves that σ_i is a best response for σ_{-i}, and therefore we do not need to compute a shortest path for player i. This leads to the *cost filter*: In profile σ, we solve several shortest path problems to obtain the bounds $\delta_\sigma(s, t)$ for all $s \in S$ and all $t \in T$. We then build the player set $M(\sigma) = \{i \in [K] \mid \mathrm{PC}_i(\sigma) > \delta_\sigma(s_i, t_i)\}$, recompute shortest paths only for players in the set $M(\sigma)$, and ignore all players in the set $[K] \setminus M(\sigma)$. After having examined all players of the set $M(\sigma)$, we can recompute the bounds $\delta_\sigma(s, t)$ to start a new iteration through the player set. If after a recomputation of the bounds no player was found in $M(\sigma)$ who could perform a selfish step, then a Nash equilibrium is found.

Though this filter can easily worsen the running time as a stand-alone solution because of its expensive computational costs, it can be used as a extension to the shortest path heuristic since we can reuse the data precomputed for the shortest path heuristic.

Detour Filter. Especially in the end phase of the selfish step algorithm, most shortest path computations performed by the selfish step algorithm are verifications of a player's best response. The aim of the detour filter is to replace a portion of these computations by a test which can be performed by looking up precomputed data. It examines paths which are known to be optimal at a certain point of time τ during the algorithm's history and which do not contain any edge whose latency has increased since τ. The idea behind the detour filter is to prove that these paths are still optimal. It can be interpreted as an extension of the cost filter, since we use a player's private cost to decide whether we recompute her path. We avoid a recomputation of the path if $PC_i(\sigma) \leq \theta_{st}$, whereby θ_{st} is a lower bound for the costs of traveling from i's origin node $s \in V$ to her destination node $t \in V$ using at least one edge whose latency decreased since the last computation of i's path.

To apply the detour filter, we need to do some bookkeeping. We introduce a counter which counts the selfish steps. With each edge $e \in E$, two timestamps are associated: The timestamps decrease(e) and increase(e) are updated to the current selfish step counter value whenever the latency of an edge decreases or increases respectively.

Using the timestamps, it is easy to compute the sets $D_\tau := \{e \in E \mid$ decrease(e) $> \tau\}$ and $I_\tau = \{e \in E \mid$ increase(e) $< \tau\}$. Furthermore, with each path a timestamp is associated which is updated whenever the path is verified as a best response. The application of the detour filter only makes sense if few latencies have changed, which is typical in the last phase of the selfish step algorithm. If too many selfish steps were performed in one iteration through the player set, the detour filter does not find better bounds than the cost filter. To avoid the expensive precomputation of the filter in such situations, we choose a parameter C which controls if the precomputation is invoked or not. We decided to do the precomputation of the detour filter if $|D_\tau| \leq C$, that is, if the latencies of at most C edges decreased during the last iteration through the player set. In our implementation, we used $C = \frac{|E|}{4}$.

When we apply the filter at a certain point of time τ', we precompute lower and upper bounds for the distance between every pair $(x, y) \in (S \times V) \cup (V \times T)$ of nodes in the graph. We denote the lower bounds by $\delta_{\sigma^{\tau'}}(u, v)$ and the upper bounds by $\Delta_{\sigma^{\tau'}}(u, v)$, where $\sigma^{\tau'}$ denotes the profile computed by the algorithm after τ' selfish steps. In our implementation, we compute the lower bounds by solving several shortest path problems, using $c_e(f_\sigma(e))$ as latency functions. As a byproduct of the computation we get a set of paths. To obtain the upper bounds, we simply compute the costs of these paths, using $c_e(f_\sigma(e) + 1)$ as latencies of the edges.

For each origin node, we first compute the set

$$A^{\tau'}(s, \tau) = \{v \in V \mid \exists e = (u, v) \in D_\tau \text{ and } \delta_{\sigma^{\tau'}}(s, u) + c_e(f_{\sigma^{\tau'}}(e) + 1) \leq \Delta_{\sigma^{\tau'}}(s, v)\}$$

of *alternative nodes*. For each node $v \in A^{\tau'}(s, \tau)$, we define the *detour cost* of using node v as

$$\alpha(v) = \min_{e=(u,v) \in D^{\tau}} \{\delta_{\sigma^{\tau'}}(s, u) + c_e(f_{\sigma^{\tau'}}(e) + 1)\}.$$

For any player i with origin node s, the detour costs provide a lower bound for the costs of traveling from s to u.

With this, we can compute the final bounds of the detour filter by

$$\theta_{st} = \min_{v \in A^{\tau'}(s,\tau)} \{\alpha(v) + \delta_{\sigma^{\tau'}}(v, t)\}.$$

In our tests, the precomputation time was dominated by the computation of the lower bounds.

To apply the filter, we have to test each path σ_i if it contains an edge whose latency has increased. If $\sigma_i \cap I_\tau = \emptyset$, we can use the improved bounds θ_{st}, otherwise we use the bounds of the cost filter. It can be shown that the detour filter only returns *true* if σ_i is a best response, but due to space limitations the proof is omitted. However, if it returns *false*, then it is unclear whether there is a path which is better than σ_i. But in this case the negative output of detour filter invokes a (non-heuristic) shortest path computation.

Edge Filter. The edge filter is based on the following intuitive idea: If a player i traveling to node t only uses edges that recently were part of a shortest path of some player whose destination node is t, it is likely that i's path is optimal, too. The edge filter works as follows: For each destination node, we maintain a set of edges E_t. We initialize the filter by setting $E_t = \emptyset$ for each $t \in T$. When we investigate a player i with current path σ_i, we only recompute her shortest path if $\sigma_i \not\subseteq E_t$. After recomputation of a shortest path to node t, we update E_t. If the optimal path of player i is σ_i^*, we set $E_t = E_t \cup \sigma_i^*$. Therefore, we quickly learn a set of promising edges for each destination node. After a while, a strong filtering is achieved. After a complete iteration through the player set, we reinitialize the filter. This is necessary since the edge filter does not react on selfish steps, that is, there may be an edge $e \in E_t$ that is no subpath of a shortest path to t of another player. In contrast to cost and detour filter, the edge filter may classify paths as optimal which are not best responses (false positives). Therefore, we switch the filter off if after an entire iteration through the player set the selfish step algorithm did not perform a selfish step and continue the selfish step algorithm without the edge filter.

4.4   Hierarchical Data Structure

In the previous subsection we presented techniques which speed up the process of confirming best responses of given players. In this section we introduce a data structure that allows us to even save these tests for many players which play best responses. The technique is based on the following observation.

Lemma 1. *Let σ be a strategy profile in a network congestion game and i, j be two players playing the strategies σ_i and σ_j in σ, with σ_j being a subpath of σ_i. If σ_i is a best response for player i to σ_{-i} then σ_j is a best response for player j to σ_{-j}.*

Proof. Since σ_i is a best response for player i to σ_{-i}, it holds that σ_i is a shortest path in the network with latencies $w_i(e) = c_e(f_{\sigma_{-i}}(e) + 1)$. But since σ_j is a subpath of σ_i it holds that σ_j is also a shortest path in the network with latencies $w_j(e) = c_e(f_{\sigma_{-j}}(e) + 1)$. Thus, σ_j is a best response of player j to σ_{-j}.□

The idea behind the data structure is to isolate for every strategy profile σ a set of strategies $L \subseteq \bigcup_i \sigma_i$ such that no strategy $\sigma' \in L$ is a subpath of a strategy $\sigma'' \in \bigcup_{j \neq i} \sigma_j$. Then, the selfish step algorithm only needs to consider players which play strategies in L, since due to Lemma 1 every player which plays a strategy that is neither in L nor a best response implicates that there must be a player which plays a strategy which is in L and is not a best response.

The data structure contains a directed weighted graph H which is built up from the network and the strategy profile σ in the following way. For every strategy σ_* which is played by at least one player in σ there is a node in H. The weight of the node is the number of players playing strategy σ_* in σ. For two strategies σ_i and σ_j and their corresponding nodes u_i and u_j there is an edge (u_j, u_i) in H if σ_j is a subpath of σ_i and in σ no player k plays a strategy σ_k such that σ_j is a subpath of σ_k and σ_k is a subpath of σ_i. Then, the nodes which have an outdegree of zero are linked in a list L'. It is easy to see that L' contains links to all nodes who represent strategies from L.

The data structure is initially built up after the shortest path allocation, and if a selfish step is performed then the data structure is updated by adding respectively removing at most one node and by updating the weights and the respective edges.

5   Experiments and Results

We evaluated our algorithms in five different traffic scenarios: A model of the autobahn network of North Rhine-Westphalia (NRW: 552 nodes, 1180 edges), a randomly generated highway network (Random: 1000, 2194) and three grid networks of different sizes (Grid 15: 225, 840; Grid 25: 625, 2400; GGrid20: 328, 1022 — in this network we randomly deleted some of the originally 400 nodes). The latency functions of the edges in the first two scenarios are conical functions [10], in the grid networks we use linear functions with randomly drawn parameters. There are 200000 players in the maps NRW, Random and Grid 15 and 100000 players in the maps GGrid 20 and Grid 25. While the topology of the maps Random and especially NRW is very close to real world highway networks, the grid networks are suitable frameworks for modeling inner city traffic networks (manhattan structure). For an overview of our results see Table 1.

In the first line of Table 1 we outline the running time of the basic selfish step algorithm during the selfish step loop without any speed up technique. In the

Table 1. Average running times of our algorithms in seconds ($\pm$ standard deviation). Each combination of map and algorithm was executed 100 times. The algorithms were tested on a 2.4 GHz Intel Core 2 Duo CPU with 1 GB RAM.

	NRW	Random	Grid 15	GGrid 20	Grid 25
Basic selfish step	75.0 ±8.1	101.6 ±11.2	16.8 ±2.5	16.4 ±2.9	172.7 ±35.6
Online heuristic	35.8 ±3.8	41.4 ±3.4	9.6 ±1.4	9.1 ±1.3	109.9 ±20.7
Shortest path heuristic	25.1 ±2.3	30.3 ±2.4	5.6 ±0.7	5.5 ±0.9	40.3 ±7.8
Hierarchical Selfish Step	20.7 ±1.4	51.5 ±3.9	11.5 ±1.1	14.9 ±1.7	157.3 ±27.3
Cost Filter	42.0 ±3.4	73.6 ±12.0	17.5 ±2.6	17.9 ±3.6	167.9 ±36.6
Detour Filter	37.6 ±2.6	57.1 ±4.5	15.7 ±1.9	14.9 ±2.0	147.9 ±28.0
Edge Filter	36.9 ±2.8	63.5 ±5.1	10.2 ±2.2	9.3 ±3.1	110.5 ±23.7
Shortest Path Heuristic, Cost Filter & Edge Filter	7.8 ±0.6	13.3 ±1.6	3.5 ±0.8	3.3 ±0.8	30.8 ±8.4
Hierarchical Selfish Step, Cost Filter & Edge filter	19.9 ±1.3	40.2 ±3.7	8.7 ±1.3	10.4 ±1.6	138.5 ±27.6

following lines we show the running times of the basic selfish step algorithm during the selfish step loop combined with the different speed up techniques or some combinations of them. We first evaluated each speed-up technique separately: We found out that the online heuristic is especially suited during the shortest path allocation. During the selfish step loop, it achieved a speed-up of the overall running time of approximately two. In this phase of the algorithm the shortest path heuristic performs best. It achieved a speed-up of a factor of up to 4 in comparison to the implementation which uses Dijkstra's algorithm (the basic selfish step).

The hierarchical selfish step algorithm is especially suited for maps with a large player set, like NRW: In this map, the algorithm reduces the number of shortest path computations from over 1 million to approximately 200000. This is why the hierarchical selfish step algorithm is more than three times faster than the basic algorithm in this scenario. As expected, the results are worse for large maps with a relatively small player set, like Grid 25. However, even in this map the hierarchical selfish step algorithm is more than 30 seconds faster than the basic selfish step.

In our evaluation of the path filters, the best results were achieved with the edge filter. It achieved a speed-up of a factor of up to 2. In most cases, cost and detour filter lead to a speed-up as well.

The best results overall were achieved by combining the shortest path heuristic with the cost filter and the edge filter. A major advantage of this combination is that we only need one precomputation to initialize all three techniques: At the beginning of each iteration through the player set, we solve several shortest path problems to compute the bounds for both shortest path heuristic and cost filter. As a byproduct of the shortest path computations, we get a predecessor tree that can be used to initialize the edge filter.

When we replace the cost filter by the detour filter in the combined solution, we get slightly worse results. The problem here is that the detour filter is incompatible with heuristic solutions. We need to switch off both cost filter and edge filter when we want the detour filter to work correctly. We also tried to combine the hierarchical selfish step algorithm with other techniques. If we use heuristic shortest path computations, the total number of shortest path computations needed to reach an equilibrium drastically increases and the running time is worse than the original hierarchical approach. If we combine the hierarchical solution with edge and cost filter, we achieve small speed-ups in most cases.

6 Conclusion

We analyzed traffic scenarios for networks, in which the latencies of the edges are not static but depend on the number of users which travel along the edge. For these dynamically behaving networks we considered the problem of computing a Nash equilibrium by using the selfish step algorithm and developed several techniques to speed up the computation. The basic observation which underlies all the techniques is that many selfish steps do not or do hardly affect the strategy choice of other players although the edge latencies behave very dynamically.

On the one hand we exploited this observation by developing heuristics which provide estimations on the lengths of shortest paths. These estimations are used by the A^*-algorithm to compute improving strategies for the players by a factor of three up to four times faster than Dijkstra's algorithm. Although the estimations are not admissible heuristics at any time during the advance of the selfish step algorithm the periodical update of the estimations still guarantees that the selfish step algorithm convergences to a Nash equilibrium.

On the other hand we improved the search for players which can perform selfish steps for a given strategy profile. One approach was to introduce efficient tests, which check for a given strategy whether it fulfills a specific condition. Based on the outcome of that test, it is decided whether or not a path computation is performed. These tests in combination with the use of the A^*-algorithm reduced the overall running time by a factor of four up to nine compared to a basic implementation of the selfish step algorithm. Another approach was to develop an intelligent rule for the selection of that player which is considered next during the advance of the selfish step algorithm. The rule uses a data structure which classifies the strategies of the players whether they are best responses for a given strategy profile or potentially not. Only strategies which were classified as "potentially not a best response" are considered for shortest path computations. This rule reduced the number of shortest path computations for a traffic scenario in which many players route through the network by a factor of five.

Naturally there remain open questions. We used an algorithm, namely the A^*, which is supposed to compute shortest paths to compute a path which is *shorter* than a given one. But one might design efficient algorithms which directly compute a shortcut for a given path, if there is a shortcut. Another crucial point is to speed up the selection of the users which have an incentive to change their strategy. For this purpose one might try to widen up the range of conditions which characterize the satisfaction of the players.

References

1. Ackermann, H., Röglin, H., Vöcking, B.: On the impact of combinatorial structure on congestion games. In: Proceedings of the 47th Annual IEEE Symposium on Foundations of Computer Science (FOCS), pp. 613–622 (2006)
2. Bast, H., Funke, S., Matijevic, D., Sanders, P., Schultes, D.: In transit to constant time shortest path queries in road networks. In: Proceedings of the 6th Workshop on Algorithms and Experiments (ALENEX), pp. 46–59 (2007)
3. Bast, H., Funke, S., Sanders, P., Schultes, D.: Fast routing in road networks with transit nodes. Science 316(5824), 566 (2007)
4. Delling, D., Wagner, D.: Landmark-Based Routing in Dynamic Graphs. In: Demetrescu, C. (ed.) WEA 2007. LNCS, vol. 4525, pp. 52–65. Springer, Heidelberg (2007)
5. Fabrikant, A., Papadimitriou, C.H., Talwar, K.: The complexity of pure Nash Equilibria. In: Proceedings of the 36th ACM Symposium on Theory of Computing (STOC), pp. 604–612 (2004)
6. Gutman, R.: Reach Based Routing: A New Approach to Shortest Path Algorithms Optimized for Road Networks. In: Proceedings of the 6th International Workshop on Algorithm Engineering and Experiments, pp. 100–111. SIAM, Philadelphia (2004)
7. Hart, P.E., Nilsson, N.J., Raphael, B.: A Formal Basis for the Heuristic Determination Of Minimum Cost Paths. IEEE Transactions on Systems Science and Cybernetics 4(2), 100–107 (1968)
8. Panagopoulou, P.N., Spirakis, P.G.: Algorithms for pure Nash equilibria in weighted congestion games. Journal of Experimental Algorithmics (11) (2006)
9. Rosenthal, R.W.: A class of games possessing pure-strategy Nash equilibria. International Journal of Game Theory 2, 65–67 (1973)
10. Spiess, H.: Conical volume-delay functions. Transportation Science 24(2), 153–158 (1990)
11. Sanders, P., Schultes, D.: Engineering fast route planning algorithms. In: Demetrescu, C. (ed.) WEA 2007. LNCS, vol. 4525, pp. 23–36. Springer, Heidelberg (2007)
12. Sanders, P., Schultes, D.: Highway hierarchies hasten exact shortest path queries. In: Brodal, G.S., Leonardi, S. (eds.) ESA 2005. LNCS, vol. 3669, pp. 568–579. Springer, Heidelberg (2005)
13. Sanders, P., Schultes, D.: Engineering highway hierarchies. In: Azar, Y., Erlebach, T. (eds.) ESA 2006. LNCS, vol. 4168, pp. 804–816. Springer, Heidelberg (2006)
14. Schultes, D., Sanders, P.: Dynamic highway-node routing. In: Demetrescu, C. (ed.) WEA 2007. LNCS, vol. 4525, pp. 66–79. Springer, Heidelberg (2007)

Experimental Study of Non-oblivious Greedy and Randomized Rounding Algorithms for Hypergraph b-Matching

(Extended Abstract)

Lasse Kliemann* and Anand Srivastav

Institut für Informatik
Christian-Albrechts-Universität Kiel
Christian-Albrechts-Platz 4
24118 Kiel
{lki,asr}@informatik.uni-kiel.de

Abstract. We consider the b-matching problem in a hypergraph on n vertices and edge cardinality bounded by ℓ. Oblivious greedy algorithms achieve approximations of $(\sqrt{n}+1)^{-1}$ and $(\ell+1)^{-1}$ independently of b (Krysta 2005). Randomized rounding achieves constant-factor approximations of $1 - \epsilon$ for large b, namely $b = \Omega(\epsilon^{-2}, \ln n)$, (Srivastav and Stangier 1997). Hardness of approximation results exist for $b = 1$ (Gonen and Lehmann 2000; Hazan, Safra, and Schwartz 2006). In the range of $1 < b \ll \ln n$, no close-to-one, or even constant-factor, polynomial-time approximations are known. The aim of this paper is to overcome this algorithmic stagnation by proposing new algorithms along with the first experimental study of the b-matching problem in hypergraphs, and to provide a first theoretical analysis of these algorithms to some extent. We propose a non-oblivious greedy algorithm and a hybrid algorithm combining randomized rounding and non-oblivious greedy. Experiments on random and real-world instances suggest that the hybrid can, in terms of approximation, outperform the known techniques. The non-oblivious greedy also shows a better approximation in many cases than the oblivious one and is accessible to theoretic analysis.

Keywords: $\mathcal{NP}$-hard problems, approximation algorithms, hypergraph matching, greedy algorithms, randomized rounding, hybrid algorithms.

1   Introduction and Previous Work

The maximum b-matching problem: definition and complexity. Let (V, E) be a hypergraph where V is a finite set, and $E \subseteq 2^V$ is a multiset. For convenience, we identify $V = [n] := \{1, \ldots, n\}$ and $E = [m]$. Elements of V are called *vertices* and elements of E (which are subsets of V) are called *hyperedges*, or just *edges*. The maximum edge cardinality is $\ell := \max_{e \in E} |e|$. Let

---

* Supported by the Deutsche Forschungsgemeinschaft, Priority Program 1307 *Algorithm Engineering*, Grant Sr7/12-1.

J. Vahrenhold (Ed.): SEA 2009, LNCS 5526, pp. 185–196, 2009.

$w : E \longrightarrow [0,1] \cap \mathbb{Q}$ be a *weight* function on the edges. We call $w(e)$ the weight of an edge $e \in E$ and denote for each set $X \subseteq E$ its weight by $w(X) := \sum_{e \in X} w(e)$.

Let $b \in \mathbb{N}_{\geq 1}$. We call a set $M \subseteq E$ a *b-matching* if no vertex is contained in more than b edges from M. MAXIMUM b-MATCHING is the problem of finding a b-matching with maximum weight. We denote the maximum weight by OPT_b. It is a classical problem in combinatorics and optimization, studied under various aspects for $b = 1$ [8,5,2,20,18,14], and for general b [12,17,22,21], and in the generalized setting of combinatorial auctions [6,10]. It is $\mathcal{NP}$-hard even if restricted to hypergraphs with $\ell \leq 3$. It is a prototype of packing integer programs as it can be written as:

$$\max\{\sum_{j=1}^{m} w_j x_j;\ Ax \leq b, x \in \{0,1\}^m\} \tag{PIP}$$

Here, $A \in \{0,1\}^{n \times m}$ is the vertex-edge incidence matrix of the hypergraph, that is $A_{ij} = 1$ if and only if vertex i is contained in edge j; and $x_j = 1$ if and only if edge j is chosen for the b-matching.

Previous approximation algorithms and results. For $0 < \alpha \leq 1$ we call an approximation algorithm an α approximation, or we say that it achieves an approximation of α, if it always delivers a solution with weight at least $\alpha\,\mathsf{OPT}_b$. Hence, the closer α is to one, the better the approximation guarantee. We will sometimes use the phrase that a b-matching M is an α approximation, meaning that $w(M) \geq \alpha\,\mathsf{OPT}_b$ holds.

Hardness upper bounds. For $b = 1$, two approximation-hardness results exist. Gonen and Lehmann [6] showed by reduction to the clique problem that no polynomial-time approximation better than $1/\sqrt{n}$ is possible for hypergraph 1-matching, unless $\mathcal{NP} = \mathcal{ZPP}$. Recently, Hazan, Safra, and Schwartz [7] proved that there is no polynomial-time $\Omega(\frac{\ln \ell}{\ell})$ approximation algorithm for hypergraph 1-matching, unless $\mathcal{P} = \mathcal{NP}$. We are not aware of any approximation-hardness results for $b \geq 2$, and indeed, for $b = \Omega(\ln n)$ there is a constant-factor approximation [22].

Algorithms. Algorithmic approaches in the existing literature mainly focus on *oblivious greedy algorithms* on the one hand and algorithms based on *randomized rounding* on the other hand. We briefly summarize both. An oblivious greedy algorithm starts with an empty set M and makes one pass over all edges in the order given by a *rank* function $\rho : E \longrightarrow \mathbb{R}_{\geq 0}$. An edge is added to M if that does not destroy the b-matching property of M. Otherwise the edge is not included. The class of functions eligible for the rank function is restricted to those which do *not* consider interaction of edges.[1] More precisely, we allow

---

[1] The term "oblivious greedy algorithm" is used by Krysta [10]. Gonen and Lehmann [6] do not use the term "oblivious", but they explain the same concept [6, Sec. 4]. (In both [10] and [6], monotonicity is also required: the rank function shall, for fixed edge e, be increasing in $w(e)$.) However, the main feature of being agnostic to edge interactions is stated rather implicitly in both works. It becomes quite clear, however, in the proof of [6, Th. 2]. We will use the term "oblivious" in the sense explained here.

$\rho(e)$ to depend only on the edge cardinality $|e|$, the edge weight $w(e)$, and the matching parameter b. Krysta [10] studied oblivious greedy algorithms for a generalized problem. Simplified for the case of hypergraph b-matching, two of his main results are: using rank functions $\rho(e) := w(e)/\sqrt{|e|}$ and $\rho(e) := w(e)/|e|$, an approximation of $(\sqrt{n}+1)^{-1}$ and $(\ell+1)^{-1}$ can be achieved, respectively. By modifying an argument due to Gonen and Lehmann [6], he showed that no oblivious greedy algorithm can achieve an approximation better than $1/\sqrt{n}$.

Algorithms based on *randomized rounding* first solve the LP relaxation of (PIP) on page 186, which can be done (theoretically) in polynomial time, and then, in a randomized manner, round the LP-optimal fractional solution $x^* \in [0,1]^m$ in order to obtain an integral vector $x \in \{0,1\}^m$, which defines a set of edges. The generic randomized rounding procedure with a parameter $\delta \in [0,1]$ is:

for $j \leftarrow 1$ **to** m **do**

$$x_j \leftarrow \begin{cases} 1 & \text{with probability } \delta x_j^* \\ 0 & \text{with probability } 1 - \delta x_j^* \end{cases}$$

Srivastav and Stangier [22] showed that for each $\epsilon > 0$ with the choice of $\delta := 1 - \frac{\epsilon}{2}$ in the randomized rounding, this technique achieves a *constant factor approximation* of $1 - \epsilon$ with a probability of at least 0.73, *provided that b is large enough*, namely $b = \Omega(\epsilon^{-2}, \ln n)$, and gave a polynomial-time derandomization.

Using a sophisticated probabilistic analysis based on correlation inequalities, Srinivasan [21] proposed a derandomized algorithm, based on randomized rounding, delivering solutions of weight $\Omega(\alpha(n,b)\, \mathsf{OPT}_b^*)$, where OPT_b^* is the optimum of the LP relaxation, and $\alpha(n,b) = (\mathsf{OPT}_b^*/n)^{1/b}$. Although this result holds for all b, it is weak for 'small' b, namely $b \ll \ln n$. That weakness lies in a crucial dependence on OPT_b^*/n. If OPT_b^*/n is small, e.g., $\mathsf{OPT}_b^*/n = 1/n^c$ for a constant $c > 0$, and we have $b \ll \ln n$, then $\alpha(n,b) = o(1)$, as $n \to \infty$, hence the approximation may become useless.

It is notable that for almost a decade no significant progress on improved polynomial-time approximation algorithms based on randomized rounding for maximum b-matching and the general integer packing was made. In view of Krysta's final results on oblivious greedy algorithms and Srinivasan's efforts to improve the randomized rounding analysis, we feel that the potential of these algorithms and their analysis is exhausted, at least when considered separately. We also believe that better approximation results for integer packing and in particular maximum b-matching cannot be obtained by further refining the probabilistic analysis for randomized rounding, unless new algorithmic ideas are invoked.

Our contribution. We present a new *non-oblivious* greedy algorithm and a hybrid algorithm *combining* randomized rounding with greedy, with focus on their experimental behavior, but provide also a theoretical foundation to some extent. In Sec. 2 we propose a family of non-oblivious greedy algorithms, called ε-*Greedy*. Roughly speaking, such an algorithm consists of an oblivious and a non-oblivious part. A parameter ε controls which of the two parts is emphasized. In Sec. 4 we describe different types of instances and their generation on which we test our algorithms. The instances are random hypergraphs with bounded ℓ or bounded VC dimension, hypergraphs based on matrices occurring in the design of finite impulse-response filters (FIR filters) in electrical engineering, and real-world instances of the three-dimensional assignment problem occurring in 3D X-ray reconstruction in prostate cancer radiation.

In Sec. 5 we compare all four algorithms (Oblivious, ε-Greedy, Randomized Rounding, Hybrid) against each other. In order to evaluate the approximation achieved, for small instances, the optimum is computed using the CBC branch-and-cut implementation of the COIN-OR Project [11]. For larger instances, we can only obtain a lower bound on the approximation, using the LP optimum OPT_b^* instead. We wish to emphasize that we focus on *worst-case* performance, that is we do not take average values, but the *minimum* over all observed approximations. Thus the approximation shown in tables is the overall worst-case for instances under consideration.

The main experimental result (Sec. 5) is that the hybrid algorithm achieves an approximation of 0.75 to 1.0 for $b > 1$, and in terms of approximation outperforms all other algorithms on all instance classes. In many cases, also ε-Greedy performs noticeably better than the oblivious greedy algorithm and is on the same level otherwise. In addition to the experimental work, we analyze ε-Greedy to some extent. We show approximation upper bounds, i.e., negative results, along with corresponding instances for some ranges of ε (Sec. 3).

Conjecture and future work. The experiments indicate that even for small b, our algorithms, especially the hybrid one, achieve good approximations, in many cases substantially better ones than the oblivious algorithm (or randomized rounding alone). It is a challenging problem to prove these observations, possibly restricted to a certain instance class, using its structural properties.

2   Algorithms

2.1 ε-Greedy

We fix some notation. For each vertex $v \in V$ denote the set of edges incident in v by $\Gamma(v) := \{e \in E; \ v \in e\}$. The *degree* (or *vertex-degree*) of v is $\deg(v) := |\Gamma(v)|$. For a set $X \subseteq E$ and a vertex $v \in V$ denote the *coverage* of v by X by $\mathrm{cover}(v, X) := |\Gamma(v) \cap X|$. Denote the *maximum coverage* of an edge $e \in E$ by a set $X \subseteq E$ by $\mathrm{maxcover}(e, X) := \max_{v \in e} \mathrm{cover}(v, X)$.

Let M be a b-matching in the following. We call $e \in E \setminus M$ a *candidate* if $M \cup \{e\}$ is again a b-matching. An equivalent formulation for being a candidate is that $\mathrm{maxcover}(e, M) < b$. Let $\mathrm{cand}(M)$ be the set of all candidates.

For each $X \subseteq E$ and any $e \in E$ define the set of *edges blocked* by e as

$$C(e, X) := \{f \in E \setminus (X \cup \{e\}); \ \mathrm{maxcover}(f, X \cup \{e\}) \geq b > \mathrm{maxcover}(f, X)\} \ .$$

So, for a candidate e we have that $C(e, M)$ is the set of edges that cannot be added to $M \cup \{e\}$ without destroying its b-matching property, but each one of them – at least alone – can be added to M. From another perspective, if all edges

---

Algorithm 1. ε-Greedy

Input: $H = (V, E)$, $w : E \longrightarrow [0, 1] \cap \mathbb{Q}$, $b \in \mathbb{N}_{\geq 1}$, $\varepsilon \in [0, 1] \cap \mathbb{Q}$.
Output: b-matching and number of iterations.

$E_0 \leftarrow E$; $M_0 \leftarrow \emptyset$; $i \leftarrow 0$;
while $|E_i| > 0$ **do**
$\quad\quad e_i \leftarrow \arg\max\{\varepsilon w(e) + (1 - \varepsilon)(|C^*(e, M_i) \cap E_i| + 1)^{-1}; \; e \in E_i\}$;
$\quad\quad M_{i+1} \leftarrow M_i \cup \{e_i\}$;
$\quad\quad E_{i+1} \leftarrow E_i \setminus (\{e_i\} \cup C(e_i, M_{i+1}))$;
$\quad\quad i \leftarrow i + 1$;
return M_i and i;

$e \in E \setminus M$ are candidates for M, then all edges from $E \setminus (M \cup \{e\} \cup C(e, M))$ are candidates for $M \cup \{e\}$.

For each $X \subseteq E$ and $e \in E$ define the set of *edges affected by e* as

$$C^*(e, X) := \{f \in E \setminus (X \cup \{e\}); \; \text{maxcover}(f, X \cup \{e\}) > \text{maxcover}(f, X)\} \; .$$

For $b = 1$ and all candidates e, we have $C(e, M) = C^*(e, M)$, but not in the general case of $b > 1$. We can now state the ε-Greedy algorithm.

This algorithm picks an edge which maximizes a function consisting of an oblivious part, namely the weight, and a non-oblivious part, namely the inverse of the number of affected edges in the remaining hypergraph (plus 1 to avoid division by zero). The chosen edge is added to the b-matching, and all edges that as a consequence of this cannot be added to M anymore are removed. So, the algorithm favors edges with high weight which affect only few other eligible edges; the parameter ε specifies which of the two criteria counts more. Note that the algorithm is oblivious for $\varepsilon = 1$, but non-oblivious for $\varepsilon \neq 1$.

Computing $C(e, M_i)$ or $C^*(e, M_i)$ takes $O(\text{edeg}(e) \cdot \ell)$ time. Hence we can bound the running time of ε-Greedy by $O(\min\{m, bn\} \cdot m \cdot \max_{e \in E} \text{edeg}(e) \cdot \ell)$, which is bounded by $O(m^3 n)$.

2.2   Hybrid Algorithm: Randomized Rounding and ε-Greedy

We use the generic scheme shown on page 187 with parameter $\delta := 1$. That is, we round up with probability x_j^* and round down with probability $1 - x_j^*$ for each $j \in [m]$. Let $X \subseteq E$ be the edge set induced by the rounded vector. This might not be a b-matching. Edges violating the b-matching property are removed in what we call a *repairing step*: we assign so-called *violation values* to the edges and then remove edges with the highest such values. This will be described in detail in the full version of our paper.

The repaired solution M, which is a b-matching, is then taken as a starting solution for a modified version of ε-Greedy. The modification lies only in the initialization: set $E_0 \leftarrow \text{cand}(M)$ and $M_0 \leftarrow M$. That is, we make ε-Greedy believe that it had computed M so far and then let it continue its work as usual.

3   Analysis: Negative Results for ε-Greedy

We have positive results for the oblivious version of ε-Greedy [3], i.e., with $\varepsilon = 1$. They extend the theory of oblivious greedy algorithms. We now give negative results for the general version. If the non-oblivious part is too strong, i.e., ε is too close to 0, then ε-Greedy can make poor choices.

Theorem 1. *For each b, there exists ε_0 such that the following holds: Let $0 < \alpha \leq 1$. There exist instances of the b-matching problem on which ε-Greedy with any $\varepsilon < \varepsilon_0$ can deliver a solution which is no better than an α approximation.*

Proof. Let $k \geq 2$ and $w := \frac{\alpha}{kb}$. We take k vertices and for each of them b edges containing that vertex; we speak of k *stacks* of singleton edges, each stack having size b before the start of the algorithm. In addition, we take one edge containing all vertices. The singleton edges receive weight w and the large edge receives weight 1. Let $\varepsilon_0 := \frac{1}{2b+5}$ and $\varepsilon < \varepsilon_0$. We show that if all stacks have size s or $s + 1$ for some $s \in \mathbb{N}$, then ε-Greedy can pick a singleton from a stack of size $s+1$. Inductively, this yields that ε-Greedy can construct a b-matching of weight $kbw = \alpha$. Choosing the large edge instead would result in weight 1, hence that solution constitutes no better than an α approximation.

Now fix $s \in \mathbb{N}$. If $s = 0$, then the large edge is no candidate anymore, and only singletons will be picked. So suppose $s \geq 1$ and for the first case also suppose that all stacks have size s. Then each singleton affects[2] s edges, namely the other $s - 1$ singletons from its stack, and the large edge. The large edge affects ks edges, namely all singletons.

For the second case suppose that all stacks have size s or $s+1$, and both sizes occur. Each singleton affects s edges:

- Those on stacks with $s + 1$ edges affect the other s singletons but *not* the large edge, since a vertex under a stack of $s + 1$ remaining singletons has smaller coverage than one under a stack of only s remaining singletons.
- Those singletons on stacks with s edges affect the other $s - 1$ singletons and the large edge.

So, in particular the algorithm sees no difference among the singletons and hence may pick any of it, provided it decides to pick a singleton and not the large edge. The large edge affects at least ks edges. Hence, in any of the two cases, a singleton from one of the higher stacks can be chosen if

$$\varepsilon + (1 - \varepsilon)\frac{1}{ks + 1} < \varepsilon w + (1 - \varepsilon)\frac{1}{s + 1} \ ,$$

which is equivalent to $1 - w < (\frac{1}{\varepsilon} - 1)(\frac{1}{s+1} - \frac{1}{ks+1})$. Since $k \geq 2$ and $s \leq b$, the right-hand side is at least $(\frac{1}{\varepsilon} - 1)(\frac{1}{s+1} - \frac{1}{2s+1}) \geq (\frac{1}{\varepsilon} - 1)\frac{1}{2b+4}$. By the choice of ε, this is at least 1. The claim follows, since $1 - w < 1$. $\qquad\square$

On the other hand, if the oblivious part is too strong, i.e., ε is too close to 1, ε-Greedy can also make poor choices. This means we have to use multiple ε values in practice.

---

[2] We always use this term relative to the *remaining* edges.

Theorem 2. *There is ε_0 such that the following holds: For each b, there exist instances on which ε-Greedy with any $\varepsilon > \varepsilon_0$ delivers a solution which is no better than a $2\ell^{-1}$ approximation.*

Proof. We give a similar construction as in the previous proof. Let $\varepsilon_0 := \frac{1}{2}$ and $\varepsilon > \varepsilon_0$. Let $k \geq 2$ and consider k vertices and for each of them a stack of b singletons of weight $w := \frac{1}{2}$ each. Then consider b large edges, each of weight 1 and each containing all vertices. If we take all kb singletons, we have a b-matching of weight $\frac{1}{2}kb$. We show that ε-Greedy may instead pick all the large edges, delivering a b-matching of weight b, so approximation is $2k^{-1} = 2\ell^{-1}$.

Let $0 \leq t < b$ large edges have already been picked for the b-matching, and no singleton. Then

- a singleton affects $(b - 1) + (b - t)$ edges;
- a large edge affects $(b - t - 1) + kb$ edges.

In order that a large edge is picked next, it must be satisfied:

$$\varepsilon w + (1 - \varepsilon)\frac{1}{(b - 1) + (b - t) + 1} < \varepsilon + (1 - \varepsilon)\frac{1}{(b - t - 1) + kb + 1} \, ,$$

which is equivalent to $w < 1 - (\frac{1}{\varepsilon} - 1)(\frac{1}{2b-t} - \frac{1}{(k+1)b-t})$. Using our knowledge on w and ε, it suffices when $\frac{1}{2} < 1 - (\frac{1}{2b-t} - \frac{1}{(k+1)b-t})$. This holds, since $0 \leq t < b$ and so for the last term we have $\frac{1}{2b-t} - \frac{1}{(k+1)b-t} \leq \frac{1}{b+1} - \frac{1}{(k+1)b} < \frac{1}{2}$. $\square$

4   Instances

We use four types of hypergraphs in our experiments. The first two are random:

- Random hypergraphs with bounded ℓ. Such a hypergraph is constructed as follows: fix the number of vertices n, the number of edges m, and the maximum edge cardinality ℓ. Then, independently and uniformly at random, draw m edges from the set of all edges of cardinality at most ℓ.

- Random hypergraphs with VC dimension at most 3. Such a hypergraph is constructed as follows: fix the number of vertices n, and for each vertex generate a random point in the two-dimensional plane. Denote these points by $V \subset \mathbb{R}^2$. Let $\mathcal{H}$ be the set of all half-planes. We then construct the set of edges as $E = \{H \cap V; \ H \in \mathcal{H}\}$. Such hypergraphs (V, E) are known to have VC dimension at most 3, see, e.g., [13]. Details of the construction will be given in the full version of our paper.

The edge weights for these two types are obtained by first drawing $w \in [0, 1]$ uniformly at random and then independently for each $e \in E$ choosing $w(e) \leftarrow 1$ or $w(e) \leftarrow w$ with probability $\frac{1}{2}$ each. Hence there occur two different edge weights (or one, if $w = 1$) in each instance.

The following two types of hypergraphs stem from applications:

- Hypergraphs based on FIR filter matrices. Matrices of a special structure with entries from $[-1, 1]$ specify a packing problem in filter design [15]. We take parts of such matrices and deterministically round the absolute values of their entries to 0 or 1. The resulting matrix is interpreted as the vertex-edge incidence matrix of a hypergraph. The resulting b-matching problem in such hypergraphs is different from the original packing problem. We consider it nonetheless, in order to have instances of a (partly) non-random structure. Edge weights are assigned randomly from $\{1, w\}$ for a fixed random w, as with the purely random hypergraphs above.

- Hypergraphs modeling 3-dimensional assignment problems (AP3). These are complete tripartite hypergraphs, i.e., $V = V_1 \times V_2 \times V_3$, $|V_1| = |V_2| = |V_3|$, and $E = \{\{v_1, v_2, v_3\}; \ v_i \in V_i, i = 1, 2, 3\}$ with a weight function $\tilde{w}$. The objective is to find a perfect 1-matching of minimum weight, i.e., a 1-matching of minimum weight such that each vertex is contained in (exactly) one edge from the matching. Such problems arise in 3D X-ray reconstruction in cancer radiation [19,4]. If we transform the weight function to $w(e) := 1 - \frac{\tilde{w}(e)}{\tilde{w}_{\max}}$, where $\tilde{w}_{\max} = \max_{e \in E} \tilde{w}(e)$, then the problem is equivalent to our (maximum-weight) 1-matching problem. We test our algorithms on AP3 instances from medicine[3] with such a transformed weight function. Edges with weights below a certain threshold are removed for the sake of reasonable running times. This has shown to be admissible in practice [19]. We also consider b-matchings for $b > 1$. These have no direct interpretation for the AP3, but could be interesting in practice as a pre-selection for a manual assignment.

5   Experimental Results

We implemented all algorithms in C++. LPs are solved with the CLP solver and optimal solutions are obtained with the CBC branch-and-cut implementation, both from the COIN-OR Project [11]. Post-processing of obtained data is done using the R System [16]. Observed worst-case approximations are given in tables, in percent and truncated to integral numbers. So, e.g., a value of 100 means that the optimum is found, and a value of 80 means that a solution of weight at least $0.8 \cdot \mathrm{OPT}_b$ is found. Columns are used for different b and different algorithms:

- "O" is based on the best result of the two oblivious greedy algorithms, which have guaranteed approximation of $(\sqrt{n} + 1)^{-1}$ and $(\ell + 1)^{-1}$, respectively.
- "G" is based on the best result of ε-Greedy for $\varepsilon = 0, \frac{1}{2}, 1$.
- "R" is based on the best of 10 trials of randomized rounding and repairing.
- "RG" is based on the best result over the same ε values as above of the hybrid algorithm. Recall that it combines randomized rounding with ε-Greedy. For

---

[3] The instances were provided by F.-A. Siebert, UKSH, Campus Kiel.

each ε, we also do 10 trials of randomized rounding here, invoke ε-Greedy on each outcome and finally take the best of these.

Instances are grouped in rows according to $\min\{\sqrt{n}+1, \ell+1\}$. The first column "#" shows the number of instances for the row, for each b. (The rough total count stated in the caption considers b as part of an instance, so it is about the sum of the first column times four.) The second column shows $a := 100 \cdot (\min\{\sqrt{n}+1, \ell+1\})^{-1}$ to give a reference to the provable approximation.

We emphasize again that the values are *worst-case* observations and no averages. For example, to create an entry in a "G" column, all instances with a fixed value of $\min\{\sqrt{n}+1, \ell+1\}$ are taken – that value determines the row – and then for each such instance the best result of ε-Greedy for $\varepsilon = 0, \frac{1}{2}, 1$ is determined and divided by OPT_b or OPT_b^*. This gives one value for each instance. Then the minimum, i.e., the worst-case, of these is taken.

Note on the number of weights. Recall that, if we use random weights, we use only two different weights for each instance. Few weights seem to give the biggest improvement of ε-Greedy over the oblivious algorithms; the hybrid however shows no such sensitivity to the number of weights. A modification of ε-Greedy is to not consider the *number* of affected edges, but their *total weight*. This generally shows improvements. In combination with using more ε values – we tested $\varepsilon = 0, 0.1, 0.2, \ldots, 1.0$ – it seems to help in case of a larger number of different weights. A detailed comparison will be given in the full version of our paper.

Now, we give results for *two different weights* (except the AP3 instances, which bring their own weights) and the ε-Greedy presented in this paper, which uses the *number* of affected edges, and the scheme of three different ε values.

Bounded ℓ: $n = 10, 20, \ldots, 50$; $m = 5, 10, \ldots, 90$; $\ell = 3, 4, 5, 6$. The whole data set comprises more than $5\,000\,000$ instances:

		$b = 1$				$b = 2$				$b = 3$				$b = 4$			
#	a	O	G	R	RG	O	G	R	RG	O	G	R	RG	O	G	R	RG
120 221	14	25	50	44	68	50	69	69	84	58	76	79	90	62	80	85	92
64 162	15	29	50	40	52	47	66	70	84	50	73	80	90	62	75	84	91
184 589	16	25	50	50	67	50	66	70	85	59	78	83	91	62	80	86	92
135 550	18	31	50	40	53	44	66	66	80	56	71	79	86	63	75	85	90
252 456	20	27	50	49	67	51	66	75	88	60	77	85	92	64	82	88	94
205 033	24	37	50	41	50	49	63	73	75	54	68	83	85	58	73	86	88
321 049	25	33	50	50	69	56	74	82	87	65	79	84	90	69	83	91	93

VC dimension at most 3: $n = 5, 10, 15$ and m up to 211. The whole data set comprises more than $9\,800\,000$ instances:

		$b = 1$				$b = 2$				$b = 3$				$b = 4$			
#	a	O	G	R	RG	O	G	R	RG	O	G	R	RG	O	G	R	RG
820 951	20	50	75	60	81	54	76	82	88	61	80	88	93	66	82	91	93
821 060	24	50	66	63	69	50	72	80	86	56	78	87	90	66	82	89	96
821 095	30	50	50	50	83	60	68	75	90	69	75	84	91	72	79	88	94

Based on FIR matrices: n up to 399 and m up to 25. The whole data set comprises more than 2 500 000 instances. The oblivious algorithms deliver approximations between 83% and 97%. ε-Greedy and the hybrid algorithm always find the optimum.

AP3 from medicine: 29 instances with n up to 204 and m up to 17 619:

		$b = 1$				$b = 2$				$b = 3$				$b = 4$		
#	a	O	G	R	RG	O	G	R	RG	O	G	R	RG	O	G	R	RG
29	25	77	92	70	95	79	94	86	96	80	94	90	97	81	95	93	98

Larger instances with bounded ℓ: $n = 1\,000$; $m = 1\,000, 2\,000, \ldots, 9\,000$; $\ell = 3, 4, 5, 6$. The LP optimum OPT_b^* is taken for comparison. The whole data set comprises more than 60 000 instances:

		$b = 1$				$b = 2$				$b = 3$				$b = 4$		
#	a	O	G	R	RG	O	G	R	RG	O	G	R	RG	O	G	R	RG
3 616	14	70	81	61	84	77	86	81	92	80	89	88	94	83	90	92	96
3 719	16	70	81	61	85	77	87	81	92	80	89	87	94	83	90	91	96
3 859	20	70	82	61	84	77	87	80	92	80	89	88	94	82	90	91	96
3 968	25	69	82	59	85	75	87	80	91	79	90	87	94	81	91	91	95

Running times. The per-instance running time is far below 1 second for most of the small instances. In order to get an idea of the running time for larger instances, we consider different densities, i.e., we let m be a function of n and then consider increasing n. We fix $b = 2$ and use random hypergraphs with bounded ℓ and two different weights. We fix the scheme of running ε-Greedy with three different values $\varepsilon = 0, \frac{1}{2}, 1$. Let first $m(n) := n \ln \ln n$. The running time of ε-Greedy appears to be quadratic in m, it follows (in seconds) the curve $\approx 1.1 \cdot 10^{-7} \cdot m^2$ on an AMD Opteron® at 2.8 G Hz. This hypothesis was obtained by linear regression based on experiments with $n = 1\,000, 2\,000, \ldots, 20\,000$. The hybrid needs more time and seems to roughly follow the cubic curve $\approx 2.7 \cdot 10^{-10} \cdot m^3$, based on experiments with $n = 1\,000, 2\,000, \ldots, 9\,000$. It took approximately 40 minutes for $n = 9\,000$ vertices. Nearly all of the time is spent solving the LP.

For denser hypergraphs, say $m(n) := n \ln n$, the situation is different. ε-Greedy needs $\approx 3.2 \cdot 10^{-7} \cdot m^2$ seconds, measured for $n = 1\,000, 1\,500, \ldots, 5\,000$. The hybrid is faster than this in the range up to 5 000 vertices, namely $\approx 1.8 \cdot 10^{-12} \cdot m^3$, however, for larger n, it shows some very high running times. This is not fully understood yet and may be related to properties of the LP solver.

Discussion. We have seen the hybrid algorithm outperforming all others in terms of approximation. Running times are different for different densities, with ε-Greedy being in favor of sparse hypergraphs.

ε-Greedy also performs well in terms of approximation: it gives substantially better approximations than oblivious greedy in most cases, and is on the same level otherwise. In many cases, ε-Greedy is similar to or better than randomized rounding alone. For instances based on FIR matrices, ε-Greedy always finds the optimum, whereas oblivious greedy can have approximations as low as 83%. Apparently, there is some structure in these instances which oblivious algorithms are unable to take advantage of, but ε-Greedy can. This is so even for a larger data set of almost 4 million instances.

6 Future Work

We wish to further speed up and perhaps parallelize the implementation of ε-Greedy, making it an even more attractive alternative to LP-based algorithms. We plan to extend the experimental basis by including more instances: e.g., other types of hypergraphs with bounded VC dimension can be constructed, and new instances for hypergraph matching arise in computational geometry in the study of symmetry detection of point sets [1,9]. We also plan to consider larger b, and especially to explore the territory around $b = \Theta(\ln n)$.

An ultimate goal is, of course, to prove the better approximation performance of our algorithms. We currently look at the outcome of experiments as if we expect a constant-factor approximation, like we have for large b. This needs not to be the case. Future work will aim at parameterizing the observed approximation ratios.

Acknowledgments. We thank our student Ole Kliemann for profiling and speeding up our implementation. We thank the anonymous referees for helpful comments and suggestions for future work. The first author thanks the Deutsche Forschungsgemeinschaft for financial support through Grant Sr7/12-1.

References

1. Alt, H., Mehlhorn, K., Wagener, H., Welzl, E.: Congruence, similarity, and symmetries of geometric objects. Discrete & Computational Geometry 3, 237–256 (1988)
2. Aronson, J., Dyer, M.E., Frieze, A.M., Suen, S.: Randomized greedy matching II. Random Structures and Algorithms 6(1), 55–74 (1995)
3. El Ouali, M., Kliemann, L., Srivastav, A.: Approximation algorithms for hypergraph b-matching. Technical report, Institut für Informatik, Christian-Albrechts-Universität Kiel (2009) (in preparation)
4. Fohlin, H., Kliemann, L., Srivastav, A.: Randomized algorithms for mixed matching and covering in hypergraphs in 3D seed reconstruction in brachytherapy. In: Alves, C.J.S., Pardalos, P.M., Vicente, L.N. (eds.) Optimization in Medicine, pp. 71–102. Springer, Heidelberg (2008)
5. Füredi, Z., Kahn, J., Seymour, P.D.: On the fractional matching polytope of a hypergraph. Combinatorica 13(2), 167–180 (1993)
6. Gonen, R., Lehmann, D.J.: Optimal solutions for multi-unit combinatorial auctions: Branch and bound heuristics. CoRR cs.GT/0202032, pp. 13–20 (2002); Presented at ACM Conference on Electronic Commerce (2000)
7. Hazan, E., Safra, S., Schwartz, O.: On the complexity of approximating k-set packing. Computational Complexity 15(1), 20–39 (2006)
8. Hochbaum, D.S.: Efficient bounds for the stable set, vertex cover and set packing problems. Discrete Applied Mathematics 6, 243–254 (1983)

9. Iwanowski, S.: Approximate Congruence and Symmetry Detection in the Plane. Ph.D thesis, Freie Universität Berlin (1990)
10. Krysta, P.: Greedy approximation via duality for packing, combinatorial auctions and routing. In: Jedrzejowicz, J., Szepietowski, A. (eds.) MFCS 2005. LNCS, vol. 3618, pp. 615–627. Springer, Heidelberg (2005)
11. Lougee-Heimer, R.: The Common Optimization INterface for Operations Research. IBM Journal of Research and Development 47(1), 57–66 (2003)
12. Lovász, L.: On the ratio of optimal integral and fractional covers. Discrete Mathematics 13 (1975)
13. Matoušek, J.: Geometric set systems. In: Proceedings of the European Congress of Mathematics, vol. 2, pp. 2–27. Birkhäuser, Basel (1998)
14. Paschos, V.T.: A survey of approximately optimal solutions to some covering and packing problems. ACM Computing Surveys 29(2), 171–209 (1997)
15. Qi, H.: Entwurf von FIR-Filtern mit extremen Wortlängenbeschränkungen. Shaker Aachen (1996)
16. R Development Core Team: R: A Language and Environment for Statistical Computing. R Foundation for Statistical Computing, Vienna, Austria (2008) ISBN 3-900051-07-0
17. Raghavan, P., Thompson, C.D.: Randomized rounding: a technique for provably good algorithms and algorithmic proofs. Combinatorica 7(4), 365–374 (1987)
18. Rödl, V., Thoma, L.: Asymptotic packing and the random greedy algorithm. Random Structures and Algorithms 8(3), 161–177 (1996)
19. Siebert, F.A., Srivastav, A., Kliemann, L., Fohlin, H., Kovács, G.: 3-dimensional reconstruction of seed implants by randomized rounding and visual evaluation. Medical Physics 34(3), 967–975 (2007)
20. Spencer, J.: Asymptotic packing via a branching process. Random Structures and Algorithms 7(2), 167–172 (1995)
21. Srinivasan, A.: Improved approximation guarantees for packing and covering integer programs. SIAM Journal on Computing 29(2), 648–670 (1999)
22. Srivastav, A., Stangier, P.: Tight approximations for resource constrained scheduling and bin packing. Discrete Applied Mathematics 79, 223–245 (1997)

Empirical Evaluation of Graph Partitioning Using Spectral Embeddings and Flow

Kevin J. Lang[1], Michael W. Mahoney[2], and Lorenzo Orecchia[3]

[1] Yahoo! Research, Santa Clara, CA, USA
[2] Stanford University, Stanford, CA, USA
[3] University of California, Berkeley, CA, USA

Abstract. We present initial results from the first empirical evaluation of a graph partitioning algorithm inspired by the Arora-Rao-Vazirani algorithm of [5], which combines spectral and flow methods in a novel way. We have studied the parameter space of this new algorithm, *e.g.*, examining the extent to which different parameter settings interpolate between a more spectral and a more flow-based approach, and we have compared results of this algorithm to results from previously known and optimized algorithms such as METIS.

1 Introduction

Graph partitioning refers to the problem of dividing an input graph into two large pieces such that the number of edges crossing the partition is minimized. There are several standard formalizations of this bi-criterion, and in this paper we focus on minimizing expansion.[1] Given an undirected, possibly weighted, graph $G = (V, E)$, the expansion $\alpha(S)$ of a set of nodes $S \subseteq V$ is defined as:

$$\alpha(S) = \frac{|E(S, \overline{S})|}{\min\{|S|, |\overline{S}|)\}},\tag{1}$$

where $E(S, \overline{S})$ denotes the set of edges having one end in S and one end in the complement $\overline{S}$, and where $|\cdot|$ denotes cardinality (or weight). The *expansion of the graph* G is then defined as:

$$\alpha(G) = \min_{S \subseteq V} \alpha(S).\tag{2}$$

It is well-known that solving (2) exactly is NP-hard. Graph partitioning is, however, of interest in many applications. For example, it has been used in divide-and-conquer algorithms; for load balancing in parallel computing applications; to segment images and, more generally, to cluster data; and to find clusters and communities in large social and information networks.

Graph partitioning is also a problem for which a wide range of algorithms have been developed, and the theoretical and/or empirical strengths and weaknesses of these algorithms have been extensively studied. Most algorithms that

---

[1] Expansion is sometimes referred to as the quotient cut objective.

J. Vahrenhold (Ed.): SEA 2009, LNCS 5526, pp. 197–208, 2009.

have been designed to find good cuts, *i.e.*, low-expansion partitions, employ one or more of the following four algorithmic ideas: *spectral methods*; *flow-based methods*; *local improvement*; and *multi-resolution*. Historically, spectral methods and flow-based methods have dominated the theoretical landscape; and variants of spectral methods, as well as local improvement techniques, often in a multi-resolution setting, have dominated applications. In particular, note that the heuristic METIS [12] is often the method of choice in applications since it finds good-quality cuts in mesh-like graphs very quickly.

Researchers also noticed that spectral and flow-based methods tend to have complementary strengths—the worst-case examples for spectral algorithms are easy graphs for flow-based methods, and vice-versa—which has lead to attempts to combine spectral and flow into a better graph partitioning algorithm. This was achieved by Arora, Rao, and Vazirani (ARV) [5], who developed the concept of "expander flows" and who introduced an algorithm that achieves an $O(\sqrt{\log n})$ worst-case approximation to expansion and several related quantities. The original version of ARV [5] yields a polynomial-time algorithm, but one that is too slow to be practical, as it must solve a large semi-definite program.

The ARV breakthrough was followed by the introduction of several related algorithms exploring the running-time versus quality-of-approximation tradeoff [3,4,13]. In particular, Orecchia, Schulman, Vazirani, and Vishnoi (OSVV) [16] developed an algorithm that performs only polylogarithmic single commodity max-flow computations to achieve an $O(\log n)$ approximation. The fastest theoretical algorithm for these single commodity flow computations has time complexity $\tilde{O}(n^{3/2})$ [10], but push-relabel methods have been shown to be faster in practice [7], potentially making the OSVV algorithm useful in applications. OSVV and other methods inspired by [5] have been reviewed in [6], where the authors of ARV pose the question of how well they will perform empirically, in particular with respect to METIS.

In this paper, we report initial results from the first empirical evaluation of an algorithm from this novel family. We have implemented the algorithm of OSVV [16], and we have compared it with several implementations of traditional graph partitioning algorithms, including METIS, on a suite of graphs designed to highlight the strengths and weaknesses of previously existing algorithms. We demonstrate that the algorithmic ideas underlying the ARV method can be implemented on medium-sized graphs to find cuts that are competitive with those returned by existing algorithms; and we demonstrate a manner in which different parameter settings interpolate between spectral and flow methods.

2 Algorithms

In this section, we describe the algorithms we used in our empirical evaluation. We compared the algorithm of OSVV [16] with two versions of METIS, two versions of the spectral method, and one purely flow-based method:

- METIS is a fast heuristic that combines a multi-resolution approach with local improvement techniques [12].

- METISRAND is a randomized variation of the basic METIS algorithm that achieves much better results.
- SPECTRAL is the classical spectral method of [1], which uses a sweep cut to round the eigenvector solution.
- SPECFLOW is a variation of SPECTRAL in which the standard sweep-cut rounding is replaced by a flow-based rounding which is guaranteed to obtain a better or equally good cut [2].
- LR is a simplified version [14] of the flow-based algorithm by Leighton and Rao [15].
- OSVV is our implementation of the the algorithm of OSVV [16], which uses ideas related to the original ARV algorithm [5]. Our implementation closely follows the theory, and the approximation guarantees of [16] still apply.

Note that our comparison includes both standard versions of the traditional algorithms (METIS and SPECTRAL), as well as modified versions (METISRAND and SPECFLOW) which in practice find much better cuts.

2.1 The Improve Algorithm

We start by describing a flow-based "improvement" procedure that will be an important building block for SPECFLOW and OSVV. The IMPROVE algorithm was originally introduced as a post-processing procedure to improve cuts returned by other methods [2]. This algorithm takes as input a bisection $(A, \overline{A})$, and it looks for a cut which optimizes a combination of low expansion and correlation with the starting bisection. The algorithm outputs a cut $(T, \overline{T})$ and a perfect matching M between A and $\overline{A}$. If the expansion of $(T, \overline{T})$ is α, then M can be routed[2] in G with congestion $1/\alpha$. This matching M can then be used as a certificate that $(T, \overline{T})$ has better expansion than all cuts strictly contained in $(A, \overline{A})$. This algorithm can be implemented by a small number of single commodity max-flow computations. We used the C++ implementation of [2], which is based on the max-flow push-relabel program hi_pr v.3.4, described in [8].

2.2 The Metis and MetisRand Algorithms

METIS is a heuristic developed by Karypis and Kumar [12] to find good balanced partitions in graphs. Although it has no theoretical guarantees, in practice it runs extremely fast. METIS makes many random choices during its execution, but the standard version of the code has a fixed random seed which makes the algorithm deterministic; our METIS results were obtained by running the unmodified pmetis program (version 4.0.1). METISRAND is a modified version of the basic program in which the random seed is left as an input to the program

---

[2] A weighted graph H can be *routed as a flow* in a graph G if every edge $e = \{u, v\} \in E(H)$ in H with weight w_e can be routed on a path from u to v in G such that the total congestion on every edge of G, i.e., the total weight of H routed across that edge in G, is less or equal to 1. In this case, we can use H as a certificate of expansion, since one can show that $\alpha(H) \leq \alpha(G)$.

and the best of $10,000$ runs is returned; in addition to small code changes to allow different seeds, we changed the matching method from the default "Sorted Heavy Edge Matching" to "Random Matching."

2.3   The Spectral and SpecFlow Algorithms

Spectral algorithms compute (exactly or approximately) the second eigenvector x of the Laplacian[3] of the graph, and then approximate the best cut in the graph by a cut defined by this vector. Recall that this eigenvector assigns to each vertex v of the graph a value x_v, and if we assume that these have been ordered, this allows one to define $n - 1$ "sweep cuts" $(S_i, \overline{S_i})$, for $1 \leq i < n$, as: $S_i = \{v \in V : x_v < x_i\}$. The eigenvector can be rounded to a cut by picking the best of these sweep cuts. Our SPECTRAL algorithm computes the smallest second, third and fourth eigenvectors of the Laplacian, applies the sweep cut rounding to each of them, and then returns the best cut found.

SPECFLOW is a randomized variant of SPECTRAL that was developed by Andersen and Lang [2] and that differs from SPECTRAL in two respects. First, rather than using each eigenvector separately, SPECFLOW uses a random combination of the lowest three non-zero eigenvectors of the Laplacian of G. This makes SPECFLOW more robust against cuts which may be hidden from a single eigenvector. Second, SPECFLOW replaces the rounding by a sweep cut with a call to IMPROVE on the bisection $(S_{n/2}, \overline{S_{n/2}})$. In [2], it is proven that the IMPROVE rounding procedure is strictly no worse than rounding by a sweep cut, and it is shown that SPECFLOW outperforms SPECTRAL on most graphs.

The computation of the second eigenvector of the Laplacian can be performed in a number of ways (we computed "exact" eigenvectors with ARPACK), but most relevant for the subsequent discussion is that it can be performed by considering exact or approximate random walks on the instance graph G. The idea underlying this approach is that random walks will mix slowly across cuts containing few edges, and conversely that if a random walk mixes slowly then there must be some cut which is constraining the spreading of the probability mass. Moreover, the second eigenvector of the Laplacian defines the slowest mixing direction, and the second smallest eigenvalue, $i.e.$, the spectral gap λ_2, characterizes the mixing time [9]. This highlights the weakness of the spectral method at finding good cuts: spectral algorithms are sensitive not only to sparse cuts but also to "large distances" in the graph—a random walk may fail to mix rapidly either because it takes a long time to overcome a sparse cut or because the graph has very long paths along which the random walk makes slow progress. The worst examples for spectral methods are based on this idea [11].

2.4   The LR Algorithm

Flow-based methods provide a very different way to find good cuts in a graph. They route a certificate graph H in G, and use this routing to provide both

---

[3] The *Laplacian* of a graph K is $L(K) = D(K) - A(K)$, where $D(K)$ is a diagonal matrix containing the degree of each vertex and $A(K)$ is the adjacency matrix.

a lower bound on expansion as well as an approximate cut by showing that if no better lower bound can be proved then by duality a good cut must exist. Leighton and Rao [15] chose H to be a scaled version of the complete graph on n vertices, and they chose a linear programming relaxation of the problem based on multi-commodity flows. The best implementations of this lead to a theoretical running time of $\tilde{O}(n^2)$, but in our empirical evaluation we used the implementation of [14], which runs faster since it only approximately solves the flow problem. Graphs on which LR is known to perform poorly include constant-degree expanders [15] and expanders with planted cuts [2].

2.5   The OSVV Algorithm

OSVV and other related algorithms subsequent to that of ARV [5] have a simple interpretation based on a modification of flow-based ideas: they strengthen the flow-based approach of Leighton and Rao [15] by removing the limitation that the graph to be routed in G be a complete graph by instead allowing it to be any graph with large spectral gap [6]. Here, we are going to give a dual interpretation of these algorithms as based on a modification of spectral ideas: OSVV strengthens the standard spectral approach by using flow-based ideas to modify the instance graph to make it more amenable to spectral methods. It does so by using the matching returned by IMPROVE to add edges to the input graph to fix the oversensitivity of spectral methods to large distances.

Before describing the OSVV algorithm, recall that the *heat kernel* of a graph $K = (V, E)$ is defined as: $H_K^\eta = \exp(-\eta L(K))$, where $L = L(K)$ is the Laplacian of K and where $\eta \geq 0$ is a learning rate. The heat kernel can be used in an alternative version of the standard spectral method to produce an approximate eigenvector of the Laplacian as follows: take a vector v picked uniformly at random from $\{+1, -1\}^V$ and consider $x = H_K^\eta v = \exp(-\eta L(K))v$. As η varies between 0 and infinity, the vector x becomes a better and better approximation to the second eigenvector of the Laplacian. Replacing the exact computation of the second eigenvector of the Laplacian with an approximation based on the heat kernel with $\eta \neq \infty$ has two potential advantages. First, the heat kernel is more robust against cuts hidden from the second eigenvector. Second, the computation of $H_K^\eta v$ is faster than that of the second eigenvector, especially for graphs with a small spectral gap.

Our OSVV algorithm takes as input a graph G, as well as parameters η, γ, and stopping condition σ. It then does the following:

1. Let $G' = \gamma G$; and $t = 0$.
2. Approximate the second eigenvector of the Laplacian of G' by performing a heat kernel computation on G'.
3. Using the bisection $(S_{n/2,t}, \overline{S_{n/2,t}})$ from the sweep cut along this approximate eigenvector, call IMPROVE with G to get a cut $(T_t, \overline{T_t})$ and a matching M_t.
4. Let $G' = G' + M_t$; and $t{+}{+}$. Until the stopping rule is satisfied, goto Step 2.
5. Return as output the cut $(T_t, \overline{T_t})$ of minimum expansion found in Step 3.

In [16], it is shown that the algorithm takes at most $O(\log^2 n)$ rounds to achieve an $O(\log n)$ approximation. Our implementation closely follows the theory [16] and has the following three parameters:

- The learning rate, η, determines how much the spectral computation of the heat kernel on G'_t is allowed to converge towards the second eigenvector of G'_t. (For higher values of η, the spectral part of the algorithm is more global, but it is also the more susceptible to the errors caused by long paths.)
- The initialization coefficient, γ, determines the weight of the instance graph G in G'. (For higher values of γ, G' depends relatively more on G, and less on the feedback matchings output by the IMPROVE algorithm, and thus the more similar the spectral computation on G' is to that of G.)
- The stopping condition, σ, is the number of iteration after which, if no improvement in the best cut found has occurred, the algorithm aborts. (A higher stopping condition can yield a better solution at the expense of time, while a lower stopping condition can make the algorithm faster but may prevent it from achieving its best expansion scores.)

Note that in addition to the instance graph G, OSVV maintains a graph G', which starts off equal to a scaled version of G and is progressively modified to be more suited to spectral methods. At every iteration t, the approximate spectral computation is performed on the current G'_t and a sweep bisection $(S_{n/2,t}, \overline{S_{n/2,t}})$ is obtained from the resulting vector. The IMPROVE algorithm is then applied to this bisection *on the input graph* G (since we are interested in cuts on G), and this will yield a cut $(T_t, \overline{T_t})$. Now, either there is a good sweep cut in the original eigenvector, which would have been found by IMPROVE, or the spectral method has been fooled by some long paths in G. To fix this problem, OSVV considers the matching M_t returned by IMPROVE at iteration t. Since the endpoints of the edges of M_t lie on opposite sides of the bisection $(S_{n/2,t}, \overline{S_{n/2,t}})$, M_t can be used to "shortcut" the long paths, and so the algorithm sets $G'_{t+1} = G'_t + M_t$. Clearly, the matching M_t can be thought as providing *iterative feedback* to the spectral method about the quality of the cut found and how to modify G'_t to explore different cuts and identify better cuts.

Note also that for large values of γ and η, OSVV becomes very spectral in flavor, as G' becomes dominated by G and the heat kernel is allowed to converge closer to the second eigenvector of G. For example, in the first iteration, the spectral computation performed by OSVV is the multiplication of a random vector by $\exp^{-\eta\gamma G}$. Hence, the higher the product $\gamma\eta$, the more the first iteration of OSVV will look like a second eigenvector computation; and similarly for subsequent iterations. Conversely, as η and γ decrease, the algorithm performs a more localized spectral computation and the feedback matchings increase in weight with respect to G, yielding an algorithm with a stronger flow-based flavor.

3   Graphs Used in Our Empirical Evaluation

In this section, we describe the graphs we used to perform our empirical evaluation of the algorithms described in Section 2. Our main testbed consists of

Table 1. Basic statistics for our main testbed of graphs, including information about the best quotient cut found during our empirical evaluation. The "cutsize" is the number of edges cut in the best quotient cut found by any method, and "smallside" is the number of nodes on the small side of the cut, in which case the balance is smallside / nodes and the quotient cut score (not displayed) would be cutsize / smallside.

graph	GM.100.6	PLANT5K	PLANT6K	WING	TOOTH	RND-A	A1.12	A3.14	A6.13	A9.10
nodes	12600	20000	20000	62032	78136	10000	10000	10000	10000	10000
edges	24974	45000	46000	121544	452591	59372	47778	52024	62561	71332
cutsize	100	5000	7055	791	3827	181	622	1723	3508	5301
smallside	6300	10000	9955	31008	39030	4945	4982	4950	4959	5000
balance	0.5000	0.5000	0.4978	0.4999	0.4995	0.4945	0.4982	0.4950	0.4959	0.5000

10 graphs, summarized in Table 1, ranging in size from $10,000$ up to $78,000$ vertices and chosen from the following five classes. These five classes were chosen to highlight the strengths and weaknesses of existing algorithms; existing graph-partitioning testbeds are less appropriate for this empirical evaluation since they tend to be easy for spectral methods, since they consist of mesh-like graphs like WING and TOOTH. Note, in addition, that the best cuts found are all very well-balanced.

3.1   Guattery-Miller Graph

This graph is based on the construction of Guattery and Miller [11] for worst-case graphs for eigenvector-based methods. It is the outer product of a double-tree (two complete binary trees connected by an edge between their roots) and a path graph; and the minimum expansion cut is obtained by separating the two trees, but the path can be made long enough so that the slow mixing along it will cause any given number of eigenvectors to cut the path instead. We include in our testbed the graph GM.100.6, in which the path has length 100 and each of the two trees is of depth 6. These parameters have been chosen so that the first 3 eigenvectors will be given by the first three modes of vibration for the path, *i.e.*, the path folded over itself once, twice, and three times. The "right" eigenvector appears in the fourth position, where it cannot be used by SPECTRAL or SPECFLOW, which only use the first 3 non-zero eigenvectors.

3.2   Expanders with Planted Bisections

Expanders with planted bisections (expander-like graphs with a distinctly good bisection planted at a random location) are known to be a worst-case inputs for LR, while they can be solved by spectral methods and by various local improvement algorithms (see, *e.g.*, [2]). We generated a family of 8 graphs each containing $20,000$ nodes, with planted bisections of size $k \cdot 1000$, for $k = 1$ through $k = 8$, by generating two $10,000$-node degree-4 expanders (each is the union of 4 disjoint random matchings of the nodes) and connecting them by a random k-matching.

As the size of the planted cut increases, it becomes harder to detect it. For example, the graph with planted cut of size 1000 is solved optimally by METISRAND, LR, SPECFLOW and all parameter choices of OSVV, while LR already fails to find the planted cut of size 2000. METISRAND, SPECFLOW and OSVV find the planted cut up through size 5000 and then all fail to find at size 6000 and larger. Thus, we include in our results PLANT5K and PLANT6K, the graphs with planted cut of size 5000 and 6000, respectively.

3.3   Finite Element Meshes

Well-shaped meshes are classic examples of "nice" low-dimensional graphs for which many graph partitioning methods have been developed. We include the finite element mesh WING and TOOTH from the archive [17].

3.4   Random Geometric Graphs

Random geometric graphs have long been standard benchmark graphs [14], and LR, SPECFLOW and METISRAND have been shown to perform well on them. We include one random geometric graph RND-A generated by picking $10,000$ random points in the unit two-dimensional disk, adding edges in increasing order of length, and stopping when the graph becomes connected. We then reduced the number of edges by removing all pairs in which one end was not among the 50 nearest neighbors of the other.

3.5   Random Geometric Graphs with Random Edges Added

Given the good results obtained by the algorithms on random geometric graphs, we make them harder by adding a number of completely random edges, as was done in [2]. Our claim that these graphs are harder than ordinary random geometric graphs is based on the empirical observation of higher-variance distributions of scores from randomized partitioning algorithms [2]. We do not know of a theoretical explanation, but intuitively the extra edges seem to cause the spectral embeddings to get twisted up, making the right answer much less obvious. Similarly, perhaps by increasing the expansion, the extra random edges also cause problems for LR.

Each graph AX.IY was obtained by first running the random geometric graph generator mentioned above. Then, $x \cdot 1000$ random edges were added (avoiding the creation of duplicates); the tag y is just an instance number. In our testbed, we include a selection of random geometric graph with random edges based on which one or more of the algorithms performs well: A1.I2, on which LR does particularly well; A3.I4, on which SPECFLOW does well; A6.I3, on which OSVV performs well; and A9.I0, for which METISRAND gave the best results.

4   Results

In this section, we describe the results of our initial empirical evaluation. Our computations were run on 4 64-bit AMD Opteron Processors, each running at

1795MHz, with 32GB of memory and 1024KB cache. We ran the deterministic SPECTRAL and METIS algorithms only once, and we report the expansion of the cut found and the time required. For LR and for each of the parameter choices of OSVV, we ran 10 trials; for SPECFLOW, we ran 1000 iterations; and for METISRAND, we ran 10,000 trials. For each of these algorithms, we report the best cut found over these runs and the total time taken. The number of trials for each algorithm was chosen in order to obtain total run times of the same order of magnitude in order to help focus the comparison on just the single criterion of the best cut found; we discuss this issue in more detail below. Finally, the timing data omit the time needed to load the description of the graph.

We explored a large number of settings of the parameters in preliminary empirical evaluations, and we determined that the interesting region of parameter space for our graphs is given by the 27 combinations of the following sets of parameters: $\eta \in \{1, 10, 100\}; \gamma \in \{0, 10, 100\}; \sigma \in \{2, 5, 10\}$. Many of our conclusions may be illustrated by considering only 3 setting of the parameters: $\eta = 100, \gamma = 100, \sigma = 10; \eta = 10, \gamma = 10, \sigma = 10;$ and $\eta = 1, \gamma = 0, \sigma = 10$. (Recall that choices of the parameters for which the product $\eta\gamma$ is larger (resp. smaller) correspond to a more spectral-like (resp. flow-like) behavior of the algorithm.) Summary statistics for the best cut results and total times for these 3 parameter settings of OSVV, compared with results from each of the other algorithms, are presented in Table 2 and Table 3.

On PLANT5K, the more spectral-based choices of parameters find the planted cut, while it seems that the more flow-based choices encounter problems similar to (but not as severe as) that of LR and are not able to detect the right cut. On PLANT6K, SPECFLOW gives the best cut overall (although no algorithm found the planted cut) and the more spectral-like choices of parameters for OSVV do marginally better than more flow-like choices. Note, though, that the entire OSVV method seems to be stuck at a quality around 10% worse than SPECFLOW, probably as a result of the constant-degree expander strongly affecting the performance of the flow part of OSVV. For meshes and random geometric graphs, all the methods do comparably. Note, though, that the

Table 2. [Best viewed in color.] Ratio of the best expansion cut score found by multiple trials of each algorithm to the best expansion cut score found overall. (See the text or the caption of Table 3 for details on the number of trials for each algorithm, and see Table 4 for results on varying the number of trials.) First and second place for each graph are highlighted in red and blue, respectively. Ratios are given to 3 decimal digits. OSVV parameters are described as OSVV-$\eta.\gamma.\sigma$.

	GM.100.6	PLANT5K	PLANT6K	WING	TOOTH	RND-A	A1.12	A3.14	A6.13	A9.10
OSVV-100.100.10	1.000	1.000	**1.098**	**1.018**	**1.003**	1.024	1.077	**1.029**	**1.009**	1.039
OSVV-10.10.10	1.000	1.000	1.102	1.069	1.033	**1.001**	**1.050**	1.039	**1.000**	**1.033**
OSVV-1.0.10	1.000	1.493	1.119	1.069	1.059	1.003	1.053	1.089	1.414	1.077
METISRAND	1.000	1.000	1.107	1.048	1.019	1.011	1.149	1.068	1.025	**1.000**
LR	1.000	2.841	2.082	1.069	1.065	1.003	**1.000**	1.163	1.072	1.075
SPECFLOW	**1.260**	1.000	**1.000**	**1.000**	**1.000**	**1.000**	1.149	**1.000**	1.037	1.081
METIS	1.640	1.526	1.130	1.169	1.208	1.322	1.445	1.330	1.190	1.059
SPECTRAL	**1.260**	**1.195**	1.207	1.253	1.111	1.517	2.624	1.878	1.414	1.661

Table 3. [Best viewed in color.] Total run time in seconds for OSVV-$\eta.\gamma.\sigma$ (10 trials), METISRAND (10000 trials), LR (10 trials), SPECFLOW (Eigensolver + 1000 flow roundings), METIS (1 try), and SPECTRAL (Eigensolver + 3 sweep roundings). Numbers are rounded to the nearest second, except for METIS and SPECTRAL, where they are rounded to the second decimal.

	GM.100.6	PLANT5K	PLANT6K	WING	TOOTH	RND-A	A1.12	A3.14	A6.13	A9.10
OSVV-100.100.10	793	367	**650**	**8167**	31847	1956	956	**735**	**1315**	1013
OSVV-10.10.10	363	304	437	2802	9923	**881**	401	370	485	**851**
OSVV-1.0.10	426	2075	3030	4201	11681	602	447	441	85	423
METISRAND	105	681	700	1049	2024	110	111	189	284	328
LR	187	660	658	8521	56378	443	509	699	1173	1637
SPECFLOW	209	636	581	4887	13254	688	639	641	724	798
METIS	0.01	0.06	0.07	0.09	0.21	0.01	0.01	0.02	0.02	0.03
SPECTRAL	7.05	**3.22**	3.27	51.48	96.02	8.98	4.40	3.10	2.32	2.46

traditional mesh WING and TOOTH slightly favors spectral methods—SPECFLOW does slightly better than other methods, and relatedly the best OSVV results are given by spectral-like and intermediate choices of parameters. As expected, on GM.100.6, all choices of parameters for OSVV find the optimal cut.

For random geometric graphs with added random edges, intermediate and flow-like parameter choices for OSVV tend to perform well for A1.12 (chosen since LR performed well on it), and the more spectral-like choices perform better for A3.14 (chosen since SPECFLOW performed well). Interestingly, for A6.13, parameters intermediate between spectral and flow decisively beat other parameter choices and all other algorithms. A similar result is seen with A9.10, for which the intermediate parameter choices nearly tie the best cut found by METISRAND and improve on the expansion found by SPECFLOW and LR by around 5%.

With respect to the stopping condition σ, (data not presented indicate that) computations behaved in expected ways, but we noticed that variations in the stopping condition seemed to impact more the quality of the score for more flow-based algorithms. Our intuition for this is that more flow-based algorithms are less aggressive in their search for sparse cuts and require more time to explore the cut space to provide their best results, while more spectral runs can achieve very good scores already in their first runs, especially if the graph is suited to the spectral method. In generally, of course, the stopping condition could be adjusted based on the relative importance of cut quality and time in the context in which the algorithm is used.

With respect to the running time, Table 3 (and data not presented) indicates that more extreme choices of parameters η and γ seem to require more running time. This is likely due to two different reasons. On the one hand, for the more spectral-like parameter settings, a larger fraction of time is spent in computing the heat kernel vector, which becomes harder as η and γ grow. On the other hand, for more flow-like parameter settings, the longer time is usually due to a larger number of iterations within the algorithm, again a result of the more conservative approach of these parameter settings. In general, if some information on the graph is available in terms of suitability to spectral or flow methods, the choice of parameters could be adjusted accordingly.

Table 4. Varying the number of trials for METISRAND and SPECFLOW. Presented is the ratio of the expansion cut score for METISRAND and SPECFLOW (as a function of the number of trials) to the best overall score. For METISRAND 10000 and SPECFLOW 1000, the score is the minimum found over all our trials, while for other number of trials, the score is an estimate of the average best score using the empirical distribution from our experiment and assuming sampling with replacement.

	Trials	GM.100.6	PLANT5K	PLANT6K	WING	TOOTH	RND-A	A1.12	A3.14	A6.13	A9.10
METISRAND	10	1.343	1.313	1.121	1.099	1.075	1.137	1.362	1.197	1.106	1.072
METISRAND	100	1.020	1.028	1.116	1.076	1.043	1.058	1.253	1.143	1.071	1.042
METISRAND	1000	1.000	1.000	1.111	1.063	1.028	1.026	1.175	1.102	1.046	1.020
METISRAND	10000	1.000	1.000	1.107	1.048	1.019	1.011	1.149	1.068	1.025	1.000
SPECFLOW	1	1.913	1.402	1.100	1.082	1.161	1.278	1.325	1.150	1.108	1.171
SPECFLOW	10	1.284	1.066	1.052	1.040	1.045	1.027	1.218	1.059	1.060	1.106
SPECFLOW	100	1.260	1.000	1.011	1.006	1.015	1.002	1.167	1.023	1.044	1.089
SPECFLOW	1000	1.260	1.000	1.000	1.000	1.000	1.000	1.149	1.000	1.037	1.081

A graph by graph inspection seems to confirm the expected behavior of the parameters of OSVV as toggling between spectral and flow methods. In addition, the results suggest that OSVV is quite robust and tends to be within 5% of the best, which is much more consistent performance than the other methods. In particular, that the performance of the intermediate choice of parameters $\eta = 10, \gamma = 10, \sigma = 10$ appears as a good global setting, performing optimally or near-optimally for all graphs except PLANT6K, for which more spectral-based methods are preferable.

Finally, we should note that Table 4 demonstrates that for METISRAND and SPECFLOW the number of trials (and thus the run time) can be decreased by a factor of 10, or in some cases 100, while still finding cuts that are only moderately worse than those found in the larger number of trials. A larger study (currently in progress) with a finer tuning of parameters, more comprehensively chosen graphs, and a more sophisticated implementation of OSVV will be necessary to fully explore these issues and validate our initial observations.

5 Conclusion

We have reported initial results from the first empirical evaluation of an algorithm from the novel family of algorithms inspired by the recent theoretical work of ARV [5]. It is important to emphasize that, prior to performing this empirical evaluation, we had no idea whether any algorithm from this family of algorithms would be at all practical in finding even moderately good cuts on graphs of any size. Thus, our primary conclusion is that a simple implementation of the algorithm of OSVV [16] performs competitively with state-of-the-art implementations of existing graph partitioning algorithms at finding good cuts on a suite of medium-sized graphs chosen to illustrate the strengths and weaknesses of these existing algorithms. Our secondary conclusion is that, as suggested by theory, different parameter choices in this algorithm can be interpreted as toggling between a more spectral-like approach and a more flow-like approach. Clearly, our initial results suggest that these methods might be a viable alternative to the classical spectral and flow methods in practical applications in large-scale data analysis and machine learning, arguing for a more comprehensive evaluation on a larger suite of larger and more realistic graphs.

References

1. Alon, N., Milman, V.: λ_1, isoperimetric inequalities for graphs and superconcentrators. J. Combin. Theory B 38, 73–88 (1985)
2. Andersen, R., Lang, K.: An algorithm for improving graph partitions. In: SODA 2008: Proceedings of the 19th ACM-SIAM Symposium on Discrete Algorithms, pp. 651–660 (2008)
3. Arora, S., Hazan, E., Kale, S.: $O(\sqrt{\log n})$ approximation to sparsest cut in $\tilde{O}(n^2)$ time. In: FOCS 2004: Proceedings of the 45th Annual Symposium on Foundations of Computer Science, pp. 238–247 (2004)
4. Arora, S., Kale, S.: A combinatorial, primal-dual approach to semidefinite programs. In: STOC 2007: Proceedings of the 39th Annual ACM Symposium on Theory of Computing, pp. 227–236 (2007)
5. Arora, S., Rao, S., Vazirani, U.: Expander flows, geometric embeddings and graph partitioning. In: STOC 2004: Proceedings of the 36th Annual ACM Symposium on Theory of Computing, pp. 222–231 (2004)
6. Arora, S., Rao, S., Vazirani, U.: Geometry, flows, and graph-partitioning algorithms. Communications of the ACM 51(10), 96–105 (2008)
7. Cherkassky, B., Goldberg, A., Martin, P., Setubal, J., Stolfi, J.: Augment or push: a computational study of bipartite matching and unit-capacity flow algorithms. Journal of Experimental Algorithmics 3, Article 8 (1998)
8. Cherkassky, B., Goldberg, A.V.: On implementing push-relabel method for the maximum flow problem. Algorithmica 19, 390–410 (1997)
9. Chung, F.: Spectral graph theory. CBMS Regional Conference Series in Mathematics, vol. 92. American Mathematical Society, Providence (1997)
10. Goldberg, A., Rao, S.: Beyond the flow decomposition barrier. Journal of the ACM 45, 783–797 (1998)
11. Guattery, S., Miller, G.: On the quality of spectral separators. SIAM Journal on Matrix Analysis and Applications 19, 701–719 (1998)
12. Karypis, G., Kumar, V.: A fast and high quality multilevel scheme for partitioning irregular graphs. SIAM Journal on Scientific Computing 20, 359–392 (1998)
13. Khandekar, R., Rao, S., Vazirani, U.: Graph partitioning using single commodity flows. In: STOC 2006: Proceedings of the 38th Annual ACM Symposium on Theory of Computing, pp. 385–390 (2006)
14. Lang, K., Rao, S.: Finding near-optimal cuts: an empirical evaluation. In: SODA 1993: Proceedings of the 4th Annual ACM-SIAM Symposium on Discrete Algorithms, pp. 212–221 (1993)
15. Leighton, T., Rao, S.: Multicommodity max-flow min-cut theorems and their use in designing approximation algorithms. Journal of the ACM 46(6), 787–832 (1999)
16. Orecchia, L., Schulman, L., Vazirani, U., Vishnoi, N.: On partitioning graphs via single commodity flows. In: STOC 2008: Proceedings of the 40th Annual ACM Symposium on Theory of Computing, pp. 461–470 (2008)
17. Walshaw, C., Cross, M.: Mesh partitioning: a multilevel balancing and refinement algorithm. SIAM Journal on Scientific Computing 22(1), 63–80 (2000)

Univariate Algebraic Kernel and Application to Arrangements

Sylvain Lazard[1], Luis Peñaranda[1], and Elias Tsigaridas[2]

[1] INRIA Nancy - Grand Est, LORIA, France
[2] INRIA Sophia-Antipolis - Méditerranée, France
FirstName.LastName@inria.fr

Abstract. We present a CGAL-based univariate algebraic kernel, which provides certified real-root isolation of univariate polynomials with integer coefficients and standard functionalities such as basic arithmetic operations, greatest common divisor (gcd) and square-free factorization, as well as comparison and sign evaluations of real algebraic numbers.

We compare our kernel with other comparable kernels, demonstrating the efficiency of our approach. Our experiments are performed on large data sets including polynomials of high degree (up to 2 000) and with very large coefficients (up to 25 000 bits per coefficient).

We also address the problem of computing arrangements of x-monotone polynomial curves. We apply our kernel to this problem and demonstrate its efficiency compared to previous solutions available in CGAL.

1 Introduction

Implementing geometric algorithms robustly is known to be a difficult task for two main reasons. First, all degenerate situations have to be handled and second, algorithms often assume a real-RAM model (a random-access machine where each register can hold a real number and each arithmetic operation has unit cost) which is not realistic in practice. In recent years, the paradigm of exact geometric computing has arisen as a standard for robust implementations [24]. In this paradigm, geometric queries, also called predicates, such as "is a point inside, outside or on a circle?", are made exactly using, usually, either (i) exact arithmetic combined, for efficiency, with interval arithmetic on doubles or (ii) interval arithmetic on arbitrary-fixed-precision floating-point numbers combined with separation bounds; on the other hand, geometric constructions, such as the circle through three points or points of intersection between two curves, may be approximated.

We address here one recurrent difficulty arising when implementing algorithms dealing, in particular, with curved objects. Such algorithms usually require evaluating, manipulating and solving systems of polynomials equations and comparing their roots. One of the most critical parts of dealing with polynomials or polynomial systems is the isolation of the real roots and their comparison.

We restrict here our attention to the case of univariate polynomials and address this problem in the context of CGAL, a C++ Computational Geometry

J. Vahrenhold (Ed.): SEA 2009, LNCS 5526, pp. 209–220, 2009.

Algorithms Library, which is an open source project and became a standard for the implementation of geometric algorithms [4].

CGAL is designed in a modular fashion following the *paradigm of generic programming*. Algorithms are typically parameterized by a *traits* class which encapsulates the geometric objects, predicates and constructions used by the algorithm. Algorithms can thus typically be implemented independently of the type of input objects. For instance, the core of a line-sweep algorithm for computing arrangements of plane curves [7] can be implemented independently of whether the curves are lines, line segments, or general curves; on the other hand, the elementary operations that depend on the type of the objects (such as, comparing x-coordinates of points of intersection) are implemented separately in traits classes. Similarly, the model of computation, such as exact arbitrary-length integer arithmetic or approximate fixed-precision floating-point arithmetic, are encapsulated in the concept of *kernel*. An implementation is thus typically separated in three or four layers, (i) the geometric algorithm which relies on (ii) a traits class, which itself relies on (iii) a kernel for elementary (typically geometric) operations. CGAL provides several predefined Cartesian kernels, for instance allowing standard Cartesian geometric operations on inputs defined with doubles and providing approximate constructions (*i.e.*, defined with double) but exact predicates. However, a kernel can also rely on (iv) a number type which essentially encapsulates the type of number (such as, double, arbitrary-length integers, intervals) and the associated arithmetic operations. A choice of traits classes, kernels and number types is useful as it gives freedom to the users and it makes it easier to compare and improve the various building blocks of an implementation.

Our Contributions. We present in this paper a CGAL-compliant algebraic kernel that provides real-root isolation of univariate integer polynomials and basic operations, *i.e.* comparisons and sign evaluations, of real algebraic numbers. This open-source kernel follows the CGAL specifications for algebraic kernels [3]. The root isolation is based on the interval Descartes algorithm [5] and uses the library RS [19]. Moreover, our kernel provides various operations for polynomials, such as gcd, which are crucial for manipulating algebraic numbers.

We compare our kernel with other comparable kernels and demonstrate the efficiency of our approach. We perform experiments on large data sets including polynomials of high degree (up to 2 000) and with very large coefficients (up to 25 000 bits per coefficient).

Finally, we apply our kernel to the problem of computing arrangements of x-monotone polynomial curves and demonstrate its efficiency compared to previous solutions available in CGAL.

Related work. Combining algebra and geometry for manipulating non-linear objects has been a long-standing challenge. Previous work includes, but it is not limited to, MAPC [14] a library for manipulating points that are defined algebraically and handling curves in the plane. More recently, the library EXACUS [2], which handles curves and surfaces in computational geometry and supports various algebraic operations, was developed and partially integrated into CGAL.

The notion of algebraic kernel for CGAL was proposed in 2004 [11]; in this work, the underlying algebraic operations were based on the SYNAPS library [15]. Several methods and algebraic kernels have been developed since then.

One kernel was developed by Hemmer and Limbach [13] following the generic programming paradigm using the C++ template mechanism. This kernel is templated by the representation of algebraic numbers and by the real root isolation method, for which two classes have been developed; one is based on the Descartes method and the other on the Bitstream Descartes method [9]. This approach has the advantage to allow, in principle, using the best instances for both template arguments.

Another kernel developed at INRIA relies on the SYNAPS library [15]. In this kernel there are several approaches concerning real root isolation, *i.e.*, methods based on Sturm subdivisions, sleeves approximations, continued fractions, and a symbolic-numeric combination of the sleeve and continued fractions methods (see [10]). Moreover, there are specialized methods for polynomials of degree less or equal than four [21].

Emiris et al. [10] presented some benchmarks of these various approaches in these two kernels as well as some tests on the kernel we present here. The authors mention that our kernel based on interval Descartes performs similarly to one approach (refer to as NCF2) based on continued fractions [20] for coefficients with (very) large bitsize but NCF2 is more efficient for small bitsize. They conclude that, first, dedicated algorithms for polynomials of degree less than (or equal to) four is always the most efficient approach and, second, that NCF2 always perform the best except for low-degree and high-bitsize polynomials, in which case the kernel based on the Bitstream Descartes method performs the best. We moderate here these conclusions.

The rest of the paper is structured as follows. In the next section we describe our algebraic kernel. In Section 3, we present various experiments on the isolation of real roots and on the comparison of algebraic numbers. In Section 4, we apply our kernel to the problem of computing arrangements and compare it to previous solutions available in CGAL. We finally conclude in Section 5.

2   Univariate Algebraic Kernel

We describe here our implementation of our univariate algebraic kernel. The two main requirements of the CGAL specifications, which we describe here, are the isolation of real roots and their comparison. We also describe our implementation of two operations, the gcd computation and the refinement of isolating intervals, that are both needed for comparing algebraic numbers.

Preliminaries. The kernel handles univariate polynomials and algebraic numbers. The polynomials have integer coefficients and are represented by arrays of GMP arbitrary-length integers [12]. We implemented in the kernel the basic functions for polynomials. An algebraic number that is a root of a polynomial F is represented by F and an isolating interval, that is an interval containing this root but no other root of F. We implemented intervals using the MPFI library [16], which represents intervals with two MPFR arbitrary-fixed-precision floating-point numbers [17]; note that MPFR is developed on top of the GMP library for multi-precision arithmetic [12].

Root isolation. For isolating the real roots of univariate polynomials with integer coefficients, we developed an interface with the library RS [19]. This library is written in C and is based on Descartes' rule for isolating the real roots of univariate polynomials with integer coefficients.

We briefly detail here the general design of the RS library; see [18] for details. RS is based on an algorithm known as *interval Descartes* [5]; namely, the coefficients of the polynomials obtained by changes of variable, sending intervals $[a, b]$ onto $[0, +\infty]$, are only approximated using interval arithmetic when this is sufficient for determining their signs. Note that the order in which these transformations are performed in RS is important for memory consumption. The intervals and operations on them are handled by the MPFI library.

Algebraic number comparison. As mentioned above, one of the main requirements of the CGAL algebraic kernel specifications is to compare two algebraic numbers r_1 and r_2. If we are lucky, their isolating intervals do not overlap and the comparison is straightforward. This is, of course, not always the case. If we knew that they were not equal, we could refine both isolating intervals until they are disjoint. Hence, the problem reduces to determining whether the algebraic numbers are equal or not.

To do so, we compute the square-free factorization of the gcd of the polynomials associated to the algebraic numbers. The roots of this gcd are the common roots of both polynomials. We calculate the intersection, I, of the isolating intervals of r_1 and r_2. The gcd has a root in this interval if and only if $r_1 = r_2$.

To determine whether the gcd has a root in interval I, it suffices to check the sign of the gcd at the endpoints of I: if they are different or one of them is zero, the gcd has a root in I and $r_1 = r_2$; otherwise, $r_1 \neq r_2$ and we can refine both intervals until they are disjoint.

Gcd computations. Computing greatest common divisors between two polynomials is not a difficult task, however, it is not trivial to do so efficiently. A naive implementation of the Euclidean algorithm works fine for small polynomials but the intermediate coefficients suffer an exponential grow in size, which is not manageable for medium to large size polynomials. We thus implemented a *modular* gcd function. We did not use some existing implementations mainly for efficiency because converting polynomials from one representation to another is substantially costly as soon as the degree and bitsize are large. Our function calculates the gcd of polynomials modulo some prime numbers and reconstructs later the result with the help of the *Chinese remainder theorem*. (See *e.g.*, [23] for details.)

Refining isolating intervals. As we mentioned before, refining the interval representing an algebraic number is critical for comparing such numbers. We provide two approaches for refinement.

Both approaches require that the polynomial associated to the algebraic number is square free. The first step thus consists of computing the square-free part of the polynomial (by computing the gcd of the polynomial and its derivative).

Our first approach is a simple bisection algorithm. It consists in calculating the sign of the polynomial associated to the algebraic number at the endpoints and midpoint of the interval. Depending on these signs, we refine the isolating interval to its left of right half.

Our second approach is a quadratic interval refinement [1]. Roughly speaking, this method splits the interval in many parts and, based on a linear interpolation, guesses in which one the root lies. If the guess is correct, the algorithm divides in the next refinement step the interval in more parts and, if not, in less.

Unfortunately, even with our careful implementation this approach turns out to be, on average, only just a bit faster than the bisection approach. Our experiments showed that the bottleneck of the refinement is the evaluation of polynomials.

3   Kernel Benchmarks

In this section, we analyze the running time of the two main functions of our algebraic kernel, that (i) isolate the roots of a polynomial and (ii) compare two algebraic numbers that is, compare the roots of two polynomials. We also compare the performance of our kernel with the one based on the Bistream Descartes method [9] and developed by Hemmer and Limbach [13] (referred to as MPII's kernel)[1] and with a kernel based on continued fractions [20] and developed on top of the SYNAPS library [15] (referred to as SYNAPS' kernel).

All tests were ran on a single-core 3.2 GHz Intel Pentium 4 with 2 Gb of RAM and 2048 kb of cache memory, using 64-bit Linux.

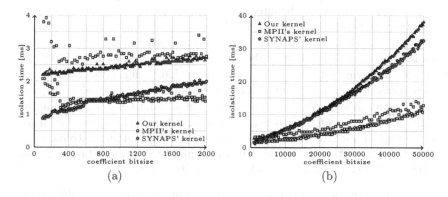

Fig. 1. Running time for isolating all the real roots of degree 12 polynomials with 12 real roots in terms of the maximum bitsize of their coefficients

---

[1] We parameterized MPII's kernel to use Bitstream Descartes as root isolator, `algebraic_real_bfi_rep` as algebraic number representation and CORE integers and rationals to represent the coefficients of the polynomials and the isolation bounds of algebraic numbers, respectively. The choice of CORE (vs. LEDA) was induced by the need of testing the kernels in the same conditions, that is, relying on GMP.

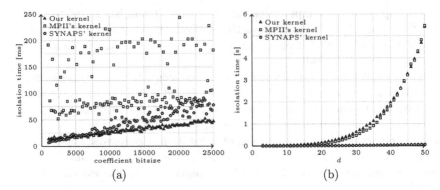

Fig. 2. Running time for isolating all the real roots of (a) degree 100 polynomials in terms of the maximum bitsize of their coefficients and (b) Mignotte polynomials of the form $f = x^d - 2(kx - 1)^2$ in terms of the degree d

Root isolation. We consider two suites of experiments in which we either fix the degree of the polynomials and vary the bitsize of the coefficients or the converse; see Figs. 1 and 2. In each experiment, we report the running time for isolating all the roots per polynomial, averaged over different trials, for our kernel, MPII's and SYNAPS' kernel.

Varying bitsize. We study here polynomials with rather low degree (12) but with no complex root and polynomials with reasonably large degree (100) with random coefficients (and thus with few real roots).

The first test sets comes from [13]. See Fig. 1. It consists of polynomials of degree 12, each one being the product of six degree-two polynomials with two roots, at least one of them in the interval $[0, 1]$; every polynomial thus has 12 real roots. We vary the maximum bitsize of all the coefficients of the input polynomial from 100 to 50 000 and average each test over 250 trials.

Secondly, we consider random polynomials with constant degree 100 and coefficients with varying bitsize. See Fig. 2(a). Note that such random polynomials have few roots: the expected number of real roots of a polynomial of degree d with coefficients independently chosen from the standard normal distribution is $\frac{2}{\pi} \ln(d) + C + \frac{2}{\pi d} + O(1/d^2)$ where $C \approx 0.625735$ [8]; this gives, for degree 100 an average of about 3.6 roots (note that this bound matches extremely well experimental observations). We vary the maximum bitsize of all the coefficients from 2 000 to 25 000 and average each test over 100 trials.

Varying degree. We consider two sets of experiments in which we study random polynomials and Mignotte polynomials (which have two very close roots).

We first consider polynomials with random coefficients of fixed bitsize for various values between 32 and 1 000. We then vary the degree of the polynomials from 100 to 2 000 and average our experiments over 100 trials (see Fig. 3). Note that the above formula gives an expected number of roots varying from 3.6 to 5.5. We observe that the running time is almost independent of the bitsize in the considered range.

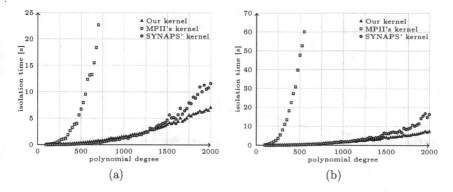

Fig. 3. Running time for isolating all the real roots of random polynomials with coefficients of bitsize (a) 32 and (b) 1000, and depending on the degree

Finally, we test Mignotte polynomials, that is polynomials of the form $x^d - 2(kx - 1)^2$. Such polynomials are known to be challenging for Descartes algorithms because two of their roots are very close to each other; the isolating intervals for these two roots are thus very small. For these tests, we used Mignotte polynomials with coefficients of bitsize 50, with varying degree d from 5 to 50. See Fig. 2(b). We averaged the running time over 5 trials for each degree. We observed essentially no difference between our kernel and MPII's one; they take roughly 0.2 and 5.5 seconds for Mignotte polynomials of degree 20 and 50, respectively. However, SYNAPS' kernel is much more efficient as the continued fractions algorithm is not so affected by the closeness of the roots.

Discussion. We observe (Fig. 1(a)) that SYNAPS' kernel is more efficient than both our and MPII's kernel in the case of polynomials of small degree (*e.g.*, twelve) and small to moderately large coefficients (up to 2 000 bits per coefficient). However, for extremely large coefficients MPII's kernel is substantially more efficient (by a factor of up to 3 for coefficients of up to 50 000 bits) than both our and SYNAPS' kernels, which perform similarly.

For polynomials of reasonable large degree, both our and SYNAPS' kernels are much more efficient that MPII's kernel; furthermore these two kernels behave similarly for degrees up to 1 500 and our kernel becomes more efficient for higher degrees (by a factor 2 for degree 2 000).

We also observe that the running time is *highly* dependent of the various settings. For instance, our kernel is up to 5 times slower when using approximate evaluation for high-degree and high-bitsize polynomials. Also, MPII's kernel is in some cases about 10 times slower when changing the arithmetic kernel to LEDA, the representation of algebraic numbers and some internal algorithms such as the refinement function. This explains why our benchmarks on both MPII's and SYNAPS' kernels are substantially better than in Emiris et al. experiments [10].

We also observe that the running time of MPII's kernel is unstable in our experiments (Figs. 1 and 2(a)); surprisingly, this instability occurs when the

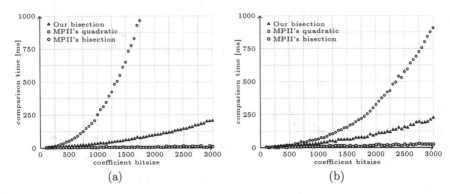

Fig. 4. Running time for comparing two distinct close roots of two almost identical polynomials of degree 20 with (a) no common roots and (b) a common factor of degree 10

experiments are performed on a 64-bits architecture, but it is stable on 32-bits architecture as shown in previous experiments [10].

Comparison of algebraic numbers. We consider three suites of experiments for comparing algebraic numbers; see Fig. 4. Recall that an algebraic number ρ is here represented by a polynomial F that vanishes at ρ and an isolating interval containing ρ but no other root of F. Recall also that the comparison of two algebraic numbers is done by (i) testing whether the intervals are disjoint; if so, report the ordering, otherwise (ii) compute the gcd of the two polynomials and test whether the gcd vanishes in the intersection of the two intervals; if so, report the equality of the numbers, otherwise (iii) refine the intervals until they are disjoint.

First, we analyze the cost of trivial comparisons that is, when the two intervals are disjoint. For that we compare the roots of two random polynomials. We observe that, as expected, the comparison time is negligible and independent of both the degree of the polynomials and the bitsize of their coefficients.

Second, we analyze the cost of comparing roots that are very close to each other but whose associate polynomials have no common root. This case is expensive because we need to refine the intervals until they do not overlap; this is, however, not the worst situation because the gcd of the two polynomials is 1 which is tested efficiently with a modular gcd. We perform these experiments as follows. We generate pairs of polynomials, one with random coefficients and the other by only adding 1 to one of the coefficients of the first polynomial. Such polynomials are such that the i-th roots of both polynomials are very close to each other. We generate such pairs of polynomials with constant degree (equal to 20) and vary the maximum bitsize of the coefficients. As the bitsize increases, the pairs of roots that are close become even closer and thus the comparison time increases. The results of these experiments are presented in Fig. 4(a), which reports the average running time for comparing two close roots. We show in this figure three curves, one corresponding to our bisection algorithm, and two corresponding the two refinement methods implemented in the MPII's kernel: the usual bisection and a quadratic refinement algorithm.

Third, we consider the, a priori, most expensive scenario in which we compare roots that are either equal or very close to each others and such that their associate polynomials have some roots in common. In this case, we accumulate the cost of computing a non-trivial gcd of the two polynomials with the cost of refining intervals when comparing two non-equal roots. In practice, we generate pairs of degree-20 polynomials each defined as the product of two degree-10 terms; one of these factors is random and common to the two polynomials; the other factor is random in one of the polynomials and slightly modified in the other polynomial where, slightly modified means, as above, that we add 1 to one of the coefficients. We then vary the maximum bitsize of the coefficients.

Discussion. We see in Fig. 4 that the MPII's quadratic refinement algorithm largely outperforms the two bisection methods. However, our bisection method is faster than MPII's one, by a factor up to 10. We also observed that the running time for comparing equal roots is negligible compared to the cost of comparing close but distinct roots. (The running time reported in Fig. 4(b) is actually the total time for comparing all pairs of roots divided by the number of comparisons of close but distinct roots.) This explains why our kernel behaves similarly in Figs. 4(a) and 4(b). Overall, it appears that comparing algebraic numbers that are very close is fairly time consuming and that the most time-consuming part of the comparison is the evaluation of polynomials performed during the interval refinements.

4   Arrangements

As an example of possible benefit of having efficient algebraic kernels in CGAL, we used our implementation to construct arrangements of polynomial functions. Wein and Fogel [22] provided a CGAL package for calculating arrangements of general curves which requires as parameter a *traits class* containing the data structures to store the curves and various primitive operations, such as comparing the relative positions of points of intersection. We implemented a traits class which uses the functions of our algebraic kernel and compared its performance with another traits classes which comes with CGAL's arrangement package and uses the CORE library [6].

In order to generate challenging data sets we proceed as follows. First we generate n random polynomials. To each of them we add 1 to the constant coefficient, m times, thus producing a data set of $n(m+1)$ univariate polynomials. Notice that the arrangement of the graphs of these polynomials is guaranteed to be degenerate, *i.e.*, there are intersections with the same x-coordinate. The arrangements generated this way have four parameters: the number n of initial polynomials, the number m of "shifts" that we perform, the degree d of the polynomials, and the bitsize τ of their coefficients. We ran experiments varying the values of the last three of these parameters and setting $n = 5$.

Fig. 5(a) shows the running time in terms of the bitsize τ for a data set where $d = 20$ and $m = 4$ (giving 25 polynomials). Fig. 5(b) shows the running time

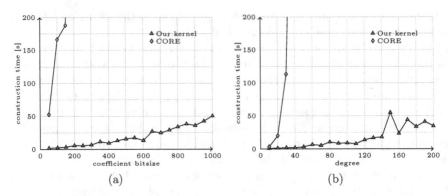

Fig. 5. Arrangements of five polynomials, shifted four times each, (a) of degree 20 and varying bitsize and (b) of bitsize 32 and varying degree

in terms of the degree d for a second data set where $\tau = 32$ and $m = 4$. We see from these experiments that running time using CORE is considerably higher than when using our kernel. We also make the following observations.

Fig. 5(a) shows that the running time depends on the bitsize. When we change the bitsize of the coefficients of the random polynomials, the size of the arrangement does not change; that means that the number of comparisons and root isolations the kernel must perform is roughly the same in all the arrangements of the test suite. The isolation time for random polynomials does not depend much on the bitsize (as shown in Fig. 2(a)), but the comparison time does. It follows that the running time increases with the bitsize.

Fig. 5(b) shows that the running time depends also on the degree of the input polynomials. As we saw in Section 3, the expected number of real roots of a random polynomial depends on its degree. The size of the arrangement thus increases with the degree of the input polynomials: each vertex is the root of the difference between two input polynomials, therefore there will be more vertices. Thus, when we increment the degree of the inputs, the number of comparisons and isolations increases; furthermore, the running time for each of these operations increases with the degree of the input.

5   Conclusion

We presented a new CGAL-compliant algebraic kernel that provides certified real-root isolation of univariate polynomials with integer coefficients based on the interval Descartes algorithm. This kernel also provides the comparison of algebraic numbers and other standard functionalities.

We compared our kernel with other comparable kernels on large data sets including, for the first time, polynomials of high degree (up to 2 000) and with extremely large coefficients (up to 25 000 bits per coefficient). We demonstrated the efficiency of our approach and showed that it performs similarly, in most cases, with one kernel based on the SYNAPS library; more precisely, our kernel is more efficient for polynomials of very large degree (greater than 1 800) and less efficient for polynomials of very small degree and with small to moderate size coefficients. Also, our kernel is a lot more efficient that the kernel developed at MPII for polynomials of large degree (greater than 200); it is however less efficient for polynomials of small degree and with extremely large coefficients.

Our tests indicate that the kernel developed at MPII appears to be less efficient than the other two for polynomials of large degree. However it should be stressed that this kernel is the only one among the three that is templeted by the number type of the coefficients. Of course this does not imply that efficiency is necessarily lost by following the generic programming paradigm, but it does imply that, from the user point of view, some substantial gain of efficiency can sometimes be made by using a kernel that does not follow this paradigm.

We also compared the performance of the kernels on the comparison of algebraic numbers. We observed in these tests that the bisection algorithm runs much faster when it is specialized on a number type since it allows for low level optimizations, confirming thus the assertion in the previous paragraph. On the other hand, it becomes evident that the bisection method is not the most efficient algorithm when a large number of refinements is needed, and MPII's quadratic refinement is the fastest method by far.

A fairly large choice of algebraic kernels and, in particular, of methods for isolating the real roots of polynomials, is now available in CGAL. This allows, in particular, to compare and improve the various methods. It appears that between the two big classes of methods, based on continued fractions and Descartes algorithms, neither is clearly much better than the other. However, some substantial differences appear between the various implementations, but, of course, it is always very difficult to benchmark implementations. For instance, we observed here that the running times are highly dependent of the various settings and architectures.

Finally, we also address the problem of computing arrangements of x-monotone polynomial curves. We apply our kernel to this problem and demonstrate its efficiency compared to previous solutions available in CGAL.

Acknowledgments

The authors are grateful to F. Rouillier, Z. Zafeirakopoulos, M. Hemmer, E. Berberich, M. Kerber, and S. Limbach for fruitful discussions and suggestions.

References

1. Abbott, J.: Quadratic interval refinement for real roots. In: International Symposium on Symbolic and Algebraic Computation (ISSAC), poster presentation (2006)
2. Berberich, E., Eigenwillig, A., Hemmer, M., Hert, S., Kettner, L., Mehlhorn, K., Reichel, J., Schmitt, S., Schömer, E., Wolpert, N.: EXACUS: Efficient and Exact Algorithms for Curves and Surfaces. In: Brodal, G.S., Leonardi, S. (eds.) ESA 2005. LNCS, vol. 3669, pp. 155–166. Springer, Heidelberg (2005)

3. Berberich, E., Hemmer, M., Karavelas, M., Teillaud, M.: Revision of the interface specification of algebaic kernel. Technical Report ACS-TR-243301-01, ACS European Project (2007)
4. CGAL, Computational Geometry Algorithms Library, http://www.cgal.org
5. Collins, G., Johnson, J., Krandick, W.: Interval Arithmetic in Cylindrical Algebraic Decomposition. Journal of Symbolic Computation 34(2), 145–157 (2002)
6. The CORE library, http://cs.nyu.edu/exact/
7. de Berg, M., Cheong, O., van Kreveld, M., Overmars, M.: Computational Geometry: Algorithms and Applications, 3rd edn. Springer, Heidelberg (2008)
8. Edelman, A., Kostlan, E.: How zeros of a random polynomial are real? Bulletin of American Mathematical Society 32(1), 1–37 (1995)
9. Eigenwillig, A., Kettner, L., Krandick, W., Mehlhorn, K., Schmitt, S., Wolpert, N.: A Descartes Algorithm for Polynomials with Bit-Stream Coefficients. In: Ganzha, V.G., Mayr, E.W., Vorozhtsov, E.V. (eds.) CASC 2005. LNCS, vol. 3718, pp. 138–149. Springer, Heidelberg (2005)
10. Emiris, I., Hemmer, M., Karavelas, M., Limbach, S., Mourrain, B., Tsigaridas, E., Zafeirakopoulos, Z.: Cross-benchmarks for univariate algebraic kernels. Technical Report ACS-TR-363602-02, ACS European Project (2008)
11. Emiris, I.Z., Kakargias, A., Pion, S., Teillaud, M., Tsigaridas, E.P.: Towards and open curved kernel. In: Proc. 20th Annual ACM Symp. on Computational Geometry (SoCG), New York, USA, pp. 438–446 (2004)
12. GMP, GNU multiple precision arithmetic library, http://gmplib.org/
13. Hemmer, M., Limbach, S.: Benchmarks on a generic univariate algebraic kernel. Technical Report ACS-TR-243306-03, ACS European Project (2006)
14. Keyser, J., Culver, T., Manocha, D., Krishnan, S.: Efficient and exact manipulation of algebraic points and curves. Computer-Aided Design 32(11), 649–662 (2000)
15. Mourrain, B., Pavone, P., Trébuchet, P., Tsigaridas, E.P., Wintz, J.: SYNAPS, a library for dedicated applications in symbolic numeric computations. In: Stillman, M., Takayama, N., Verschelde, J. (eds.) IMA Volumes in Mathematics and its Applications, pp. 81–110. Springer, New York (2007), http://synaps.inria.fr
16. MPFI, multiple precision interval arithmetic library, http://perso.ens-lyon.fr/nathalie.revol/software.html
17. MPFR, library for multiple-precision floating-point computations, http://mpfr.org/
18. Rouillier, F., Zimmermann, Z.: Efficient isolation of polynomial's real roots. J. of Computational and Applied Mathematics 162(1), 33–50 (2004)
19. Rs, a software for real solving of algebraic systems. F. Rouillier, http://fgbrs.lip6.fr
20. Tsigaridas, E.P., Emiris, I.Z.: On the complexity of real root isolation using Continued Fractions. Theoretical Computer Science 392, 158–173 (2008)
21. Tsigaridas, E.P., Emiris, I.Z.: Real algebraic numbers and polynomial systems of small degree. Theoretical Computer Science 409(2), 186–199 (2008)
22. Wein, R., Fogel, E.: The new design of CGAL's arrangement package. Technical report, Tel-Aviv University (2005)
23. Yap, C.: Fundamental Problems of Algorithmic Algebra. Oxford University Press, Oxford (2000)
24. Yap, C.: Robust geometric computation. In: Goodman, J.E., O'Rourke, J. (eds.) Handbook of Discrete and Computational Geometry, 2nd edn., ch. 41, pp. 927–952. Chapman & Hall/CRC, Boca Raton (2004)

Fast Algorithm for Graph Isomorphism Testing[*]

José Luis López-Presa[1] and Antonio Fernández Anta[2,**]

[1] DIATEL, Universidad Politécnica de Madrid, 28031 Madrid, Spain
[2] LADyR, GSyC, Universidad Rey Juan Carlos, 28933 Móstoles, Spain

Abstract. In this paper we present a novel approach to the graph isomorphism problem. We combine a direct approach, that tries to find
a mapping between the two input graphs using backtracking, with a
(possibly partial) automorphism precomputing that allows to prune the
search tree. We propose an algorithm, *conauto*, that has a space complexity of $O(n^2 \log n)$ bits. It runs in time $O(n^5)$ with high probability if either one of the input graphs is a $G(n, p)$ random graph, for
$p \in [\omega(\ln^4 n/n \ln \ln n), 1 - \omega(\ln^4 n/n \ln \ln n)]$. We compare the practical
performance of conauto with other popular algorithms, with an extensive
collection of problem instances. Our algorithm behaves consistently for
directed, undirected, positive, and negative cases. Additionally, when it
is slower than any of the other algorithms, it is only by a small factor.

1 Introduction

The Graph Isomorphism problem (GI) tests whether there is a one-to-one mapping between the vertices of two graphs, preserving the arcs. This is of both theoretical and practical interest. In practice, it has applications in many fields, like
pattern recognition, computer vision, information retrieval, data mining, VLSI
layout validation, and chemistry. Its main theoretical interest comes from the
fact that, while GI is clearly in NP, it is not known if it is in P or NP-complete.

Previous work. As could be expected, GI has been extensively studied[1]. On
the theoretical side, there is much work trying to place GI into a complexity
class. There is strong evidence that GI is not NP-complete since, otherwise, the
polynomial time hierarchy would collapse to its second level ($\Sigma_2^p = \Pi_2^p = $ AM)
[4,16] and because it would be the only NP-complete problem to be polynomial-
time equivalent in its decision and counting versions [10]. Recently, Arvind and
Kurur [1] have shown that GI is in SPP ("Stoic PP"). GI is known to be solvable
in polynomial time for some restricted classes of graphs, like trees or planar
graphs [9]. However there are graph families that are specially hard, like certain
families of strongly regular graphs (SRG) and projective planes. As far as we

---

[*] Partially supported by grants MICINN TIN2008-06735-C02-01, CAM
S-0505/TIC/0285, and MEC PR2008-0015.
[**] Done in part while on leave at Alcatel-Lucent Bell Laboratories, Murray Hill, NJ.
[1] This review of the literature is necessarily incomplete. The reader can see the surveys
of Reed and Corneil [15], Fortin [7], Goldberg [9], and Gati [8]. See also [14].

J. Vahrenhold (Ed.): SEA 2009, LNCS 5526, pp. 221–232, 2009.
© Springer-Verlag Berlin Heidelberg 2009

know, the best bound for general graphs up to now is due to Babai and Luks [3], whose canonical labeling (see below) algorithm runs in $\exp(n^{1/2+o(1)})$ time. GI has also been studied on random graphs $G(n, p)$. For $p = 1/2$, Babai et al. [2] proposed a canonical labeling algorithm that labels all graphs in expected linear time. Recently, Czajka and Padurangan [6] have given a linear time algorithm that canonically labels a $G(n, p)$ random graph with high probability[2], for $p \in [\omega(\ln^4 n/n \ln \ln n), 1 - \omega(\ln^4 n/n \ln \ln n)]$.

GI algorithms use mainly two approaches. The *direct* approach tries to find an isomorphism between the two input graphs directly with a classical backtracking algorithm, possibly using heuristics to prune the search tree. Examples of direct algorithms are Ullman's [18] or vf2 [5]. The major drawback of these algorithms is that they are slow when the graphs being tested have many automorphisms, since they usually do not detect them. The *canonical labeling* approach applies some function $C()$ to each graph, which returns a *certificate* (canonical labeling) of the graph, such that $C(G) = C(H)$ if and only if graphs G and H are isomorphic. Nauty [11,12] is a canonical labeling algorithm that is currenly considered the fastest GI algorithm. The main problem of nauty, and any other complete canonical labeling algorithm, is that it needs to compute the whole automorphism group (which is hard). Not surprisingly, Miyazaki [13] has found a family of graphs with exponential lower time bounds for nauty.

Contributions. We propose an algorithm for GI that combines the best of the two approaches. Our algorithm, which we call *conauto*, is a direct algorithm since it tries to find a mapping between the two input graphs using backtracking. However, to drastically prune the search tree, it looks for automorphisms in the graphs, as canonical labeling algorithms do, but without necessarily computing the whole automorphism group. We show that our algorithm has a space complexity of $O(n^2 \log n)$ bits when run with n-node graphs. Additionally, using results of Czajka and Padurangan [6], we show that conauto runs in time $O(n^3)$ w.h.p. if either one of the input graphs is a $G(n, p)$ random graph, for $p \in [\omega(\ln^4 n/n \ln \ln n), 1 - \omega(\ln^4 n/n \ln \ln n)]$.

We claim that conauto is very practical. To back this claim we compare it with other algorithms, namely nauty [12] and vf2 [5]. The former is included because it is considered to be the fastest practical GI algorithm, while the latter is included as a modern example of a direct algorithm. The comparison is done by running programs implementing the algorithms on an extensive benchmark that we have built [14], with positive and negative isomorphism cases, and directed and undirected graphs from several families. The benchmark used combines simple graph families, like random graphs, with other families that are known to be hard to handle by most GI algorithms, like some SRG families or the point-line graphs of Desarguesian projective planes. The comparison concludes that, when conauto is not able to handle a family of graphs (it cannot finish in 10,000 seconds), none of the other two can, while there are families that are handled easily by conauto and not by the others. Additionally, when it is slower than any of the other algorithms, it is only by a small factor. In general,

---

[2] W.h.p., probability at least $1 - O(n^{-c})$, for some $c > 0$ and large enough n.

conauto behaves more consistently in all cases (directed versus undirected, and positive versus negative). It is worth mentioning that an early version of conauto was recoded by Johannes Singler and included in the LEDA C++ class library of algorithms [17]. As noted in [17], both implementations (LEDA's and ours) of that early version of conauto have a very uniform behavior, but the LEDA implementation was found to be slower than ours. The version of conauto we present in this paper has a more complete search for automorphisms and uses them more exhaustively than the one included in LEDA.

Paper structure. In Section 2 we give basic definitions and notation. In Section 3 we describe the theoretic concepts on which the algorithm is based, while the algorithm is presented in Section 4. In Section 5 the asymptotic complexity of the algorithm is evaluated, and in Section 6 its practical performance is compared with nauty and vf2.

2 Definitions and Notation

A *directed graph* $G = (V, R)$ consists of a finite non-empty set V of vertices and a binary relation $R \subseteq V \times V$. An *arc* $(u, v) \in R$ is considered to be directed from u to v. R can be represented by an *adjacency matrix* $Adj(G) = A$ with size $|V| \times |V|$ in the following way.

$$A_{uv} = \begin{cases} 0 \text{ if } (u, v) \notin R \wedge (v, u) \notin R \\ 1 \text{ if } (u, v) \notin R \wedge (v, u) \in R \\ 2 \text{ if } (u, v) \in R \wedge (v, u) \notin R \\ 3 \text{ if } (u, v) \in R \wedge (v, u) \in R \end{cases}$$

Let $V_1 \subseteq V$, the *available degree* of v in V_1 under G, denoted by $ADg(v, V_1, G)$, is the 3-tuple (D_3, D_2, D_1) where $D_i = |\{u \in V_1 : A_{vu} = i\}|$ for $i \in \{1, 2, 3\}$. Extending the notation, we use $ADg(V_1, V_2, G) = d$ to denote that $\forall u, v \in V_1, ADg(u, V_2, G) = ADg(v, V_2, G) = d$, for $V_1, V_2 \subseteq V$. Let $ADg(V_1, V_2, G) = (D_3, D_2, D_1)$, then we define $Neigh(V_1, V_2, G) = D_3 + D_2 + D_1$ (i.e. the number of neighbors *each vertex* of V_1 has in V_2), and the predicate $Lnkd(V_1, V_2, G) = (Neigh(V_1, V_2, G) > 0)$. We say that $(D_3, D_2, D_1) \prec (E_3, E_2, E_1)$ when the first 3-tuple precedes the second one in lexicographic order. This notation will be used to order the available degrees of both vertices and sets.

Definition 1. *Let* $G = (V_G, R_G)$ *and* $H = (V_H, R_H)$. *An isomorphism of* G *and* H *is a one-to-one mapping* $m : V_G \longrightarrow V_H$ *such that for all* $u, v \in V_G$ $(v, u) \in R_G \iff (m(v), m(u)) \in R_H$.

Graphs G and H are *isomorphic*, written $G \simeq H$, if there is at least one isomorphism of them. An *automorphism* of G is an isomorphism of G and itself.

Like other GI algorithms, conauto relies on vertex classification. This is performed using the available degree of the vertices, and refining the successive partitions in an iterative process. A *partition* of a set S is a sequence $\mathcal{S} = (S_1, ..., S_r)$ of disjoint nonempty subsets of S such that $S = \bigcup_{i=1}^{r} S_i$. The sets S_i are called the *cells* of partition S. The empty partition is denoted $\emptyset$. If $S = (S_1, ..., S_r)$ and $T = (T_1, ..., T_s)$ are partitions of two disjoint sets S and T, the *concatenation* of S and T, denoted $S \circ T$, is the partition $(S_1, ..., S_r, T_1, ..., T_s)$. Clearly, $\emptyset \circ S = S = S \circ \emptyset$.

Partitions may be refined by two means: vertex and set refinements. A vertex refinement classifies the vertices in each cell using the adjacency type they have with a *pivot vertex*. This way, each cell may be split into up to four subcells. A set refinement classifies the vertices in each cell using their available degree with respect to a *pivot set* (cell). Let $V_1, V_2 \subseteq V$, $SetPart(V_1, V_2, G)$ is the *set partition* of V_1 by V_2, which is a partition $(S_1, ..., S_r)$ of V_1 such that $\forall i, j \in \{1, ..., r\}, i < j$ implies $ADg(S_i, V_2, G) \succ ADg(S_j, V_2, G)$. If $V_2 = \{v\} \not\subseteq V_1$ we have the *vertex partition* of V_1 by v, denoted $VtxPart(V_1, v, G)$. Let $V_1 \subseteq V$, $S = (S_1, ..., S_r)$ be a partition of V_1, and $P = S_x$ for some $x \in \{1, ..., r\}$ be a pivot set, then

1. The *vertex refinement* of S by the pivot vertex $v \in P$, denoted $VtxRef(S, v, G)$, is the partition $T = T_1 \circ ... \circ T_r$ such that $\forall i \in \{1, ..., r\}, T_i = \emptyset$ if $\neg Lnkd(S_i, V_1, G)$, and $T_i = VtxPart(S_i \setminus \{v\}, v, G)$ otherwise.

2. The *set refinement* of S by P, denoted $SetRef(S, P, G)$ is the partition $T = T_1 \circ ... \circ T_r$ such that $\forall i \in \{1, ..., r\}, T_i = \emptyset$ if $\neg Lnkd(S_i, V_1, G)$, and $T_i = SetPart(S_i, P, G)$ otherwise.

Let $G = (V_G, R_G)$ and $H = (V_H, R_H)$ be two graphs. Let $S = (S_1, ..., S_r)$ and $T = (T_1, ..., T_s)$ be partitions of $V_1 \subseteq V_G$ and $V_2 \subseteq V_H$ respectively, S and T are *compatible* under G and H, denoted $Comp(S, T, G, H)$, if $r = s$, and $\forall i \in \{1, ..., r\}, |S_i| = |T_i|$ and $ADg(S_i, V_1, G) = ADg(T_i, V_2, H)$.

A sequence of partitions starts with an initial partition (e.g., the degree partition) and each subsequent partition is obtained by applying some refinement to the previous one. A set refinement is labeled SET, and a vertex refinement is labeled VTX (from *vertex*) when the pivot set has only one vertex, and BTK (from *backtrack*) when it has more than one. More formally, a *sequence of partitions* for a graph $G = (V, R)$ is a tuple $(\mathsf{S}, \mathsf{R}, \mathsf{P})$, where $\mathsf{S} = (S^0, ..., S^t)$, are the partitions, $\mathsf{R} = (R^0, ..., R^{t-1})$ indicate the type of each refinement applied, and $\mathsf{P} = (P^0, ..., P^{t-1})$ are the pivot sets used. For all $i \in \{0, ..., t\}$, let $S^i = (S_1^i, ..., S_{r_i}^i)$, $V^i = \bigcup_{j=1}^{r_i} S_j^i$. Then the following statements must hold:

1. $\forall i \in \{0, ..., t-1\}, R^i \in \{\text{VTX}, \text{SET}, \text{BTK}\}$, and $P^i \in \{1, ..., |S^i|\}$.
2. $\forall i \in \{0, ..., t-1\}, R^i = \text{SET} \Rightarrow S^{i+1} = SetRef(S^i, S_{P^i}^i, G)$.
3. $\forall i \in \{0, ..., t-1\}, R^i \neq \text{SET} \Rightarrow S^{i+1} = VtxRef(S^i, v, G)$ for some $v \in S_{P^i}^i$.
4. $\forall x \in \{1, ..., r_t\}, \neg Lnkd(S_x^t, V^t, G) \vee |S_x^t| = 1$.

For convenience, for any $l \in \{1, ..., t-1\}$, we refer to the tuple (S^l, R^l, P^l) as *level l*. Level t is identified by S^t, since R^t and P^t are not defined. Note that, at each refinement step, from the definitions of vertex and set refinements, the relative order of the vertices is preserved, and the vertices with no links are discarded. It is hence possible to define a (partial) order of the vertices of a graph, induced by a sequence of partitions, in the following way. Let $\mathsf{Q} = (\mathsf{S}, \mathsf{R}, \mathsf{P})$ be a sequence of partitions for graph $G = (V, R)$. $\forall i \in \{0, ..., t\}$, let $S^i = (S_1^i, ..., S_{r_i}^i)$, and $V^i = \bigcup_{j=1}^{r_i} S_j^i$. Q induces a (partial) *order* $\prec_\mathsf{Q}$ in V as follows.

1. $\forall i \in \{0, ..., t\}, \forall x, y \in \{1, ..., r_i\}, x < y \Rightarrow \forall u \in S_x^i, \forall v \in S_y^i, u \prec_Q v.$
2. $\forall i \in \{0, ..., t-1\}, \forall x \in \{1, ..., r_i\}, \forall u \in (S_x^i \setminus V^{i+1}), \forall v \in (S_x^i \cap V^{i+1}), u \prec_Q v.$

An *Order* $<_Q$ *induced by a sequence of partitions* Q is any total order that extends the order $\prec_Q$. The ith vertex with respect to $<_Q$ is denoted $\omega_Q(i)$.

Let $G = (V_G, R_G)$ and $H = (V_H, R_H)$ be two graphs. Let $Q_G = (S_G, R_G, P_G)$, and $Q_H = (S_H, R_H, P_H)$ be two sequences of partitions for graphs G and H respectively. Q_G and Q_H are said to be *compatible* if $|S_G| = |S_H| = t, |R_G| = |R_H| = t - 1, |P_G| = |P_H| = t - 1$, and they satisfy all the following. Let $R_G = (R_G^0, ..., R_G^{t-1})$, $R_H = (R_H^0, ..., R_H^{t-1})$, $P_G = (P_G^0, ..., P_G^{t-1})$, $P_H = (P_H^0, ..., P_H^{t-1})$, $S_G = (S^0, ..., S^t)$, and $S_H = (T^0, ..., T^t)$. Then

1. $\forall i \in \{0, ..., t - 1\}, R_G^i = R_H^i,$ and $P_G^i = P_H^i.$
2. $\forall i \in \{0, ..., t\}, Comp(S^i, T^i, G, H).$
3. Let $S^t = (S_1^t, ..., S_r^t), T^t = (T_1^t, ..., T_r^t),$ then $\forall x, y \in \{1, ..., r\}, ADg(S_x^t, S_y^t, G) = ADg(T_x^t, T_y^t, H).$

As will be seen, finding compatible sequences of partitions for two graphs gives an isomorphism between them, by just mapping the vertices in any of the orders induced by the sequences.

3   Theoretical Foundations

The algorithm conauto solves GI by trying to find compatible sequences of partitions for the input graphs. The following theorem shows that this in fact solves GI. All the proofs can be found in [14].

Theorem 1. *Two graphs G and H are isomorphic if and only if there are two compatible sequences of partitions Q_G and Q_H for graphs G and H respectively.*

Basically, conauto first constructs a sequence of partitions for one of the graphs, and then tries to find a compatible one for the other. Reproducing in the second sequence a refinement labeled SET or VTX is direct, since there is only one possible pivot set or vertex. However, a refinement labeled BTK implies several potential pivot vertices, what may lead to backtracking. The rest of this section explores how a limited automorphism search in the first graph can avoid some of this backtracking, transforming BTK into VTX for some refinements.

Two vertices $u, v \in V$ of a graph $G = (V, R)$ are *equivalent*, denoted $u \equiv v$, if there is an automorphism π of G such that $\pi(u) = v$. A vertex $w \in V$ is *fixed* by π if $\pi(w) = w$. When two vertices are equivalent, they belong to the same *orbit*. The set of all the orbits of a graph is called the *orbit partition*. Our algorithm performs a partial computation of the orbit partition incrementally, starting from the singleton partition. Since only a limited search for automorphisms is done, it is possible to stop before the orbit partition is really found. Then, only a semiorbit partition is obtained. A *semiorbit partition* of G is any partition $O = \{O_1, ..., O_k\}$ of V, such that all vertices in O_i are equivalent, for all i.

Lemma 1. *At any level l of a sequence of partitions Q_G, all the vertices in a cell with no remaining links are mutually equivalent.*

Using this lemma, some equivalences are detected using only one sequence of partitions. However, conauto generates two sequences of partitions to detect most equivalences. From Theorem 1 and the definition of automorphism, it follows that two compatible sequences of partitions for a graph G define an automorphism of G. Let l be a backtracking level of a sequence of partitions Q_G (i.e., $R^l = \text{BTK}$), let $S^l_{P^l}$ be the pivot cell and $p \in S^l_{P^l}$ the pivot vertex used for the vertex refinement at level l. Consider any $p' \in S^l_{P^l}, p \neq p'$. Let Q'_G be a sequence of partitions compatible with Q_G, generated using p' as pivot instead of p at level l. Note that Q_G and Q'_G are equal up to level l. Let $<_{\mathsf{Q}_G}$ be an order induced by Q_G on the vertices of V, and let $<_{\mathsf{Q}'_G}$ be an order induced by Q'_G on the same set of vertices V. Then,

Lemma 2. *The mapping π induced by $<_{\mathsf{Q}_G}$ and $<_{\mathsf{Q}'_G}$, defined as $\pi(\omega_{\mathsf{Q}_G}(i)) = \omega_{\mathsf{Q}'_G}(i), \forall i \in \{1, ..., |V|\}$, is an automorphism of G.*

Let k be such that $p = \omega_{\mathsf{Q}_G}(k)$, then $\pi(p) = p' = \omega_{\mathsf{Q}'_G}(k)$; $\forall j \in \{k, ..., |V|\}$, $\omega_{\mathsf{Q}_G}(j) \equiv \omega_{\mathsf{Q}'_G}(j)$; and π fixes vertices $\omega_{\mathsf{Q}_G}(1), ..., \omega_{\mathsf{Q}_G}(k-1)$. Two vertices $u, v \in V$ of a graph $G = (V, R)$ are *equivalent at level l*, denoted $u \equiv_l v$, if there is an automorphism of G that permutes them, and fixes all the vertices in $V \setminus V^l$ (i.e., those discarded in previous levels). Note that p and p' are equivalent at level l.

Lemma 3. *If $u \equiv_l v$, then $u \equiv_i v$, $\forall i \in \{0, ..., l-1\}$.*

Let $u \equiv_l v$, if $u \equiv_l p$, then $v \equiv_l p$, and if $u \not\equiv_l p$, then $v \not\equiv_l p$. This implies that when $u \equiv_l v$, their semiorbits can be merged at level l. Let us now extend the sequence of partitions to include a semiorbit partition.

Definition 2. *An extended sequence of partitions E for a graph $G = (V, R)$ is a tuple (Q, O), where Q is a sequence of partitions, denoted as $SeqPart(\mathsf{E})$, and O is a semiorbit partition of G, denoted as $Orbits(\mathsf{E})$.*

We observe now that when all the vertices in a pivot set used at a backtracking level l ($R^l = \text{BTK}$) are proved to be equivalent, R^l can be set to VTX, eliminating the backtracking point. This follows from the fact that automorphisms are preserved under isomorphisms, as stated in the following lemma.

Lemma 4. *If the vertices of a pivot set in a sequence of partitions Q_G for graph G are equivalent, then in a compatible sequence of partitions Q_H for graph H, the vertices in the corresponding pivot set must also be equivalent.*

The only information conauto stores about automorphisms is the semiorbit partition. Hence, with an extended sequence of partitions, it knows that two vertices are equivalent (but it does not know all the vertices that are fixed by an automorphism that permutes them). Nevertheless, for each two vertices u and v that belong to the same semiorbit in a semiorbit partition, there is at least one automorphism that fixes all the vertices that belong to singleton semiorbits, and permutes u and v.

Algorithm 1. Test whether G and H are isomorphic (*conauto*)

$Iso(G, H)$: boolean
 1 **if** degree partitions do not match **then return** FALSE
 2 $Q_G \leftarrow GenSeqOfPart(G)$; $Q_H \leftarrow GenSeqOfPart(H)$
 3 $E_G \leftarrow FindAuto(G, Q_G)$; $E_H \leftarrow FindAuto(H, Q_H)$
 4 **if** $BtkAmount(SeqPart(E_G)) \leq BtkAmount(SeqPart(E_H))$
 5 **then return** $Match(0, G, H, SeqPart(E_G), Orbits(E_H))$
 6 **else return** $Match(0, H, G, SeqPart(E_H), Orbits(E_G))$

Algorithm 2. Generate a sequence of partitions for a graph G

$GenSeqOfPart(G)$: sequence of partitions
 1 Start with the degree partition
 2 **while** there are non-singleton cells with links **do**
 3   **if** there is a singleton cell $\{v\}$ with links **then** label VTX; Refine by vertex v
 4   **else** label SET; Refine by set exhaustively
 5   **if** no set refinement succeeded **then** relabel BTK; Refine by vertex
 6 label FIN
 7 **return** the computed sequence of partitions

Algorithm *conauto*

In this section we present the algorithm *conauto* (Algorithm 1) which applies the previous theoretical discussion. If both graphs have the same vertex degrees, first it generates a sequence of partitions for each graph, and then tries to eliminate potential backtracking points looking for vertex equivalences at these backtracking points. Then, it chooses the graph with less backtracking levels ($BtkAmount()$, i.e., number of levels l with $R^l = $ BTK) in its sequence of partitions as the target, and tries to find a compatible sequence of partitions for the other graph. If one such sequence of partitions is found, it returns TRUE. Otherwise it returns FALSE.

Algorithm 2, *GenSeqOfPart*, starts from the degree partition of the vertex set, and generates a sequence of partitions iteratively as follows.

1. If there are singleton cells in the partition, one of them is chosen as the pivot set, and a vertex refinement is performed to obtain the next partition in the sequence (Line 3).
2. Otherwise, the algorithm performs set refinements using different cells in the partition as pivot sets, until one of them is able to split at least one cell (maybe itself), or all of them have been tried unsuccessfully (Line 5).
3. If no cell meeting the conditions of Cases 1 and 2 has been found, then some cell is chosen as the pivot set, and a vertex in that cell is used as the pivot vertex to generate the new partition performing a vertex refinement (Line 6).

The search for automorphisms is performed by Algorithm 3. First it uses algorithm *ProcCellsWithNoLinks* to apply Lemma 1. Then traverses the sequence

---

Algorithm 3. Look for automorphisms

$FindAuto(G, \mathsf{Q})$: extended sequence of partitions
 1 $\mathsf{O} \leftarrow$ the singleton partition of V
 2 $ProcCellsWithNoLinks(\mathsf{O})$
 3 **for each** level l labeled BTK, in decreasing order of l **do**
 4   **for** each non-pivot vertex v in the pivot cell **do**
 5   Generate an alternative sequence of partitions using v
 6   **if** the sequences of partitions are compatible **then**
 7              $ProcCompSeqsOfPart(\mathsf{O})$
 8   **if** all the vertices in the pivot cell are equivalent **then**
 9   relabel original partition VTX
10 **return** (Q, O)

Algorithm 4. Find a sequence of partitions compatible with the target

---

$Match(l, G, H, \mathsf{Q}_G, \mathsf{O}_H)$: boolean
 1 **if** partition labeled VTX **then**
 2   success $\longleftarrow$ Ref. by vertex are compat. **and** $Match(l + 1, G, H, \mathsf{Q}_G, \mathsf{O}_H)$
 3 **else if** partition labeled SET **then**
 4   success $\longleftarrow$ Ref. by set are compat. **and** $Match(l + 1, G, H, \mathsf{Q}_G, \mathsf{O}_H)$
 5 **else if** partition labeled BTK **then**
 6   **for each** vertex v in the pivot cell, while not success **do**
 7   **if** v may not be discarded according to O_H **then**
 8   success $\longleftarrow$ Ref. by vertex are compat. **and** $Match(l + 1, G, H, \mathsf{Q}_G, \mathsf{O}_H)$
 9 **else** (i.e. partition labeled FIN)
10   success $\leftarrow$ adjacencies in both partitions match
11 **return** success

---

of partitions upwards looking for vertex equivalences among the vertices in the pivot sets at the levels labeled BTK, applying Lemma 2. This way, Lemma 3 will be applicable, so the automorphisms already found may be used when processing previous partitions in the sequence. The generation of an alternative sequence of partitions is performed in a straightforward way, avoiding bactracking. If this alternative sequence of partitions is compatible with the original one, then new vertex equivalences have been found, and they are used to iteratively compute the semiorbit partitions of the graphs using algorithm $ProcCompSeqsOfPart$.

When, at a backtracking point, all the vertices in the pivot cell are found to be equivalent, that level is relabeld from BTK to VTX. Recall that, from Lemma 4, this equivalence must hold for the other graph, so only one vertex in the corresponding pivot cell will need to be tested during the search for an equivalent sequence of partitions.

Algorithm 4 ($Match$) is a recursive algorithm that receives a level l to process in the sequence of partitions, the graphs G and H to test, the sequence of partitions Q_G for graph G, and the semiorbit partition O_H previously obtained for graph H. It returns TRUE if it is able to find a sequence of partitions for graph H that is compatible with Q_G, and FALSE otherwise. Algorithm 4 starts with a partition that is compatible with the original (e.g., both start with the degree partition). Then, if the current level is labeled VTX, it applies a vertex refinement to the current partition. If the new partition generated is compatible with the original, it recursively calls itself to process the next partition in the sequence. Levels labeled SET are processed in a similar way, but applying a set refinement. If the current level is labeled BTK, it applies Lemma 2 to prune the search space. More sophisticated automorphism management may help here, but we have discarded for now that possibility in favor of simplicity. Hence, vertex equivalence will only be applied when all the previously fixed vertices belong to singleton semiorbits. At the last level (labeled FIN), Condition 3 from the definition of compatibility between sequences of partitions is tested.

The algorithm conauto directly applies the theoretical results from the previous section. Hence, the following theorem.

Theorem 2. *Two graphs G and H are isomorphic iff $Iso(G, H)$ returns* TRUE.

5   Complexity Analysis

Algorithm conauto requires to store the adjacency matrices and the sequences of partitions for each of the graphs. The matrices need $O(n^2)$ words for graphs of n vertices. (We assume words of $O(\log n)$ bits, since they need to store vertex identifiers.) Each partition may be represented using $O(n)$ words. It is not hard to see that a sequence of partitions has at most $2n$ partitions. Then, a sequence of partitions requires $O(n^2)$ words. Since at most three sequences have to be stored at any time (those of the graphs and a temporary sequence to find automorphisms), the sequences of partitions take $O(n^2)$ words. This yields a total amount of space required by conauto of $O(n^2)$ words, or $O(n^2 \log n)$ bits.

Regarding time, a careful analysis of each type of refinement gives that generating a new partition in a sequence takes at most time $O(n^2)$. Then, a sequence of partitions is built in time $O(n^3)$. In order to find automorphisms at most $O(n^2)$ sequences are created. Hence, creating a target sequence of partitions requires time $O(n^5)$. Now, the time to find a sequence of partitions compatible with the target directly depends on the number of backtracking points in the target sequence. If there are no backtracking points it is just the time to generate a sequence, $O(n^3)$ time. In general, let α be the number of backtracking points; then the time complexity is $O(n^{\max(\alpha+3,5)})$.

Finally, let us consider a random graph $G(n, p)$ for $p \in [\omega(\ln^4 n/n \ln \ln n), 1 - \omega(\ln^4 n/n \ln \ln n)]$. Sort the degrees of the neighbors of a vertex into its degree vector. Czajka and Pandurangan [6] have shown that, with high probability, no two vertices have the same degree vector, and that a canonical labeling for the graph is obtained from the lexicographic ordering of the degree vectors. If no two vertices have the same degree vector, conauto will generate a sequence of partitions without backtracking points, first obtaining the degree partition and then by repeatedly applying set refinements. Then, our algorithm will finish in time $O(n^5)$ with high probability if any of the graphs is a random graph $G(n, p)$.

6   Performance Evaluation

In this section we compare the performance of an implementation of conauto with the two other programs of reference: *nauty* and *vf2*. The tests have been carried out in a Pentium III at 1.0 GHz with 256 MB of main memory, under Linux RedHat 9.0. All the programs have been compiled with the same compiler, GNU's gcc, and using the same optimization flags. The execution time considered is the real time (not CPU time) consumed by the programs, excluding loading time (the time needed by the programs to load from disk the graphs being tested). The CPU time limit for each program run was set to 10,000 seconds. If a program was unable to finish within this CPU time limit for a pair of graphs of some size, all its tests for that and bigger sizes were discarded. Some of the curves obtained have been omitted due to space restrictions. They can be found, with a detailed description of the benchmark used in the evaluation, at [14].

The first graphs considered are random graphs $G(n, 0.1)$ (only isomorphic cases). As expected, all algorithms run very fast with these graphs, finishing in less than a second even for graphs of 1,000 nodes. However, vf2 is one order of magnitude worse than the other programs, nauty being the fastest. The second family of graphs are 2D-meshes. In this case, for undirected graphs all algorithms behave similarly, finishing in, at most, a few seconds (for 1,000 nodes). A difference in behavior is observed for directed graphs. While conauto behaves as with undirected 2D-meshes, the time of nauty increases and the time of vf2 decreases, both in about one order of magnitude. The next family of graphs considered are Paley graphs, a subclass of SRGs. In this case all programs run in reasonable time (at most tens of seconds). It may be worth to note that vf2 is more than 2

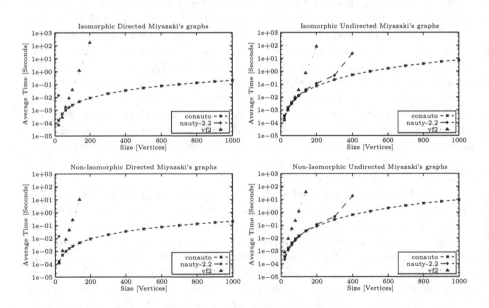

Fig. 1. Performance of conauto with Miyazaki's graphs

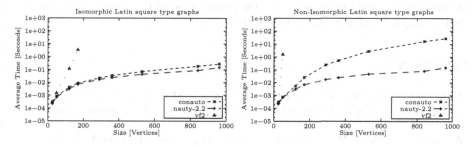

Fig. 2. Performance of conauto with Latin square graphs

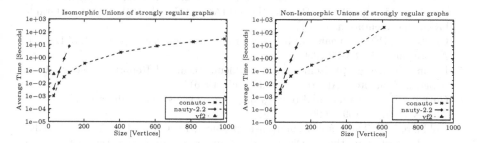

Fig. 3. Performance of conauto with unions of strongly regular graphs

orders of magnitude slower than conauto and nauty. For triangular graphs and lattice graphs, also subclasses of SRGs, we observe a symmetric phenomenon: all programs run fast (at most a few seconds) and vf2 is about one order of magnitude faster.

The first family of graphs in which a substantial difference in behavior can be observed are Miyazaki's graphs (see Figure 1). These are known to be very hard graphs for nauty [13] (e.g., with the directed version, it is not able to label graphs of 40 vertices in 10,000 s.). As can be seen in the figure, this family of graphs is only handled fast by conauto, which always finishes in a few seconds. The other algorithms cannot go beyond 400-node graphs (200 nodes if directed). A second interesting family are Latin square graphs, which are SRGs. For this family vf2 is not able to finish beyond graphs of 200 nodes (see Figure 2). Additionally, while nauty has the same low running time for positive and negative cases, conauto shows good (similar to nauty) running times for positive cases but about 2 orders of magnitude more for negative cases. The third interesting family of graphs are those obtained as unions of SRGs with the same parameters (29, 14, 6, 7) (see the results in Figure 3). These graphs are already known to make nauty exponential in time (cf. [13]). For vf2, they are so hard, that it can only finish within time with graphs of one component. On the other hand, conauto runs reasonably fast for positive cases, and faster than the others for the negative cases. However, it can not find an answer for graphs above 600 vertices for non-isomorphic pairs of graphs. The hardest family we have in our benchmark are point-line graphs of

Desarguesian projective planes. For this family none of the programs is able to deal with graphs of more than 200 vertices.

References

1. Arvind, V., Kurur, P.P.: Graph isomorphism is in SPP. In: FOCS 2002 (2002)
2. Babai, L., Kučera, L.: Canonical labeling of graphs in linear average time. In: FOCS 1979 (1979)
3. Babai, L., Luks, E.M.: Canonical labeling of graphs. In: STOC 1983 (1983)
4. Boppana, R., Hastad, J., Zachos, S.: Does co-NP have short interactive proofs? Information Processing Letters 25, 27–32 (1987)
5. Cordella, L.P., Foggia, P., Sansone, C., Vento, M.: An improved algorithm for matching large graphs. In: Proc. of the 3rd IAPR-TC-15 Int'l Workshop on Graph-based Representations (2001)
6. Czajka, T., Pandurangan, G.: Improved random graph isomorphism. J. of Discrete Algorithms 6(1), 85–92 (2008)
7. Fortin, S.: The graph isomorphism problem. Technical Report TR 96-20, U. of Alberta, Edmonton, Alberta, Canada (July 1996)
8. Gati, G.: Further annotated bibliography on the isomorphism disease. J. of Graph Theory 3(2), 95–109 (2006)
9. Goldberg, M.: The graph isomorphism problem. In: Gross, J.L., Yellen, J. (eds.) Handbook of graph theory. Discrete Mathematics and its Applications, ch. 2.2, pp. 68–78. CRC Press, Boca Raton (2003)
10. Mathon, R.: A note on the graph isomorphism counting problem. Information Processing Letters 8(3), 131–132 (1979)
11. McKay, B.D.: Practical graph isomorphism. Congr. Numer. 30, 45–87 (1981)
12. McKay, B.D.: The nauty page. CS Dept., Australian National U. (2004), http://cs.anu.edu.au/~bdm/nauty/
13. Miyazaki, T.: The complexity of McKay's canonical labeling algorithm. In: Finkelstein, L., Kantor, W.M. (eds.) Groups and Computation II. DIMACS Series in Discrete Mathematics and Theoretical Computer Science, vol. 28 (1997)
14. López Presa, J.L.: Efficient Algorithms for Graph Isomorphism Testing. Ph.D thesis, ETSI Telecomunicación, U. Rey Juan Carlos, Madrid, Spain (March 2009), http://www.diatel.upm.es/jllopez/tesis/thesis.pdf
15. Read, R.C., Corneil, D.G.: The graph isomorphism disease. J. of Graph Theory 1, 339–363 (1977)
16. Schöning, U.: Graph isomorphism is in the low hierarchy. J. of Computer and System Sciences 37(3), 312–323 (1988)
17. Singler, J.: Graph isomorphism implementation in LEDA 5.1. Technical report, Algorithmic Solutions Software GmbH (December 2005)
18. Ullmann, J.R.: An algorithm for subgraph isomorphism. J. ACM 23(1) (1976)

Algorithms and Experiments for Clique Relaxations—Finding Maximum s-Plexes

Hannes Moser*, Rolf Niedermeier, and Manuel Sorge**

Institut für Informatik, Friedrich-Schiller-Universität Jena,
Ernst-Abbe-Platz 2, D-07743 Jena, Germany
{hannes.moser,rolf.niedermeier,manuel.sorge}@uni-jena.de

Abstract. We propose new practical algorithms to find degree-relaxed variants of cliques called s-plexes. An s-plex denotes a vertex subset in a graph inducing a subgraph where every vertex has edges to all but at most s vertices in the s-plex. Cliques are 1-plexes. In analogy to the special case of finding maximum-cardinality cliques, finding maximum-cardinality s-plexes is NP-hard. Complementing previous work, we develop combinatorial, exact algorithms, which are strongly based on methods from parameterized algorithmics. The experiments with our freely available implementation indicate the competitiveness of our approach, for many real-world graphs outperforming the previously used methods.

1 Introduction

Finding maximum-cardinality cliques in graphs now for a long time is a major challenge for algorithmic graph theory and corresponding algorithm engineering efforts (cf. DIMACS challenge [5]). The corresponding MAXIMUM CLIQUE problem is NP-hard and neither effective approximation nor parameterized approaches exist that allow for efficient algorithms with provable performance bounds. Hence, the use of heuristic approaches always has been an important tool for practical solutions of MAXIMUM CLIQUE. The concept of cliques, however, has been criticized for its overly restrictive nature asking for *complete* subgraphs. A more relaxed concept of a dense subgraph has been introduced by Seidman and Foster [14] with the notion of s-plexes. A 1-plex is the same as a clique. For $s \geq 2$, an s-*plex* of a graph $G = (V, E)$ is a vertex set $S \subseteq V$ such that in the induced subgraph $G[S]$ every vertex has degree at least $|S| - s$. Unfortunately, finding maximum-cardinality s-plexes turns out to be computationally basically as hard as clique detection is [2, 8]. Thus, recently the development of practical (heuristic) algorithms for s-plex detection has received quite some interest [2, 9, 15]. In this work, we contribute novel tools for the efficient detection of maximum-cardinality s-plexes. Other than previous work [2, 9, 15] (where [15] deals with s-plex *enumeration*), our algorithms draw on methods from parameterized algorithmics [10].

* Supported by the DFG, project AREG, NI 369/9.
** Supported by the DFG, project PIAF, NI 369/4.

The MAXIMUM s-PLEX problem for an integer $s \geq 1$ is defined as follows.

Input: A graph $G = (V, E)$ and a nonnegative integer k.
Question: Is there an s-plex $S \subseteq V$ of size at least k?

Clearly, in our experiments we actually choose to maximize the value of k. Recent work on clique finding has exploited the close connection (indeed, duality) between MAXIMUM CLIQUE and the MINIMUM VERTEX COVER problem [1, 4]. We follow the same spirit here and make use of the duality between MAXIMUM s-PLEX and the MINIMUM d-BOUNDED-DEGREE DELETION problem (d-BDD for short). The latter problem is defined as follows.

Input: A graph $G = (V, E)$ and a nonnegative integer k.
Question: Is there a vertex set $S \subseteq V$ of size at most k whose deletion makes $G[V \setminus S]$ a graph of maximum degree d?

Clearly, we are interested in minimizing the value k. The point now is that an n-vertex graph has an s-plex of size k iff its complement graph has a solution set for d-BDD of size $n - k$ with $d := s - 1$. We exploit this close connection by making use of fixed-parameter tractability results for d-BDD [6, 8] and adding some new ones.

Our contributions. On the theoretical side, we provide an improved depth-bounded search tree for 1-BDD (the search tree has size $O(2.31^k)$ instead of previously $O(3^k)$ [8]) and an algorithm for 1-BDD based on iterative compression (exponential factor 2^k). Note that, by duality, these algorithms can be used for finding 2-plexes. Moreover, we present several very effective heuristics (still yielding optimal solution sets) which help to significantly boost the performance of the underlying fixed-parameter algorithms in applications. We perform a number of computational studies, comparing with previous work [2, 9] on exact solutions for s-plex finding which mainly rely on integer linear programming and branch-and-bound. For several real-world graphs, we mostly achieved speedups by orders of magnitude when compared to the previous work. Concerning some dense synthetic instances, we are most of the time slightly slower than approaches based on integer linear programming.

Preliminaries. In this paper, all graphs are simple and undirected. For a graph $G = (V, E)$ and a vertex set $S \subseteq V$, we write $G[S]$ to denote the graph induced by S in G, that is, $G[S] := (S, \{e \in E \mid e \subseteq S\})$. For a vertex $v \in V$, we write $G - v$ instead of $G[V \setminus \{v\}]$ and for a vertex set $S \subseteq V$ we write $G - S$ instead of $G[V \setminus S]$. We define $N(v) := \{u \in V \mid \{u, v\} \in E\}$, $N[v] := N(v) \cup \{v\}$; the *degree* of a vertex v is $|N(v)|$. If every vertex in G has degree at most d, then we say that G has *maximum degree* d. A vertex set $S \subseteq V$ is a *d-bdd-set* if $G - S$ has maximum degree d.

A parameterized problem is *fixed-parameter tractable* if it can be solved in $f(k) \cdot n^{O(1)}$ time, where f is a computable function depending only on the parameter k, not on the input size n [10]. We also employ search trees for our fixed-parameter algorithms. Search tree algorithms work in a recursive manner.

The number of recursion calls is the number of nodes in the according tree. This number is governed by linear recurrences with constant coefficients. These can be solved by standard mathematical methods [10]. If the algorithm solves a problem instance of size s and calls itself recursively for problem instances of sizes $s - d_1, \ldots, s - d_i$, then $(d_1, \ldots, d_i)$ is called the *branching vector* of this recursion. It corresponds to the recurrence $T_s = T_{s-d_1} + \cdots + T_{s-d_i}$ for the asymptotic size T_s of the overall search tree.

Due to the lack of space, some details are deferred to a full version of the paper.

2   Algorithms

Before coming to some new (mostly fixed-parameter) algorithms, we start with surveying algorithmic approaches that have been developed so far.

Known Approaches. Balasundaram et al. [2] presented a 0/1 integer linear program for MAXIMUM s-PLEX, generalizing a known formulation for the special case MAXIMUM CLIQUE. In addition, they carried out a polyhedral study of the problem and discussed a branch-and-cut implementation as the basis of computational tests. In follow-up work, McClosky and Hicks [9] described combinatorial algorithms for MAXIMUM s-PLEX, both of heuristic (without provable guarantees on the solution quality) and exact nature. Their heuristic algorithms are based on certain upper and lower bounds for vertex coloring and their exact algorithms are based on adapting known algorithms for MAXIMUM CLIQUE.

As mentioned before and already undertaken for the special cases of MAXIMUM CLIQUE and MINIMUM VERTEX COVER (cf. [1, 4]), an alternative route to solving MAXIMUM s-PLEX is to do a "detour" via d-BDD in the complement graph. This is our approach, which, thus, can also be seen as work on d-BDD. Concerning d-BDD, Nishimura et al. [11] presented a depth-bounded search tree yielding a solving algorithm running in $O((d + k)^{k+3} \cdot k + n \cdot (d + k))$ time. Subsequently, an improved simple search tree algorithm running in $O((d + 2)^k \cdot (d + k)^2 + n \cdot (d + k))$ time was described [8]. Finally, very recently, an intricate combinatorial data reduction algorithm has been developed [6]. More specifically, it was shown that MINIMUM d-BOUNDED-DEGREE DELETION with a solution set of size k possesses a problem kernel[1] containing at most $(d^3 + 4d^2 + 6d + 4) \cdot k$ vertices, which is computable in $O(n^{5/2} \cdot m + n^3)$ time.

Concerning implementations and experimental work, only the investigations of Balasundaram et al. [2], McClosky and Hicks [9], and Wu and Pei [15] have been accompanied by computational studies. Hence, it is one of the goals of our work to study the practical potential of the new approaches that are based on combinatorial algorithms that avoid polyhedral methods.

---

[1] Intuitively, a problem kernel is an equivalent problem instance whose size can be upper-bounded by a function independent of the size of the original input instance but only depending on the parameter k (see [10] for details).

Our main algorithm uses a bounded search tree and polynomial-time data reduction rules interleaving with the search tree. In general, the branching strategy of the search tree algorithm chooses a vertex v of degree at least $d + 1$, and then branches into the subcases of deleting v and every possibility of deleting all but d neighbors of v. In this case we say that the strategy "branches on v and $N(v)$". In practice, it is favorable to delete many vertices in each branching step, that is, v should be a vertex of high degree. Most parts of the subsequent descriptions of new algorithmic approaches refer to this.

Conditional application of BDD-Rule. By preliminary experiments, we found out that the direct application of the aforementioned problem kernel of at most $O(d^3 \cdot k)$ vertices is only effective for very few real-world graphs. Therefore, we turned our attention to use the corresponding data reduction rule (called *BDD-rule*) as an interleaving step in a search tree approach. However, applying the rule in every search tree node is not practical. We only apply it in a search tree node if there is a high probability that it will successfully reduce the graph.

Guided branching. The aforementioned problem kernel is based on a $(d + 2)$-approximate solution[2] X (hence, $|X| \leq (d + 2) \cdot k$). With this size bound on X, by applying the BDD-rule, the size bound for the reduced graph can be derived. This means that the interleaving of this kernel with the search tree algorithm can only be effective if X is small compared to $V \setminus X$ (more precisely, if $|V \setminus X| > (d + 1)^2 \cdot |X|$). That is why it is beneficial when the branching strategy tends to branch on vertices in X (thereby deleting more vertices in X) such that after few branching steps X gets small enough. However, in order to decrease the size of X more efficiently, it can be useful to branch on v and only a subset of $N(v)$. To this end, among the vertices of maximum degree, the vertex v to branch on is chosen such that $|N[v] \cap X|$ is maximized and the algorithm only branches on v and $N(v) \cap X$. Since $|X|$ is an upper bound on the size of an optimal solution, this branching strategy can also help in speeding up the search process (by using this upper bound in the search tree to detect branches that cannot lead to a minimum solution) even if interleaving with the BDD-rule is not effective.

Edge-count rule. The *edge-count rule* tests whether the given d-BDD instance is a no-instance. The rule counts how many edges can be deleted from the graph $G = (V, E)$ by at most k vertex deletions based on the vertex degree distribution of the graph. If the number of such edges is too small, then the graph cannot be turned into a graph with maximum degree d by at most k vertex deletions. The number of edges m' that can be deleted by at most k vertex deletions is computed as follows: sort the vertices of G by non-decreasing degree and sum up the degrees of the first k vertices in that order. Then, test whether $m - m' > \frac{d \cdot n}{2}$. If so, then (G, k) is a no-instance. Due to its simplicity, this rule can be implemented to run very efficiently.

---

[2] This $(d + 2)$-approximate solution can be computed by greedily finding a maximal collection of vertex-disjoint copies of stars with $(d + 1)$ leaves.

Improved search tree for $d = 1$. For the practically relevant special case $d = 1$, we give a more refined branching strategy with an improved search tree size of $O(2.31^k)$. We refrain from conceivable further asymptotic improvements (which appear likely when using even further refined branching strategies) in order to keep the algorithm easy to implement and efficient by avoiding the overhead incurred by more complicated strategies.

We start with considering a vertex v of degree $t > 1$. Clearly, v either needs to be deleted or all but one of its neighbors to achieve maximum degree one. Let $N(v) = \{u_1, \ldots, u_t\}$. If not deleting v, branch into the following $t + 1$ subcases:

1. Delete $N(v)$.
2. For each $u_i \in N(v)$, $1 \leq i \leq t$, delete $(N(v) \setminus \{u_i\}) \cup (N(u_i) \setminus \{v\})$.

The correctness of this branching can be seen as follows. First, clearly in each subcase v eventually gets maximum degree one. Second, the branching covers all possibilities how v can be made a maximum-degree-one vertex: one can keep at most one vertex from $N(v)$, the rest has to be deleted. If u_i is the neighbor that shall not be deleted, then clearly all vertices from $N(v) \setminus \{u_i\}$ have to be deleted (otherwise, v would have degree greater than one) and all neighbors of u_i except for v (that is, $(N(u_i) \setminus \{v\})$ have to be deleted (otherwise, u_i would have degree greater than one). Finally, the case of deleting all of $N(v)$ also needs to be considered since, otherwise, one would overlook the situation that all of v's neighbors have to be deleted for reasons lying outside the neighborhood of v. One obtains a branching into $t+2$ cases with the corresponding branching vector

$$(1, t, t - 1 + |N(u_1) \setminus N[v]|, \ldots, t - 1 + |N(u_t) \setminus N[v]|).$$

It is not hard to check[3] that the worst-case branching vector occurs for $t = 2$ and $|N(u_1) \setminus N[v]| = |N(u_2) \setminus N[v]| = 1$, meaning $(1, 2, 2, 2)$ with the branching number 2.31. In analogy to the general result [8], this gives the following.

Theorem 1. MINIMUM   1-BOUNDED-DEGREE   DELETION   *is   solvable   in* $O(2.31^k \cdot k^2 + kn)$ *time.*

Theorem 1 is a pure worst-case result. In the implementation, it is clearly favorable to first branch on high-degree vertices (large t-values), making the approach typically much more efficient than the theoretical bound predicts. Without proof, we mention in passing that 1-BDD can be also solved in $O(2^k \cdot k^{5/2} + n + m)$ time using the technique of iterative compression; however, here we focus on the more practical search tree algorithm as described above.

3   Implementation, Algorithmic Tricks, and Experiments

Implementation. Our implementation is written in the functional programming language Objective Caml[4]. A reason for this choice was that we could

---

[3] We omit some details here; basically, one can argue that for $t = 2$ cases where $|N(u_1) \setminus N[v]| = 0$ are actually easier (often avoiding branching at all) and $t > 2$ gives branching vectors with smaller branching numbers.

[4] See, http://caml.inria.fr/

make use of a purely functional graph data structure. This data structure makes the implementation of a search-tree based algorithm much easier, since we do not have to care about undoing changes to the data structure that were applied in other search tree branches. Moreover, it is a stated (and usually achieved) goal of the Objective Caml developers that Objective Caml code runs at most twice as slow as code generated by a decent C compiler. This speed difference is not a major factor for our considerations, since we are interested in the relative performance of algorithms. Moreover, since we are dealing with exponential-time algorithms, algorithmic improvements usually lead to time savings that cannot be bounded by any constant factor, so this effect seems small in comparison.

Our implementation is open source and it is freely available.[5] In Figure 1, we give the pseudocode of the basic search tree algorithm to compute a minimum d-bdd-set of size at most k for a graph. The data reduction rules in lines 3–7 remove parts of the graph that can be omitted from further consideration (line 3), high-degree vertices (lines 4–5), and some neighbors of degree-1 vertices (lines 6-7). The simple correctness proofs for these rules are omitted here. Note that the rules not only have to delete vertices from the graph G, but also from the d-bdd-set X (see "guided branching" in Section 2), in order to preserve the invariant that X is a d-bdd-set for G. Concerning the BDD-rule (lines 8–10), we changed the condition from $|V \setminus X| > (d+1)^2 \cdot |X|$ (which guarantees success of the BDD-rule application, see Section 2) to $|N(X)| > (d+1) \cdot |X|$ (which makes the success of the BDD-rule probable in practice, even if the condition $|V \setminus X| > (d+1)^2 \cdot |X|$ is not met). In lines 12–15 we perform several tests whether the instance resulting by the application of the data reduction rules is a no-instance. In line 15 we test whether the instance has already bounded degree d. Then, in line 16 the algorithm selects a vertex to branch on. The branching is then performed in lines 18–21. Then, in lines 22–24 the algorithm either returns that the input instance is a no-instance or returns the best solution that it has found.

Algorithmic Tricks. Concerning the initial $(d + 2)$-approximate solution X needed for the guided branching, it turns out that a greedy solution, computed by simply taking a vertex of highest degree into the solution until the remaining graph has bounded degree d, very often is smaller than a $(d + 2)$-approximate solution, although this method does not provably guarantee an approximation factor of $d + 2$. Such a greedy solution is also computed at the beginning of the computation (before invoking the search tree algorithm), and its size is taken as the initial value of k. Note that our implementation contains many algorithmic tweaks that are not covered by the basic description in Figure 1. For instance, the effect of the guided branching can be improved by recomputing X from time to time in the course of the branching process. Moreover, it improves performance significantly if one updates the value of k if a branch has found a solution that is smaller than the initial k. For $d = 1$, we implemented the improved branching described in Section 2 instead of the branching shown in Figure 1.

In the following, we comment about some particularities of our search tree implementation. One of the most important issues was the computation of the

---

[5] http://theinf1.informatik.uni-jena.de/splex/

Algorithm. bddsolve (G, X, k)
Input: A graph $G = (V, E)$, a d-bdd-set X for G, and an integer $k \geq 0$.
Output: A minimum-size d-bdd-set S for G with $|S| \leq k$, or "no-instance".

```
 1  S ← ∅
 2  repeat
 3      Remove each vertex v from G for which ∀_{w∈N[v]} deg(w) ≤ d.
 4      while ∃v ∈ V : deg(v) > d + k              ▷ High-degree rule
 5          G ← G − v; X ← X \ {v}; S ← S ∪ {v}; k := k − 1.
 6      while ∃v ∈ V : v has at least d + 1 degree-1 neighbors  ▷ Degree-1 rule
 7          G ← G − v; X ← X \ {v}; S ← S ∪ {v}; k := k − 1.
 8      if |N(X)| > (d + 1) · |X| then
 9          call BDD-rule to obtain vertex sets A and B ⁶
10          G ← G − (A ∪ B); X ← X \ A; S ← S ∪ A; k ← k − |A|
11  until none of the rules applies.
12  if k < 0 then return "no-instance"
13  l := greedily computed lower bound of the size of a minimum d-bdd-set.
14  if k < l or edge-rule tells "no-instance" then return "no-instance"
15  if maximum degree of G is d then return S
16  Among all max.-deg. vertices, choose a vertex v where |N[v] ∩ X| is maximum.
17  if |N(v) ∩ X| > d then                        ▷ Branch on v and N(v) ∩ X
18      call bddsolve (G − v, X \ {v}, k − 1)
19      for all size (|N(v) ∩ X| − d)-subsets C ⊆ N(v) ∩ X do
20          call bddsolve (G \ C, X \ C, k − |C|)
21  else branch analogously to lines 18–20 on v and N(v).
22  if all recursive calls of bddsolve returned "no-instance" then
23      return "no-instance"
24  else return S ∪ S', where S' is a smallest set returned by the bddsolve calls.
```

Fig. 1. Pseudocode of the basic algorithm to compute a minimum d-bdd-set.

complement graph, which has to be performed before executing the **bddsolve** algorithm (Figure 1). For sparse graphs, the complement graph is dense and, surprisingly, in practice the amount of time and memory to compute it exceeds often the time and memory needed for finding a maximum s-plex. Therefore, we implemented a wrapper that simulates a complement graph, rather than actually computing it. This wrapper, of course, is theoretically slower than the original graph data structure, since the data structure calls have to be translated by the wrapper. However, in practice, this method turns out to be almost always much more efficient than computing the complement graph directly.

For the graphs we considered, it turned out that applying the data reduction rules (see lines 3–10 in Figure 1) in every search tree node yields the best results. In particular, the degree-one rule and the high-degree rule are mostly very

---

⁶ The BDD-rule [6] returns two disjoint vertex sets A and B such that there exists a minimum-cardinality d-bdd-set S with $A \subseteq S$ and $S \cap B = \emptyset$.

effective. To be able to apply these rules more quickly, it seems to be reasonable to implement a data structure that provides fast access to vertices with a particular degree. However, this results in an increase of memory usage, and since the data structure has to be updated very frequently, many operations take more time. For instance, the deletion of a vertex, which is one of the most frequently called routines, needs about twice the time in our experiments. Moreover, we noticed an increased garbage collection overhead. Summarizing, such a data structure slows down the algorithm; surprisingly, for the degree-one and the high-degree rule a simple sweep over all vertices gives a faster implementation.

Experiments. All experiments were run on AMD Athlon 64 3700+ machines with 2.2 GHz, 1 M L2 cache, and 3 GB main memory running under the Debian GNU/Linux 4.0 operating system with the Objective Caml 3.09.2 compiler. The experiments of Balasundaram et al. [2] were performed on Dell Precision PWS690 machines with a 2.66 GHz Xeon Processor, 3 GB main memory, implemented using ILOG CPLEX 10.0. The processor speeds are comparable, so we compare the running times directly without applying a correction factor. The experiments of McClosky and Hicks [9] were run on a 2.2 GHz Dual-Core AMD Opteron processor with 3 GB main memory. We assume that their implementation uses one core only and compare the running times directly. Note that for both papers [2, 9] the corresponding source code is not publicly available.

Balasundaram et al. [2] performed experiments with two groups of graphs. One group can be characterized as social networks, which are derived from real-world data. The second group of graphs contains various graphs using the *Sanchis generator* [13] and clique instances from the second DIMACS challenge [5]. Balasundaram et al. [2] used an integer linear programming formulation combined with branch & cut methods. One of their exact algorithms, called BC(MIS), generates cuts based on a greedily computed independent set. For the real-world graphs, they use a variant called IPBC, which iterates over all vertices and searches an s-plex only in the vicinity of each iterated vertex (using the BC(MIS) approach). In the following, we compare our approach with the BC(MIS) and IPBC algorithms and also with the exact algorithm "Oster-Plex" by McClosky and Hicks [9], which is an adapted version of an algorithm for finding maximum-cardinality cliques by Östergård [12]. The experiments of McClosky and Hicks [9] cover almost all social networks and the instances from the DIMACS challenge.

Social Networks. This group contains a set of Erdős collaboration networks [7] (ERDŐS graphs), collaboration networks in computational geometry [3] (GEOM graphs), and text-mining networks based on Reuters news [3] (DAYS graphs). Due to space constraints, we omit the DAYS graphs; our results for ERDŐS and GEOM graphs also hold for the DAYS graphs in the qualitative sense.

ERDŐS graphs: Each vertex in an Erdős graph represents a scientist, and two vertices are adjacent if the corresponding scientists have published together. The graphs, obtained from [7], are named "ERDOS-x-y", where x represents the last two digits of the year for which the network was constructed, and y the maximum

Table 1. Running time and number of search tree nodes for ERDŐS and GEOM graphs compared with the running times of the IPBC and OsterPlex algorithm. Note that the OsterPlex experiments [9] were aborted after one hour.

s	graph	IPBC seconds [2]	OsterPlex seconds [9]	search tree algorithm			
				no guided branching		guided branching	
				seconds	nodes	seconds	nodes
2	ERDOS-97-1	2.9	0	0.9	179	0.3	311
	ERDOS-97-2	2123	1253	12.7	187	8.6	502
	ERDOS-98-1	2.2	0	1.1	201	0.4	358
	ERDOS-98-2	2251	1514	33.1	181	9.8	398
	ERDOS-99-1	4.2	0	1.2	212	0.4	357
	ERDOS-99-2	2442	1757	44.1	194	11.0	414
3	ERDOS-97-1	7.2	19	25.5	118620	0.7	10295
	ERDOS-97-2	32773	≥ 3600	620	596753	14.6	54695
	ERDOS-98-1	17.8	20	11.7	51965	1.0	13637
	ERDOS-98-2	45448	≥ 3600	762	694455	26.3	120605
	ERDOS-99-1	15.6	21	12.6	56704	1.7	28753
	ERDOS-99-2	40164	≥ 3600	1425	969064	36.8	132981
2	GEOM-0	12147	397	5.2	0	5.2	0
	GEOM-1	946	1118	0.3	20	0.3	20
	GEOM-2	487	1145	0.2	17	0.1	32
3	GEOM-0	20948	≥ 3600	5.2	0	5.2	0
	GEOM-1	1027	≥ 3600	1.0	5065	0.4	887
	GEOM-2	489	≥ 3600	0.2	1225	0.1	3

distance from each vertex to Erdős in the graph. As Balasundaram et al. [2] and McClosky and Hicks [9], we consider $x \in \{97, 98, 99\}$ and $y \in \{1, 2\}$.

GEOM graphs: Each vertex represents an author in computational geometry. For each pair of authors the number of joint publications is available. Given a threshold t, two authors are adjacent if they have more than t joint publications. The graphs, obtained from [3], are named "GEOM-t", where $t \in \{0, 1, 2\}$.

We compared the IPBC algorithm [2] and the OsterPlex algorithm [9] with our methods. We discovered experimentally that the guided branching has a strong effect on the running time for these instances, while the BDD-rule and the edge-count rule had only minuscule effects. Therefore, we performed experiments with and without guided branching. The resulting running times for the ERDŐS and GEOM graphs are given in Table 1. For the ERDŐS graphs, our method without guided branching outperforms the approach of Balasundaram et al. [2] by one or two orders of magnitude. With guided branching, the running time is improved by three orders of magnitude. To our surprise, the BDD-rule (almost) does not apply at all. The reason is that X (see "guided branching" in Section 2) is rather big, and we apply the high-degree rule first (see Figure 1), which reduces the graph so effectively that the condition for applying the BDD-rule is (almost) never met. When switching off the high-degree rule, almost all reduction is then performed by the BDD-rule. The OsterPlex algorithm [9] is mostly faster than

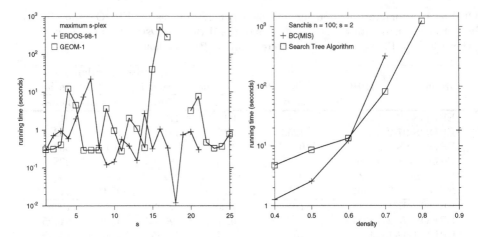

Fig. 2. Left: running times of our approach for $1 \leq s \leq 25$ on the ERDOS-98-1 and the GEOM-1 graph. Missing data points are due to exceeded running time limit of 60 minutes. Right: running times of our approach (search tree algorithm) compared with the running times of the BC(MIS) approach [2].

the IPBC algorithm [2], and for some instances it has running times comparable to our approach with guided branching, while in general it is about two orders of magnitude slower than our approach.

For the GEOM graphs, we observe similar speedups of up to three orders of magnitude (see Table 1). Interestingly, for some instances our approach does not branch at all; it immediately finds a solution using the data reduction rules. Since the data reduction rules are very effective and few branchings take place, the effect of the guided branching is not as pronounced as for the ERDŐS graphs.

Since the preceding experiments indicate that the running time of our approach does not increase too much with increasing s (recall that $s = d + 1$), we performed experiments on two of the real-world graphs for $1 \leq s \leq 25$. The results are shown in Figure 2. For most values of s, the instances can be solved within some seconds, only some take around one hour or more. Interestingly, there is a peak of the running time around $s = 19$. We conclude that our approach seems to be able to find maximum s-plexes for a wide range of the parameter s for these types of graph.

Sanchis and DIMACS Graphs. The Sanchis generator [13] produces graphs with known maximum clique size with a specified number of vertices n and edges m, and a construction parameter r. As Balasundaram et al. [2], we fixed the maximum clique size at $\lceil n/5 \rceil$, and the construction parameter to $\lfloor 0.75(n/c-1) \rfloor$. The number of edges is determined by the density d, that is, we compute the number of edges as $m := \lfloor dn(n-1)/2 \rfloor$. We performed experiments for $n \in \{100, 200\}$ and $d \in \{0.4, 0.5, 0.6, 0.7, 0.8, 0.9\}$.

Balasundaram et al. [2] used Sanchis graphs to study how the efficiency of their methods depends on the number of graph vertices, the density of the graph, and on the value s defining s-plexes. Their methods perform best on sparse graphs, and become less effective on dense graphs. Likewise, small graphs can be solved quickly, while larger graphs become more difficult to solve. Balasundaram et al. [2] performed experiments with the BC(MIS) algorithm for $s \in \{1, 2\}$. They observed that the case $s = 2$ is generally more difficult to solve than $s = 1$.

We observe the same general behavior as for the BC(MIS) algorithm, that is, dense Sanchis graphs are harder to solve than sparse ones, and graphs with many vertices are harder to solve than graphs with few vertices. We can observe that, especially on sparse instances, our approach is about one order of magnitude slower than the BC(MIS) algorithm (see Figure 2). However, the available data seems to indicate that the running time of our approach increases not as quickly with increasing density as the BC(MIS) algorithm does, but this needs to be checked more carefully with further experimentation, also including higher values of s. For Sanchis graphs with more vertices, there are too few instances where the BC(MIS) algorithm gave exact results within a three-hour running time limit in order to do a similar comparison, and likewise our approach did mostly not terminate within that time.

Finally, we briefly report about our findings concerning instances from the DIMACS challenge. Here, we compare with the BC(MIS) algorithm [2] and the OsterPlex algorithm [9]. Summarizing, out of the 32 considered instances we could solve 23 instances for $s = 1$ and 14 instances for $s = 2$, while BC(MIS) could solve 20 instances for $s = 1$ and 16 instances for $s = 2$ within a running time limit of three hours. For the instances that neither BC(MIS) nor our approach could solve exactly within three hours, we observe that our lower/upper bounds seem to be worse than the ones computed by BC(MIS). Compared to the OsterPlex algorithm, we could solve within one hour all but five instances for $s = 2$, which OsterPlex can solve within that time. Summarizing, BC(MIS) and OsterPlex are at least as good as our approach for these instances. In this respect, it would be interesting to study whether the OsterPlex and the BC(MIS) algorithms could be efficiently combined with ours.

4   Conclusion and Outlook

In some analogy to previous work on maximum-cardinality clique finding [1, 4], we demonstrated that an exact combinatorial approach provides competitive algorithms for finding maximum-cardinality s-plexes. Clearly, due to the NP-hardness of the problem, there are limitations concerning the range of practical feasibility. On the one hand, we believe that there is still some room for further tuning our algorithms and implementations (which in future work also should be compared with other approaches in an experimental study that is based on the *same* platform); on the other hand, we think that at some point more restrictions such as the one of "isolation" (see [8]) have to be imposed in order to gain practical algorithms. Our focus was on finding s-plexes of maximum size; studies concerning efficient approximation algorithms are left open.

References

[1] Abu-Khzam, F.N., Fellows, M.R., Langston, M.A., Suters, W.H.: Crown structures for vertex cover kernelization. Theory Comput. Syst. 41(3), 411–430 (2007)

[2] Balasundaram, B., Butenko, S., Hicks, I.V., Sachdeva, S.: Clique relaxations in social network analysis: The maximum k-plex problem (February 2008) (manuscript), http://iem.okstate.edu/baski/files/kplex4web.pdf

[3] Batagelj, V., Mrvar, A.: Pajek datasets (2006), http://vlado.fmf.uni-lj.si/pub/networks/data/ (accessed, January 2009)

[4] Chesler, E.J., et al.: Complex trait analysis of gene expression uncovers polygenic and pleiotropic networks that modulate nervous system function. Nat. Genet. 37(3), 233–242 (2005)

[5] DIMACS. Maximum clique, graph coloring, and satisfiability. Second DIMACS implementation challenge (1995), http://dimacs.rutgers.edu/Challenges/ (accessed, November 2008)

[6] Fellows, M.R., Guo, J., Moser, H., Niedermeier, R.: A generalization of Nemhauser and Trotter's local optimization theorem. In: Proc. 26th STACS, Germany, pp. 409–420. IBFI Dagstuhl, Germany (2009)

[7] Grossman, J., Ion, P., Castro, R.D.: The Erdős number project (2007), http://www.oakland.edu/enp/ (accessed, January 2009)

[8] Komusiewicz, C., Hüffner, F., Moser, H., Niedermeier, R.: Isolation concepts for enumerating dense subgraphs. In: Lin, G. (ed.) COCOON 2007. LNCS, vol. 4598, pp. 140–150. Springer, Heidelberg (2007)

[9] McClosky, B., Hicks, I.V.: Combinatorial algorithms for the maximum k-plex problem (January 2009) (manuscript), http://www.caam.rice.edu/~bjm4/CombiOptPaper.pdf

[10] Niedermeier, R.: Invitation to Fixed-Parameter Algorithms. Oxford University Press, Oxford (2006)

[11] Nishimura, N., Ragde, P., Thilikos, D.M.: Fast fixed-parameter tractable algorithms for nontrivial generalizations of Vertex Cover. Discrete Appl. Math. 152(1-3), 229–245 (2005)

[12] Östergård, P.R.J.: A fast algorithm for the maximum clique problem. Discrete Appl. Math. 120(1-3), 197–207 (2002)

[13] Sanchis, L.A., Jagota, A.: Some experimental and theoretical results on test case generators for the maximum clique problem. INFORMS J. Comput. 8(2), 103–117 (1996)

[14] Seidman, S.B., Foster, B.L.: A graph-theoretic generalization of the clique concept. Journal of Mathematical Sociology 6, 139–154 (1978)

[15] Wu, B., Pei, X.: A parallel algorithm for enumerating all the maximal k-plexes. In: Washio, T., Zhou, Z.-H., Huang, J.Z., Hu, X., Li, J., Xie, C., He, J., Zou, D., Li, K.-C., Freire, M.M. (eds.) PAKDD 2007. LNCS (LNAI), vol. 4819, pp. 476–483. Springer, Heidelberg (2007)

A Design-for-Yield Algorithm to Assess and Improve the Structural and Energetic Robustness of Proteins and Drugs

Giuseppe Nicosia and Giovanni Stracquadanio

Department of Mathematics and Computer Science
University of Catania - Viale A. Doria 6, 95125, Catania, Italy
{nicosia,stracquadanio}@dmi.unict.it

Abstract. Robustness is a property that pervades all aspects of nature. The ability of a system to adapt to perturbations due to internal and external agents, aging, wear, or to environmental changes is one of the driving forces of evolution. At the molecular level, understanding the robustness of a protein has a great impact on the *in-silico* design of polypeptide chains and drugs. The chance of computationally checking the ability of a protein to preserve its structure in the native state may lead to the design of new compounds that can work in a living cell more effectively. Inspired by the well known *robustness analysis framework* used in Electronic Design Automation, we introduce a formal definition of robustness for proteins and a dimensionless quantity, called *yield*, to quantify the robustness of a protein. Then, we introduce a new *robustness-centered* protein design algorithm called *Design-For-Yield*. The aim of the algorithm is to discover new conformations with a specific functionality and high yield values. We present extensive characterizations of the robustness properties of many peptides, proteins, and drugs. Finally, we apply the DFY algorithm on the *Crambin* protein (1CRN) and on the *Oxicitin* drug (DB00107). The obtained results confirm that the algorithm is able to discover a *Crambin-like* protein that is 23.61% more robust than the wild type. Concerning the Oxicitin drug a new protein sequence and the corresponding protein structure was discovered with an improved robustness of 3% at the global level.

1 Introduction

The in silico design of complex systems has become an effective, fast and reliable working flow reducing the time required for the design and manufacturing of new systems and products. In many engineering fields, for example the Electronic Design Automation (EDA) field, the in silico design approach is the *de-facto* standard for the design of new devices and circuits [1]. In particular, computational approaches are widely applied in various production phases including design, simulation, optimization and testing. It is important to have a measure for estimating how robust the designed device is towards "perturbations" that may occur during the production or working phases. For example, in EDA, a

J. Vahrenhold (Ed.): SEA 2009, LNCS 5526, pp. 245–256, 2009.
© Springer-Verlag Berlin Heidelberg 2009

computational methodology, called *Monte-Carlo Analysis*, is widely used to provide a statistical measure of circuit *robustness*. The quantitative measure for the robustness of a circuit is called the *yield* of the circuit. In order to minimize the manufacturing costs, the interest is directed towards the obtainment of circuits with a specific behaviour, that at the same time maximize the yield [2].

In general, the concept of robustness pervades all aspects of nature. An organism is said to be robust if it is able to adapt and resist to environmental changes, to fight against internal and external agents or because its character persists during the evolutionary process [3]. From this point of view, the notion of *persistence* plays a crucial role in the definition of robustness. The persistence of some particular properties of a system defines how robust the system is. In particular, robustness can be seen as a measure of property persistence under perturbations caused by changes in the system composition, system topology, or the environment in which the system is put [3].

According to Krakauer [4], robustness is viewed differently in stability theory because it considers multiple perturbations in multiple dimensions, instead of focusing on a single perturbation [5]. There, the concept of persistence emerges for various reasons not necessarily related to robust design, including constancy of the environment, developmental constraints or evolutionary constraints.

In Biology, a system is considered robust to mutations if it continues to function after genetic changes in its parts. The concept of robustness is pervasive on different levels of biological organization, from macromolecules to genetic networks or to whole organisms [6]. In any case, a strict connection between robustness and evolvability has been shown [7,8,9]. According to Kirschner and Gerhart [10], if a system can be protected from lethal mutations, then the accumulation of variability may permit it to move to a state within the same neutral conformation, such that fewer subsequent mutations are needed to produce a major innovation.

In the last twenty years, computational approaches have been largely applied in biochemistry. In particular, the prediction of the three dimensional structure of polypeptides is probably the best-known field [11,12]. Many efforts have been put in defining effective and efficient folding algorithms [13,14,15] to computationally design new proteins. Despite the complexity of the problem, many good approaches are available and the in silico design of new proteins [16] has become an emerging field in *drug design*. From an engineering point of view, a plethora of methods and algorithms for the design of new proteins are available but, to our knowledge, there is no universally accepted approach to estimate, in silico, the robustness of wild type and synthetic (computer designed) protein structures. However, the estimation of protein robustness is a key point in protein design. Physical mutations may occur at any stage of the synthesis process, and in any stage of the many bio-chemical processes occurring in a living organism. Since, in protein science, functionality follows structure, the estimation of the yield of a protein structure is crucial to measure how well it maintains its functionality under structural mutations. It should now be clear why a method to assess the robustness of a protein is extremely useful in the design of low toxic and cost-effective drugs.

In the present article we draw inspiration from the current *state-of-the-art* in EDA, to formally define the concept of *energetic yield* for proteins. Then we introduce new statistical methods to estimate the robustness of the structure of a protein. In particular, we define two new algorithms, the *Protein Monte-Carlo Sampling* (PMCS) algorithm and the Design-for-Yield (DFY) algorithm. They use three different methods to perturb the structure of the protein. The presented experimental studies are focused on two goals. Firstly, we assess the effectiveness of the methodology for finding *sensitive regions* and *sensitive residues* in proteins. Then, we also show how the DFY algorithm is able to make a protein more robust through the systematic mutation of faulty and sensitive residues of the structure. This enables the design of new proteins with the same specific function but with a more robust conformation. Finally, the DFY algorithm is applied to drug molecules. Our experiments show that it is possible to systematically maximize, in-silico, the robustness of these molecules.

2   Methods

A biological system is robust if it continues to function after perturbation [7]. Hence, the robustness of a system Ω is the study of the persistence of a certain property ϕ of Ω under perturbations introduced by internal or external agents. In our protocol, a perturbation is defined as a function $\tau = \alpha(\Omega, \sigma)$. Here α applies a stochastic noise σ to the system Ω. The result of the perturbation is called the *trial sample*, τ, while the α function is called α-perturbation (or α-analysis). We assume that the noise is defined by a random distribution. In order to simulate a statistically meaningful perturbation phenomenon, we generate an ensemble T of perturbed systems. The element τ of the ensemble T is considered robust to a perturbation (mutation) of the stochastic noise σ for a given property ϕ, if the following *robustness condition* is verified:

$$\rho(\Omega, \tau, \phi, \epsilon) = \begin{cases} 1 \; if \; | \; \phi(\Omega) - \phi(\tau) \; | \le \epsilon \\ 0 \qquad \qquad otherwise \end{cases} \tag{1}$$

Here Ω is the *reference system*, ϕ is a property of the system, τ is a *trial sample* of the ensemble T, and ϵ is a *robustness threshold*. The ρ function does not make any assumption on the property function ϕ, hence, the property function is not necessarily strictly related to features and properties of the given system. However, the property function ϕ implicitly assumes that the property under investigation is quantifiable. The robustness of a system Ω in respect to a property ϕ is the number of robust trials of T over the total number of trials ($| \; T \; |$). In EDA, this measure is denoted as the *yield* of the system. Formally, we can define our *yield function* Γ as follows:

$$\Gamma(\Omega, T, \phi, \epsilon) = \frac{\sum_{\tau \in T} \rho(\Omega, \tau, \phi, \epsilon)}{|T|} \tag{2}$$

The function $\Gamma(\Omega, T, \phi, \epsilon)$ may be used to assess the yield of a general system, and is not only restricted to proteins and drugs as in this manuscript.

It may be of interest to note that the yield does not decrease as ϵ values increase. Choosing a meaningful threshold value is crucial. Although it is good practice to set a strict threshold value, it is important not to restrict the analysis to a small set of feasible trials (in this work, feasible protein structures), to avoid that a large quantity of plausible systems are excluded. In the EDA domain, the threshold value is typically set by expert designers by taking into account the manufacturing system, the physical properties of materials and the adherence to the original design. In the area of robust protein design, the setting of an ad-hoc ϵ value is not a trivial task. Setting different threshold values according to the protein family is a plausible approach. In this work we performed extensive computational experiments to detect general and reasonable threshold values: 1.0 $kcal/mol$ for local analysis (local robustness) and 5.0 $kcal/mol$ for global analysis (global robustness). It is important to remark that the yield of a system is strictly related to the perturbation (the type of mutation that is applied). We use the term α-yield to refer to the yield value of the system Ω perturbed by the α perturbation, and the term α-analysis to indicate the whole process assessing the robustness of the ensemble of systems generated by the perturbation α.

We studied the thermodynamic robustness of proteins using the $ECEPP/3$ potential energy model, with explicit solvent terms according to the model proposed by Ooi [17]. All the potential energy calculations have been conducted using the Simple Molecular Mechanics for Proteins (SMMP) [18].

3   Algorithms for Robustness Analysis

We introduce an ad-hoc Monte-Carlo algorithm, called Protein Monte-Carlo Sampling (PMCS), derived from classical Monte-Carlo Sampling algorithms, to study the robustness of proteins under various molecular deformations. Following the classical algorithms, PMCS takes a protein in input and generates n trial conformations of the protein by randomly perturbing the protein structure in the native state. The protein structure can be described by using angles. In particular we use an internal coordinates representation (torsion angles). Each residue type requires a fixed number of torsion angles to fix the 3D coordinates of all atoms. To apply the definition of perturbation given in the previous section, we need to define the random noise. In our experiments, we choose a classical normal distribution. However, other distributions could also be used. We consider three different α-analyses: a Global analysis, a Local analysis and a Residue analysis. In the Global analysis procedure, deformations are applied to the whole structure of the protein: all the angles of the protein are mutated. With this procedure we aim at analyzing strong and dramatic mutations, which in practice may occur due to changes of the cellular environment or to an error regarding the synthesis process. The Local analysis procedure perturbs an individual's dihedral angle in order to find sensitive points in the structure. By identifying the most sensitive angle of the structure, this analysis is especially helpful for the design of *de-novo* optimization algorithms based on potential energy functions. Finally, the Residue analysis procedure perturbs all the angles of a residue. This analysis is especially indicated for identifying key residues in the polypeptide chain. In particular, the identification of sensitive amino acids allows us to define a new class of algorithms that focus on yield-optimization.

The Design-for-Yield Algorithm (DFY). In this paragraph we introduce a new algorithm based on the robust protein design principle, called Design-for-Yield. The aim of the design-for-yield optimization algorithm is to discover new proteins with a specific function with improved robustness. In particular, we are interested in finding mutants of wild type proteins with the same functionality but greater robustness to perturbations. From a mathematical point of view, the algorithm tries to discover structures with potential energy distribution that is well centered on the native energy value (the nominal value in EDA terminology), and with the tightest spread. The basic idea is to find the most sensitive regions of the structure and replace it with the other amino acids that maximize the yield. Obviously it is possible to iteratively replace each amino acid of a sequence and calculate its yield value. However, there are two important constraints on this protein redesign process. Firstly, each amino acid mutation must be neutral [19]. Since we want to preserve the functionality of the protein, it is important to consider only mutations that preserve it. This is mandatory especially in drug-design (e.g., we do not want to deal with high toxic drugs). Secondly, the mutant protein has to fold correctly. This constraint can be checked by evaluating its potential energy. Positive values imply that a protein is not in the native state. Since the functionality of a protein is structure-dependent, it is important to check that a mutant's structure does not differ more than $1\mathring{A}$ from the wild type, in terms of $RMSD_{C_\alpha}$. Now we are ready to describe the DFY algorithm consisting of two main procedures. The first procedure is responsible for the evaluation of the neutrality of the mutant. In our work, we used the SDM server (http://mordred.bioc.cam.ac.uk/~sdm/sdm.php)[20]. SDM takes in input a wild type protein, the position to be mutated and the amino acid to be added. The server outputs a Boolean value stating whether the mutant is neutral or not.

The second procedure uses the PMCS algorithm to estimate the robustness of the obtained protein. Since we are working at the residue level, we perform a residue analysis to understand whether the mutation has made the protein's structure more robust. The pseudo-code of the algorithm is shown in figure 1. The algorithm takes in input a wild type protein C and a robustness threshold ϵ to be used for the residue analysis. Firstly, the algorithm performs a residue analysis on the wild type protein to discover the most sensitive residue. Then the algorithm mutates the protein by changing the most sensitive amino acid with the other 19 remaining amino acids, and queries the SDM server to establish if the mutant is neutral. If it is neutral, then it undergoes a full regularization of the structure and it will be used for the residue analysis. Otherwise it will be discarded. From the set of neutral mutants, the mutant with the highest residue yield value is returned.

1: **procedure** DFY(C, ϵ)
2:     $trials \leftarrow PMCS_{residue}(C)$
3:     $s \leftarrow \min(Yield_{residue}(trials, \epsilon))$ ▷ returns the position of the most sensitive residue
4:     $mutants \leftarrow []$
5:     **for** $a \in Aminoacid \wedge a \neq a_s$ **do**
6:         $M_c^a \leftarrow C[s] \leftarrow a$ ▷ Mutate the s-th residue with the amino acid a
7:         $neutral \leftarrow SDM(M_c^a)$ ▷ Returns true if the mutation is neutral
8:         **if** neutral = true **then**
9:             $trials_{M_c^a} \leftarrow PMCS_{residue}(M_c^a)$
10:            $mutants \leftarrow [M_c^a, Yield_{residue}(trials)]$
11:        **end if**
12:    **end for**
13:    **return** $\max(mutants)$ ▷ Return the mutant with the highest residue yield
14: **end procedure**

Fig. 1. The pseudo-code of the DFY algorithm: the procedure takes in input a conformation C and a robustness threshold ϵ, and it returns the mutant that maximize the residue yield

Test set and parameter settings. To validate our methodology, we performed several experiments on a large set of peptides, proteins and drugs; the benchmark contains a paradigmatic peptide (PDB Id. 1PLW), the three basic classes $(\alpha, \alpha + \beta, \beta)$ (PDB IDs: 1CRN, 1IGD, 1BDD, 1GAB,1E0L), a protein with two of its mutants (PDB IDs: 1AML, 1BJB, 1BJC), the Enzyme (1ST7) and two Drugs (PDB Ids: 1NPO, 1GCN). We generated 10^4 trials for the global analysis perturbation method and $(200\times$ Na) trials (Na is the number of torsion angles of the given protein) for the local and residue analysis methods. For each kind of analysis, we studied the energetic distributions due to global, local and residue mutations using energy histograms. Each histogram was computed by sampling the energy landscape with bins of 1.0 $kcal/mol$ and considering a radius of 1000 bins around the potential energy value of the native structure for the global yield, and a radius of 50 bins for the local and the residue yields.

4   Results

The experiments reported in Table 1 show some interesting energetic robustness properties. Firstly, from our global analysis using normally distributed perturbations we note that the protein conformations on each dihedral angle of non-small proteins, unfold. These kind of mutations may occur, for example, in the synthesis due to a bias in the process. In these cases, a misfolded protein or a totally different protein from the designed one may be produced. Although this is evident for small and medium size proteins, this is not necessarily true for peptides or proteins that are fully exposed to a solvent. For the Met-enkephalin (PDB code 1PLW) or the Amyloid of the Alzheimer's disease (PDB code 1AML, 1BJB and 1BJC), the protein robustness is close to the maximum even in global

Table 1. A description of the proteins taken into account in our yield analysis. For each protein, we report the PDB code, the number of residues, the number of angles (Na), the corresponding class, the potential energy value of the native state (E_0), and the yield values for the three robustness analyses.

PDB & Na	Class	E_0(kcal/mol)	Analysis	Yield	PDB & Na	Class	E_0(kcal/mol)	Analysis	Yield
1PLW(5) 24	-	-24.835	global local residue	99.80 99.98 100.00	1AML(40) 224	α	-276.133	global local residue	2.86 97.13 76.36
1CRN(46) 235	$\alpha + \beta$	-225.219	global local residue	0.72 64.71 14.00	1BJB(28) 161	α	-235.491	global local residue	17.62 98.32 85.02
1IGD(61) 356	$\alpha + \beta$	-584.261	global local residue	0.00 80.75 22.54	1BJC(28) 159	α	-263.719	global local residue	75.31 98.10 87.38
1BDD(60) 357	α	-659.484	global local residue	0.00 88.29 46.14	DB00107 1NPO(9) 48	Drug	-65.783	global local residue	95.40 99.76 98.06
1GAB(53) 324	α	-419.262	global local residue	0.00 87.34 42.76	DB00040 1GCN(29) 175	Drug	-273.502	global local residue	73.30 97.57 86.57
1E0L(37) 221	β	-233.022	global local residue	0.00 87.44 43.93	1ST7(86) 527	Enzyme	-890.776	global local residue	0.00 74.78 23.56

analysis (see table 1). This could be justified by the fact that small proteins maintain a well-defined structure due to very small coil regions, which, typically, do not connect *structural motif*. Instead for fully exposed proteins, it is seams clear that the solvent force plays a central role in the definition of the structure. From the local analysis it clearly appears that, although there are angles that are responsible for the complete misfolding of a protein, the yield of the structure achieves high values. From an other point of view, this kind of analysis reveals that a protein undergoing perturbations on a single dihedral angle has a very low probability of misfolding. However, from a de-novo point-of-view, a single angle perturbation is not a viable way to find the native structures of protein. To this end, mutations of more angles during the folding process seem to be more reasonable approach. Also the residue analysis led to interesting findings. In particular it reveals that robustness of a protein varies according to its class. The $\alpha + \beta$ class seems to be the most sensitive to residue mutation with an average yield of $\sim 18.27\%$; the α and β classes report an average yield of 44.45% and 43.93% respectively; for the Amyloid A4 (1AML) and its mutants (1BJB and 1BJC) the average yield is 82.92% with the mutants achieving a higher yield than the ancestor. The box-error-plot of the residue analysis shows some sensitive residues for the robustness of the proteins (for the 1CRN protein see Figure 2 plot (b), (d) and (f)). In particular, there are mutations of a single amino acid that give greater unstableness than others of many orders of magnitude. This amino acid can be classified as the main actor of the protein folding process. By inspecting the Amyloids sequences, it is interesting to note that the three proteins are identical for the first 28 residues, while the mutants are different on the last residues. Probably the second α helix of the 1AML, with its surrounding coil region, makes the protein less robust. For this set of proteins,

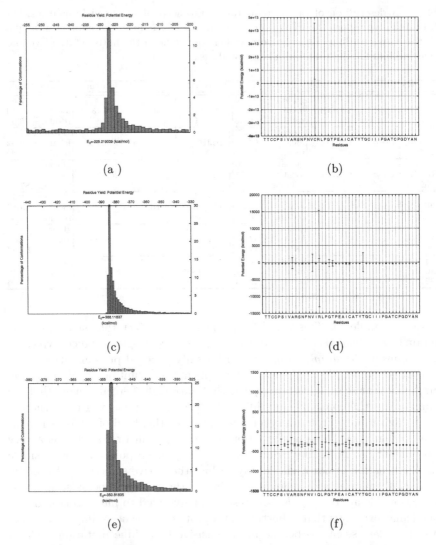

Fig. 2. Robustness analysis of the Crambin and of the most robust synthetic mutants. the potential energy distribution of 1CRN obtained by the residue robustness analysis (a); the potential energy variation of 1CRN obtained by the residue robustness analysis (b); the potential energy distribution of 1CRN-I(16) obtained by the residue robustness analysis (c); the potential energy variation of 1CRN-I(16) obtained by the residue robustness analysis (d); the potential energy distribution of 1CRN-Q(17) obtained by the residue robustness analysis (e); the potential energy variation of 1CRN-Q(17) obtained by the residue robustness analysis (f).

the most sensitive residues have the same order of magnitude of the sensitive residues of the other Amyloids sequences. This is in contrast with the results reported for the $\alpha + \beta$ class (see Table 1 and Figure 3). The characterization

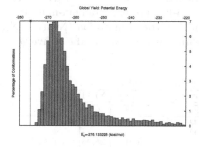 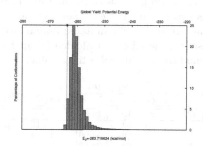

Fig. 3. 1BJC, the case of a robust peptide. The potential energy histograms show that the 1BJC mutant has strict energy spread well centered on the native potential energy value (E_0); the 1AML shows a larger spread and the histogram shows a great variability of the potential energy of the trials.

of the robustness of proteins seems to confirm that yield analysis is an effective approach to obtain more robust proteins. In order to validate this assumption, we apply the DFY algorithm on the less robust protein in terms of residue yield value, the Crambin (see Table 1). The most sensitive residue of the 1CRN is the 16th residue, a Cysteine (see Figure 2 plot (b)). We systematically change this amino acid with the remaining 19 amino acids, and, by using the SDM server [20], we identify three *disruptive mutations* (for Asparagine, Proline, Threonine) and *sixteen neutral mutations*. From the set of the neutral mutants in Table 2, the mutant with the Isoleucine at the sixteenth position, 1CRN-I(16), reports a yield of 36.04%. This means it is more robust that the wild type 1CRN of 22.04%. The robustness of the whole structure is obvious by inspecting figure 2. It is possible to note that the energy histogram of the mutant, 1CRN-I16, is well centered around the native state E_0 and it is smoother than the wild type Crambin. Moreover, the yield optimization provides a structure with a lower potential energy variation. In Table 2, we report the yield values of the mutant 1CRN-I(16) (the reference structure with yield 36.04%) with the second most sensitive residue, the Arginine at the seventeenth position, changed. For the second time, it has been possible to improve the robustness of the protein by mutating the second most sensitive residue, the Arginine, with a glutamine. This mutation achieves a yield of 37.61%. Starting with a yield of 14.00% for the wild type Crambin, we obtained a yield of 36.04% by mutating the most sensitive residue, Cysteine (16), with Isoleucine. Then, by mutating the second most sensitive residue Arginine (17), with the glutamine, a mutant with high yield, similar structure ($RMSD_{C_\alpha} = 1.121\text{Å}$) and same functionality has been obtained. The designed algorithm, using a sort of learning cascade, was able to find out a Crambin-like protein that is $\sim 23.61\%$ more robust than the wild type. The robustness analysis on the two drugs (see Table 1) confirms that both are robust to all the three types of perturbations. The global yield varies from 73.30% for the 1GCN to 95.40% of the 1NPO, and higher yield values are reported both for local and residue yield. In particular, for the 1NPO, we obtained

Table 2. Yield analysis of the mutants of the Crambin, the Reference Structure (RS) in this experiment. For each new synthetic Crambin-like protein (crambin-mutant, or mutant) we report the mutated amino acid (aa), the type of mutation (Neutral (N), Disease Associated (D), Unfolded (U)), the potential energy value in the native state (E_0), and the residue robustness value.

		1CRN-I(16)					1CRN-Q(17)		
AA	Mut.	E_0 (kcal/mol)	Yield	$RMSD_{C_\alpha}$ (Å)	AA	Mut.	E_0 (kcal/mol)	Yield	$RMSD_{C_\alpha}$ (Å)
C	RS	−225.219	14.00	-	R	RS	−388.118	36.04	1.184
A	N	−365.964	25.58	1.015	A	D	-	-	-
D	N	−399.786	34.99	1.113	C	D	-	-	-
E	N	−369.268	24.52	0.646	D	N	−322.001	25.96	1.374
F	N	−388.038	30.34	1.337	E	D	-	-	-
G	N	−364.802	32.17	0.956	F	D	-	-	-
H	N	−394.265	30.42	0.911	G	N	−350.270	32.14	1.185
I	N	−388.118	**36.04**	1.184	H	U	-	-	-
K	N	−394.239	30.46	1.239	I	U	-	-	-
L	N	−391.632	30.18	1.143	K	N	−343.820	23.37	**1.111**
M	N	−388.424	32.11	1.282	L	U	-	-	-
N	D	-	-	-	M	D	-	-	-
P	D	-	-	-	N	D	-	-	-
Q	N	−392.754	30.79	0.606	P	D	-	-	-
R	N	−418.925	35.04	1.006	Q	N	−350.818	**37.61**	1.121
S	N	−361.138	23.84	0.624	S	U	-	-	-
T	D	-	-	-	T	D	-	-	-
V	N	−358.409	26.13	**0.611**	V	N	−314.261	23.37	1.370
W	N	−384.837	32.07	0.840	W	N	−344.361	30.08	1.150
Y	N	−397.363	27.49	1.196	Y	N	−283.419	27.95	1.215

Table 3. Mutation of the most sensitive residue (Glutamine, 4 residue) of 1NPO, the Reference Structure (RS) in this experiment. For each mutant we report the mutated amino acid (AA), the type of mutation (Neutral (N), Disease Associated (D)) the potential energy value in the native state, and the global yield value.

AA	Mut.	E_0 (kcal/mol)	Yield	$RMSD_{C_\alpha}$ (Å)	AA	Mut.	E_0 (kcal/mol)	Yield	$RMSD_{C_\alpha}$ (Å)
Q	RS	−65.783	95.40	-					
A	N	−44.661	96.01	**0.039**	L	N	−44.843	**98.03**	2.305
C	N	−45.590	97.04	2.102	M	N	−45.449	97.46	2.126
D	N	−67.121	96.10	0.146	N	N	−69.997	95.47	0.145
E	N	−62.733	97.61	2.124	P	N	−53.425	97.18	0.477
F	N	−48.942	96.56	0.143	R	N	−81.161	96.98	0.145
G	D	-	-	-	T	N	−53.020	95.70	0.034
H	N	−53.560	97.49	2.080	V	N	−44.980	97.17	2.170
I	N	−49.900	97.19	2.105	W	N	−52.621	94.57	0.369
K	N	−52.106	97.94	2.085	Y	N	−56.144	97.05	0.148

a local yield of 99.76% and a residue yield of 98.06%. The application of our DFY algorithm to the 1NPO led to a new mutant through the replacement of the most sensitive amino acid, a glutamine at the fourth position, with a serine 3. Since the Oxicitin is a peptide, we do not limit our investigation to the residue analysis but we take into account also the global one. This choice was made because in our preliminary study it had turned out that global analysis is more meaningful for peptides. The mutant has a global yield value of ∼ 3% better than that of the wild type protein (Table 3). This result proves that the DFY algorithm is also able to improve robust proteins.

5   Conclusions

In the present research work, we introduced a computational framework and a dimensionless quantity, the yield, for studying and improving protein robustness. The extensive studies on a set of proteins and drugs show some well-defined properties. Proteins are robust to local mutations, but become more sensitive to residue-level or global mutations. In particular for global mutations, only small peptides and proteins with strong secondary structure, like α-helix, maintain a good robustness. The robustness principle was the starting point for the DFY algorithm. DFY systematically mutates the most sensitive residue of a protein in order to discover new mutants with the same functionality and an improved robustness. It is important to remark that the suggested methodology is absolutely general and transparent to the problem domain. The claimed universality is due to our definitions of system, robustness and yield. We made no assumption on the nature of the system, the properties or the features to be analyzed. The robustness design principle could be applied to any kind of system definable in mathematical terms, from biological and electronic circuits to algorithms. The general applicability of this approach opens new frontiers towards in-silico automatic design of molecular, synthetic and information processing systems.

Supplementary information. Supplementary data are available at: http://www.dmi.unict.it/~stracquadanio/protein-robustness.html

References

1. Milor, L., Sangiovanni-Vincentelli, A.L.: Minimizing production test time to detect faults in analog circuits. IEEE Transactions on Computer-Aided Design of Integrated Circuits and Systems 13(6), 796–813 (1994)
2. Graeb, H.E.: Analog Design Centering and Sizing. Springer, Heidelberg (2007)
3. Jen, E.: Robust Design: A Repertoire of Biological, Ecological, and Engineering Case Studies. Oxford University Press, USA (2005)
4. Krakauer, D.C.: Robustness in Biological Systems-A Provisional Taxonomy. In: Complex Systems Science in Biomedicine, pp. 183–205. Springer, US (2006)
5. Tartaglia, G.G., Cavalli, A., Vendruscolo, M.: Prediction of Local Structural Stabilities of Proteins from Their Amino Acid Sequences. Structure 15(2), 139–143 (2007)
6. Wagner, A.: Distributed robustness versus redundancy as causes of mutational robustness. BioEssays 27(2), 176–188 (2005)
7. Wagner, A.: Robustness and evolvability in living systems. Princeton University Press, Princeton (2007)
8. Wagner, A.: Neutralism and selectionism: A network-based reconciliation. Nature Reviews Genetics 9, 965–974 (2008)
9. Wagner, A.: Robustness and evolvability: A paradox resolved. Proc. Roy. Soc. London Series. B 275, 91–100 (2008)
10. Kirschner, M., Gerhart, J.: Evolvability. Proceedings of the National Academy of Sciences 95(15), 8420–8427 (1998)

11. Cutello, V., Narzisi, G., Nicosia, G.: A multi-objective evolutionary approach to the protein structure prediction problem. J. of the Royal Society Interface 3(6), 139–151 (2006)
12. Floudas, C.A.: Computational methods in protein structure prediction. Biotechnol. Bioeng. 97(2), 207–213 (2007)
13. Klepeis, J.L., Floudas, C.A.: ASTRO-FOLD: A Combinatorial and Global Optimization Framework for Ab Initio Prediction of Three-Dimensional Structures of Proteins from the Amino Acid Sequence. Biophysical J. 85(4), 2119–2146 (2003)
14. Klepeis, J.L., Pieja, M.J., Floudas, C.A.: Hybrid Global Optimization Algorithms for Protein Structure Prediction: Alternating Hybrids. Biophysical J. 84(2), 869–882 (2003)
15. Nicosia, G., Stracquadanio, G.: Generalized pattern search algorithm for peptide structure prediction. Biophysical J. 95(10), 4988–4999 (2008)
16. Floudas, C.A., Fung, H.K., McAllister, S.R., Mönnigmann, M., Rajgaria, R.: Advances in protein structure prediction and de novo protein design: A review. Chemical Engineering Science 61(3), 966–988 (2006)
17. Ooi, T., Oobatake, M., Nemethy, G., Scheraga, H.A.: Accessible surface areas as a measure of the thermodynamic parameters of hydration of peptides. Proc. Natl. Acad. Sci. USA 84(10), 3086–3090 (1987)
18. Eisenmenger, F., Hansmann, U.H.E., Hayryan, S., Hu, C.K.: [SMMP] A modern package for simulation of proteins. Computer Physics Communications 138(2), 192–212 (2001)
19. Bemporad, F., Gsponer, J., Hopearuoho, H.I., Plakoutsi, G., Stati, G., Stefani, M., Taddei, N., Vendruscolo, M., Chiti, F.: Biological function in a non-native partially folded state of a protein. EMBO J. 27(10), 1525 (2008)
20. Worth, C.L., Bickerton, G.R., Schreyer, A., Forman, J.R., Cheng, T.M., Lee, S., Gong, S., Burke, D.F., Blundell, T.L.: A structural bioinformatics approach to the analysis of nonsynonymous single nucleotide polymorphisms (nsSNPs) and their relation to disease. J. Bioinform. Comput. Biol. 5(6), 1297–1318 (2007)

Multi-level Algorithms
for Modularity Clustering

Andreas Noack and Randolf Rotta

Brandenburg University of Technology, 03013 Cottbus, Germany
{an,rrotta}@informatik.tu-cottbus.de

Abstract. Modularity is a widely used quality measure for graph clusterings. Its exact maximization is prohibitively expensive for large graphs. Popular heuristics progressively merge clusters starting from singletons (coarsening), and optionally improve the resulting clustering by moving vertices between clusters (refinement). This paper experimentally compares existing and new heuristics of this type with respect to their effectiveness (achieved modularity) and runtime. For coarsening, it turns out that the most widely used criterion for merging clusters (modularity increase) is outperformed by other simple criteria, and that a recent multi-step algorithm is no improvement over simple single-step coarsening for these criteria. For refinement, a new multi-level algorithm produces significantly better clusterings than conventional single-level algorithms. A comparison with published benchmark results and algorithm implementations shows that combinations of coarsening and multi-level refinement are competitive with the best algorithms in the literature.

1 Introduction

A *graph clustering* partitions the vertex set of a graph into disjoint subsets called *clusters*. *Modularity* was introduced by Newman and Girvan as formalization of the common requirement that the connections within graph clusters should be dense, and the connections between different graph clusters should be sparse [1]. It is by far not the only quality measure for graph clusterings [2,3], but one of the most widely used measures, and has been successfully applied for detecting meaningful groups in a wide variety of real-world systems.

The problem of finding a clustering with maximum modularity for a given graph is NP-hard [4], and even recent exact algorithms scale only to graphs with a few hundred vertices [4,5,6]. In practice, modularity is almost exclusively optimized with heuristic algorithms.

Simple yet reasonably effective are *coarsening heuristics*, which iteratively merge cluster pairs starting from singleton clusters [7,8]. Various strategies for selecting the merged cluster pairs were developed [9,10,11], but the proposals have not been coherently organized or combined, and the published evaluation results are largely incomparable due to the use of different (and often small) graph collections. Therefore, Sect. 3 systematically describes major design alternatives for coarsening algorithms, including two new prioritizing criteria for merges, and Sect. 5 compares them experimentally.

J. Vahrenhold (Ed.): SEA 2009, LNCS 5526, pp. 257–268, 2009.
© Springer-Verlag Berlin Heidelberg 2009

The clusterings produced by coarsening are often improved with *refinement algorithms*, which iteratively reassign vertices to different clusters [12,11,13]. While *single-level* refinement only moves individual vertices, *multi-level* refinement reassigns entire clusters from several levels of the coarsening hierarchy. This proved to be very effective for minimum cut partitioning problems [14,15], but has not previously been adapted to modularity clustering. Section 4 introduces single-level and multi-level refinement heuristics, and Sect. 5 compares them experimentally.

To demonstrate that simple combinations of coarsening and refinement are among the most effective and efficient heuristics for modularity clustering, Sect. 6 provides a comparison with published benchmark results and implementations of various algorithms from the literature.

2   Graph Clusterings and Modularity

Graphs and Clusterings. A *graph* (V, f) consist of a finite set V of *vertices* and a function $f : V \times V \to \mathbb{N}$ that assigns an *edge weight* to each vertex pair. For simplicity, graphs are assumed to be undirected, i.e., $f(u, v) = f(v, u)$ for all $u, v \in V$. The *degree* $\deg(v)$ of a vertex v is the total weight $\sum_{u \in V} f(u, v)$ of its edges. The degrees and weights are naturally generalized to sets of vertices, e.g., $f(V, V) = \sum_{u \in V, v \in V} f(u, v)$. Note that $\deg(V) = f(V, V)$, and generally $\deg(U) \geq f(U, U)$ for all $U \subseteq V$. A *graph clustering* $\mathcal{C} = \{C_1, \ldots, C_k\}$ partitions the vertex set V into disjoint non-empty subsets C_i.

Modularity. Modularity is a widely used quality measure for graph clusterings. It was defined by Newman and Girvan [1,16] as

$$Q(\mathcal{C}) := \sum_{C \in \mathcal{C}} \left(\frac{f(C, C)}{f(V, V)} - \frac{\deg(C)^2}{\deg(V)^2} \right).$$

Intuitively, the first term is the *actual* fraction of intra-cluster edge weight. The second term specifies the *expected* fraction of intra-cluster edge weight in a null model where the end-vertices of $\frac{1}{2} \deg(V)$ edges are chosen at random, and the probability that an end-vertex of an edge attaches to a particular vertex v is $\frac{\deg(v)}{\deg(V)}$ [17]. In this null model, the edge weight $f(u, v)$ between each vertex pair $(u, v) \in V^2$ is binomially distributed with the expected value $\frac{\deg(u) \deg(v)}{\deg(V)}$.

Merging two clusters C and D increases the modularity by

$$\Delta Q_{C,D} := \frac{2f(C, D)}{f(V, V)} - \frac{2 \deg(C) \deg(D)}{\deg(V)^2},$$

and moving a vertex v from its current cluster C to another cluster D increases the modularity by

$$\Delta Q_{v \to D} := \frac{2f(v, D) - 2f(v, C-v)}{f(V, V)} - \frac{2 \deg(v) \deg(D) - 2 \deg(v) \deg(C-v)}{\deg(V)^2}.$$

3   Coarsening Algorithms

Coarsening algorithms compute clusterings by iteratively merging either one cluster pair or several disjoint cluster pairs, as detailed in the first paragraphs, and choose the merged cluster pairs according to certain priority criteria, which are discussed in the final paragraph. Implementation details and runtime bounds of the algorithms can be found in the extended version of this paper [18].

Single-Step. Single-Step coarsening starts with single-vertex clusters, and iteratively merges the cluster pair with the highest priority, until this merge would decrease the modularity.

Multi-Step. To prevent extremely unbalanced cluster growth, Schuetz and Caflisch introduced Multi-Step coarsening, which iteratively merges the l *disjoint* cluster pairs with the highest priority (unless the merge decreases the modularity) [11]. Single-Step coarsening corresponds to the special case of $l = 1$ (at least conceptually, the implementation differs). To make the parameter l independent of the graph size, we specify it as percentage of the number of modularity-increasing cluster pairs, and call it *merge fraction*.[1]

Merge Prioritizers. A merge prioritizer assigns to each cluster pair (C, D) a real number called *merge priority*, and thereby determines the order in which the coarsening algorithms merge cluster pairs. Because the coarsening algorithms use only the order of the priorities, two prioritizers can be considered as *equivalent* if one can be transformed into the other by adding a constant or multiplying with a positive constant.

The *Modularity Increase (MI)* $\Delta Q_{C,D}$ resulting from the merge of the clusters C and D is an obvious and widely used merge prioritizer [7,8,11,13].

The *Weight Density (WD)* is defined as $\frac{f(C,D)}{\deg(C)\deg(D)}$, and is equivalent (in the above sense) to $\frac{\Delta Q_{C,D}}{\deg(C)\deg(D)}$. Unlike the classic notion of density $\frac{f(C,D)}{|C||D|}$, it normalizes the actual edge weight $f(C, D)$ not with the potential unit edge weight but with the expected edge weight (up to a constant factor) in the null model (see Sect. 2). The weight density has not previously been used as merge prioritizer, which is surprising given that the modularity measure was originally introduced to formalize the requirement of intra-cluster density and inter-cluster sparsity [1], and its optimal clusterings indeed fulfill this requirement [20,21].

The *Significance (Sig)*, another new prioritizer, is defined as $\frac{\Delta Q_{C,D}}{\sqrt{\deg(C)\deg(D)}}$, and is thus a natural compromise between Modularity Increase and Weight Density. A further motivation is its relation to the (im)probability of the edge weight $f(C, D)$ in the null model described in Sect. 2. Under this null model, both the expected value and the variance (at least for large $\deg(V)$) of the edge

---

[1] Recently, Schuetz and Caflisch provided the empirical formula $l_{opt} := \alpha\sqrt{f(V,V)}$ for good values of l [19]. It does not outperform our formula for unweighted graphs (see Sect. 6), and is unsuitable for weighted graphs, because scaling all edge weights with a positive constant changes l_{opt} but not the optimal clustering.

weight between C and D are $\frac{\deg(C)\deg(D)}{\deg(V)}$, and the Significance is equivalent to the number of standard deviations that separate the actual edge weight from the expected edge weight.

Danon et al. (DA) observed that the Modularity Increase $\Delta Q_{C,D}$ tends to prioritize pairs of clusters with large degrees, and proposed the prioritizer $\frac{\Delta Q_{C,D}}{\min(\deg(C),\deg(D))}$ to avoid this bias [9]. It equals Significance if $\deg(C) = \deg(D)$, and is another compromise between Modularity Increase and Weight Density.

Wakita and Tsurumi found that coarsening by Modularity Increase tends to merge clusters of extremely uneven sizes [10]. To suppress unbalanced merges, they proposed the merge prioritizer $\min\left(\frac{\text{size}(C)}{\text{size}(D)}, \frac{\text{size}(D)}{\text{size}(C)}\right)\Delta Q_{C,D}$, where size$(C)$ is either the number of vertices in C (prioritizer *HN*) or the number of other clusters to which C is connected by an edge of positive weight (prioritizer *HE*).

4   Refinement Algorithms

Refinement algorithms perform a local search by iteratively moving individual vertices to different clusters (including newly created clusters) such that the modularity increases. This section describes three simple variants of greedy refinement, and proposes, for the first time in modularity clustering, to apply refinement on more than one level of the coarsening hierarchy. Excluded from consideration are algorithms with several tunable parameters or explicit randomness, like simulated annealing [20,22,23] or extremal optimization [24].

Complete Greedy. Complete Greedy refinement repeatedly performs the best vertex move, until no further modularity-increasing vertex moves are possible. Here the *best* vertex move is a move with the largest modularity increase $\Delta Q_{v\rightarrow D}$ over all vertices v and all target clusters D.

Fast Greedy. Fast Greedy refinement repeatedly iterates through all vertices and moves each vertex to its best cluster, until no improvement is found for any vertex. It has been previously proposed by Schuetz and Caflisch [11] and Ye et al. [13].

Adapted Kernighan-Lin. Kernighan-Lin refinement extends Complete Greedy refinement with a basic capability to escape local maxima. The algorithm was originally proposed by Kernighan and Lin for minimum cut partitioning [25], and was adapted to modularity clustering by Newman [12] (though with a limitation to two clusters). In its inner loop, the algorithm iteratively performs the best vertex move, with the restriction that each vertex is moved only once, but without the restriction that each move must increase the modularity. After all vertices have been moved, the inner loop is restarted from the best found clustering. Preliminary experiments indicated that it is much more efficient and rarely less effective to abort the inner loop when the best found clustering has not improved in the last $k := 10\log_2 |V|$ vertex moves [26].

Multi-Level Refinement. The refinement algorithms in the previous paragraphs are unlikely to move an entire group of densely interconnected vertices to another cluster, because this would require a series of sharply modularity-decreasing vertex moves. However, such a vertex group may well have been merged into a single cluster at some stage of the coarsening, and thus can be easily reassigned by moving entire clusters of this coarsening level, instead of individual vertices. This is the basic idea of multi-level refinement, which has already proved to be very effective for minimum cut partitioning problems [14,15].

As a prerequisite for Multi-Level refinement, intermediate results of the coarsening algorithm are stored as *coarsening levels*. Starting with the initial singleton clusters, a clustering is recorded as coarsening level whenever the number of clusters has decreased by a certain percentage since the previous level; this percentage is provided as a parameter called *reduction factor*. Each coarsening level is considered as a graph where each cluster at the respective state of coarsening is collapsed into a single vertex. The refinement algorithm (e.g., any algorithm from the previous paragraphs) is applied successively to every coarsening level, from the coarsest level to the original graph. The conventional Single-Level refinement, which executes a refinement algorithm only on the original graph, is the special case of Multi-Level refinement with a reduction factor of 100%.

Several recent algorithms for modularity clustering are related to Multi-Level refinement, but differ in crucial respects. Djidjev's method is not itself a multi-level algorithm, but a divisive method built on an existing multi-level algorithm for minimum cut partitioning [27]. Blondel et al. use local search on multiple levels to coarsen graphs, but do not refine the results of the coarsening [28]. Ye et al.'s algorithm performs local search on multiple coarsening levels, but only moves vertices of the original graph instead of coarse vertices (clusters) [13].

5   Experiments

This section experimentally compares the effectiveness (achieved modularity) and efficiency (runtime) of the heuristics in the previous sections.

Setup. The heuristics were implemented in C++ and compiled with GCC 4.2.3. The source code is available at www.informatik.tu-cottbus.de/~rrotta/.

To compare the effectiveness of the heuristics, the mean modularity over a fixed set of graphs is measured; higher means indicate more effective algorithms. (Thus only the relative values of the means are interpreted, the absolute values are not intended to be meaningful.) Generated graphs are not used because they do not necessarily permit generalizations to graphs from real applications. Instead the graph set contains 58 real-world graphs from various sources listed in the extended version [18]. The available graphs were classified by their application domain, and graphs of diverse size that fairly represent all major domains were selected, with a preference for common benchmarks like Zachary's karate club network [29]. The graphs range from a few to 75k vertices and 352k edges.

All runtimes were measured on a 3.00GHz Intel Pentium 4 processor with 1GB main memory, and exclude the time required for reading the graph.

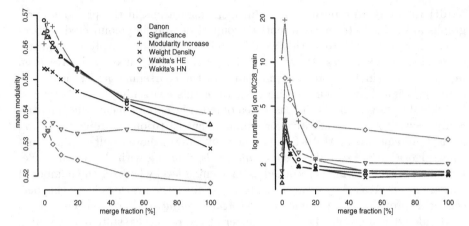

Fig. 1. Mean modularity on all graphs (left) and runtime on the graph 'DIC28_main' (right) by merge fraction and merge prioritizer

Coarsening Algorithms. Figure 1 compares the merge prioritizers for Single-Step coarsening (represented by a merge fraction of 0%) and Multi-Step coarsening with merge fractions of 2%, 5%, 10%, 20%, 50%, and 100%. No refinement was used. The relative runtimes shown for the graph 'DIC28_main' are typical for larger graphs.

Concerning the merge prioritizers, Wakita's HE and HN are much less effective than the others, and not more (usually even less) efficient.

Concerning the algorithms, Multi-Step coarsening is generally less effective and less efficient than the simpler Single-Step coarsening. Only for the Modularity Increase prioritizer, Multi-Step coarsening with merge fractions of 2% and 5% is slightly more effective, but still similar to Single-Step coarsening with Danon and Significance. Apparently the other merge prioritizers do not benefit from Multi-Step's tendency to balance cluster sizes because, unlike Modularity Increase, they have no strong bias towards merging large clusters.

Refinement Algorithms. Figure 2 compares the refinement algorithms for Single-Level refinement (reduction factor 100%) and Multi-Level refinement with reduction factors of 5%, 10%, 20%, and 50%. As coarsener the Single-Step algorithm with the Significance prioritizer was chosen, because it proved to be effective and efficient in the previous paragraph.

Multi-Level refinement with a reduction factor of 50% turns out to be more effective than Single-Level refinement, and similarly efficient. Reduction factors below 50% do not considerably improve the modularity, but significantly increase the runtime for Fast Greedy.

Fast Greedy refinement is about as effective as Complete Greedy, and just slightly less effective than Kernighan-Lin, but much faster.

Combining Coarsening and Refinement. Concerning Single-Level vs. Multi-Level refinement, Fig. 3 shows that Multi-Level refinement is consistently more effective for all merge prioritizers.

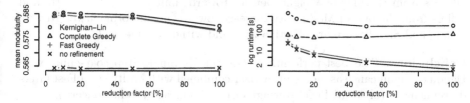

Fig. 2. Mean modularity on all graphs (left) and runtime on the graph 'DIC28_main' (right) by reduction factor and refinement method

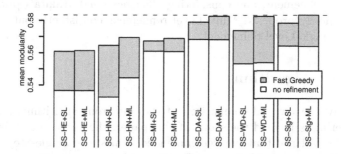

Fig. 3. Mean modularity by merge prioritizer. Left bars for reduction factor 100% (Single-Level), right bars 50% (Multi-Level). Both use Single-Step coarsening.

Concerning Single-Step vs. Multi-Step coarsening, Fig. 4 shows that for the best merge prioritizers, both are similarly effective with Multi-Level refinement, while Single-Step coarsening is more effective when excluding refinement (as detailed in Fig. 1). Clearly, Multi-Level refinement benefits from the uniform cluster growth enforced by Multi-Step coarsening. Overall, Single-Step coarsening is still preferable because of its greater simplicity and efficiency.

Concerning the merge prioritizers, Figs. 3 and 4 show that Modularity Increase is only competitive without refinement (ignoring efficiency), and Weight Density is only competitive with Multi-Level refinement. Here Multi-Level refinement

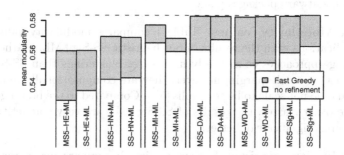

Fig. 4. Mean modularity by merge prioritizer. Left bars show merge fraction 5% (Multi-Step), right bars 0% (Single-Step). Both use reduction factor 50%.

benefits from the bias of Weight Density towards uniform cluster growth, and suffers from the bias of Modularity Increase towards nonuniform cluster growth. Danon and Significance are effective with and without refinement.

Conclusions. The best-performing algorithm is Single-Step coarsening with Danon or Significance as merge prioritizer combined with Multi-Level Fast Greedy refinement (or Multi-Level Kernighan-Lin, if efficiency is no concern).

With the best merge prioritizers, Single-Step coarsening outperformed the more complex Multi-Step coarsening. The Danon and Significance merge prioritizers clearly outperformed the much more widely used Modularity Increase (especially with refinement, and considering efficiency) and Wakita's prioritizers. Finally, the newly proposed Multi-Level refinement consistently outperformed the popular Single-Level refinement.

6   Related Algorithms

An exhaustive review and comparison of the numerous algorithms for modularity clustering is beyond the scope of this paper; the purpose of this section is to provide evidence that our recommended heuristic – Single-Step coarsening by Significance with Multi-Level Fast Greedy refinement (SS-Sig+ML) – is competitive with the best existing methods.

Basic Approaches. Algorithms for modularity clustering can be categorized into the following four types: *Subdivision* heuristics try to divide the graph, for example by iteratively removing edges [1] or by recursively splitting the graph using eigenvectors [12]. *Coarsening* (or agglomeration) heuristics iteratively merge clusters starting from singletons. Cluster pairs can be selected based on random walks [30,31], increase of modularity [8,11,13], or other criteria [10,9,32]. *Local search* heuristics move vertices between clusters, with Kernighan-Lin-style and greedy search being the most prominent examples. Other approaches include Tabu Search [33], Extremal Optimization [24], and Simulated Annealing [20,22,23]. Finally, *mathematical programming* approaches model modularity maximization as a linear or quadratic programming problem which can be solved with existing software packages [5,4,6].

Published Modularity Values. Table 1 compares modularity values from various publications with the results of our heuristic SS-Sig+ML. Mathematical programming approaches consistently find better clusterings than SS-Sig+ML, though by a very small margin; however, they are computationally much more expensive and do not scale to large graphs [5,6]. Compared to the best algorithms in the three other classes, the results of SS-Sig+ML are very competitive, and for large graphs significantly better.

Published Implementations. In order to directly compare our heuristics with existing algorithms, a range of publicly available implementations was retrieved from authors' websites and through the *igraph* library of Csárdi and Nepusz [44].

Table 1. Best published modularity values for four algorithm classes, compared to the modularity values for our heuristic SS-Sig+ML. Where possible, missing values were substituted with results from published implementations, which are shown in italics.

graph	size	subdivision	coarsening	local search	math prog	SS+ML
karate [29]	34	[12] .419	[13] .4198	[24] .4188	[5] .4197	.41978
dolphins [34]	62	[12] *.4893*	[30] *.5171*	[20] *.5285*	[6] .5285	.52760
polBooks [35]	105	[12] *.3992*	[19] *.5269*	[28] *.5204*	[5] .5272	.52693
afootball [36]	115	[37] .602	[13] .605	[28] *.6045*	[5] .6046	.60028
jazz [38]	198	[12] .442	[9] .4409	[24] .4452	[5] .445	.44467
celeg_metab [24]	453	[12] .435	[11] .450	[24] .4342	[5] .450	.44607
email [39]	1133	[12] .572	[9] .5569	[24] .5738	[5] .579	.57744
Erdos02 [40]	6927	[12] *.5969*	[31] .6817	[20] *.7094*		.71626
PGP_main [41]	11k	[12] .855	[9] .7462	[24] .8459		.88418
cmat03_main [42]	28k	[12] .723	[13] .761	[24] .6790		.81432
ND_edu [43]	325k		[8] .927	[28] .935		.95090

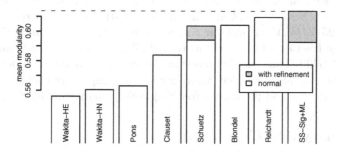

Fig. 5. Mean modularities from the published implementations and our recommended heuristic SS-Sig+ML on unweighted graphs

Because some of these implementations cannot process graphs with weighted edges, only 23 of the 58 graphs in the graph collection could be used in the experiments. In some of these graphs, negligible differences in edge weights and small amounts of self-edges were removed.

The included coarsening heuristics are the fast greedy joining of Clauset et al. [8], the algorithms of Wakita and Tsurumi [10], the recent multi-step greedy algorithm of Schuetz and Caflisch [11] (with parameter $l = 0.25\sqrt{f(V,V)/2}$, as recommended by Schuetz and Caflisch in [19]), and the algorithm of Pons and Latapy [30] based on short random walks (here of length 4, the default value). The examined local search heuristics are simulated annealing of Reichardt and Bornholdt [20] (here with at most 120 clusters) and the recent hierarchical algorithm of Blondel et al. [28].

The results are shown in Figs. 5 and 6. Compared to our recommended heuristic, Single-Step coarsening by Significance with Multi-Level Fast Greedy refinement (SS-Sig+ML), only Reichardt and Bornholdt's implementation produces clusterings of similarly high modularity, but it is much slower, and only Blondel et al.'s implementation is faster, but it produces worse clusterings.

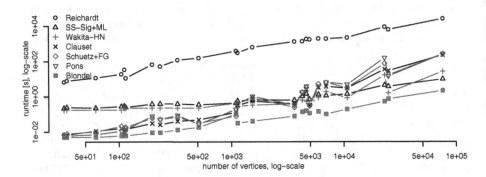

Fig. 6. Runtime of the published implementations on unweighted graphs, log-log scaled. The runtime of SS-Sig+ML includes the Fast Greedy refinement.

A minor problem raised in Sect. 3 is the choice of the parameter l for Multi-Step coarsening. Schuetz and Caflisch's implementation with their recommended choice $l = 0.25\sqrt{f(V,V)/2}$ and our corresponding implementation MS5-MI+SL with a merge fraction of 5% perform very similarly, with mean modularities of 0.6037 and 0.6069. Thus the difference in the formulas for the parameter l does not affect the conclusions about Multi-Step coarsening, in particular its inferiority to the simpler and parameter-free Single-Step coarsening by Significance.

7   Summary and Conclusion

Various coarsening and refinement heuristics for modularity clustering can be organized into a design space with four dimensions: merge fraction (including Single-Step and Multi-Step coarsening), merge prioritizer, refinement algorithm, and reduction factor (including Single-Level and Multi-Level refinement). In an experimental comparison of achieved modularities and required runtimes, some widely used or rather complex design choices – for example, Multi-Step coarsening, merge prioritization by Modularity Increase, or Single-Level refinement – were outperformed by newly proposed or simpler alternatives – particularly Single-Step coarsening by Significance with Multi-Level Fast Greedy refinement. In a comparison with published implementations and benchmark results, this heuristic required less runtime than algorithms that achieved similar modularities, and achieved higher modularities than algorithms with similar or better runtimes.

References

1. Newman, M.E.J., Girvan, M.: Finding and evaluating community structure in networks. Physical Review E 69, 026113 (2004)
2. Gaertler, M.: Clustering. In: Brandes, U., Erlebach, T. (eds.) Network Analysis: Methodological Foundations. LNCS, vol. 3418, pp. 178–215. Springer, Heidelberg (2005)

3. Schaeffer, S.E.: Graph clustering. Computer Science Review 1(1), 27–64 (2007)
4. Brandes, U., Delling, D., Gaertler, M., Görke, R., Hoefer, M., Nikoloski, Z., Wagner, D.: On modularity clustering. IEEE Transactions on Knowledge and Data Engineering 20(2), 172–188 (2008)
5. Agarwal, G., Kempe, D.: Modularity-maximizing graph communities via mathematical programming. The European Physical Journal B 66(3), 409–418 (2008)
6. Xu, G., Tsoka, S., Papageorgiou, L.G.: Finding community structures in complex networks using mixed integer optimisation. The European Physical Journal B 60(2), 231–239 (2007)
7. Newman, M.E.J.: Fast algorithm for detecting community structure in networks. Physical Review E 69, 066133 (2004)
8. Clauset, A., Newman, M.E.J., Moore, C.: Finding community structure in very large networks. Physical Review E 70, 066111 (2004)
9. Danon, L., Díaz-Guilera, A., Arenas, A.: Effect of size heterogeneity on community identification in complex networks. Journal of Statistical Mechanics: Theory and Experiment, P11010 (2006)
10. Wakita, K., Tsurumi, T.: Finding community structure in mega-scale social networks. Preprint arXiv:cs/0702048 (2007)
11. Schuetz, P., Caflisch, A.: Efficient modularity optimization by multistep greedy algorithm and vertex mover refinement. Physical Review E 77, 046112 (2008)
12. Newman, M.E.J.: Modularity and community structure in networks. Proceedings of the National Academy of Sciences 103(23), 8577–8582 (2006)
13. Ye, Z., Hu, S., Yu, J.: Adaptive clustering algorithm for community detection in complex networks. Physical Review E 78(4), 046115 (2008)
14. Hendrickson, B., Leland, R.W.: A multi-level algorithm for partitioning graphs. In: Proceedings of the ACM/IEEE Supercomputing Conference, SC 1995 (1995)
15. Karypis, G., Kumar, V.: A fast and high quality multilevel scheme for partitioning irregular graphs. SIAM Journal on Scientific Computing 20(1), 359–392 (1998)
16. Newman, M.E.J.: Analysis of weighted networks. Physical Review E 70, 056131 (2004)
17. Newman, M.E.J.: Finding community structure in networks using the eigenvectors of matrices. Physical Review E 74, 036104 (2006)
18. Noack, A., Rotta, R.: Multi-level algorithms for modularity clustering. Preprint arXiv:0812.4073 (2008)
19. Schuetz, P., Caflisch, A.: Multistep greedy algorithm identifies community structure in real-world and computer-generated networks. Physical Review E 78, 026112 (2008)
20. Reichardt, J., Bornholdt, S.: Statistical mechanics of community detection. Physical Review E 74, 016110 (2006)
21. Noack, A.: Modularity clustering is force-directed layout. Physical Review E 79, 026102 (2009)
22. Medus, A., Acuña, G., Dorso, C.O.: Detection of community structures in networks via global optimization. Physica A 358(2-4), 593–604 (2005)
23. Massen, C.P., Doye, J.P.K.: Identifying communities within energy landscapes. Physical Review E 71, 046101 (2005)
24. Duch, J., Arenas, A.: Community detection in complex networks using extremal optimization. Physical Review E 72, 027104 (2005)
25. Kernighan, B., Lin, S.: An efficient heuristic procedure for partitioning graphs. Bell System Technical Journal 49(2), 291–307 (1970)
26. Rotta, R.: A multi-level algorithm for modularity graph clustering. Master's thesis, Brandenburg University of Technology (2008)

27. Djidjev, H.N.: A scalable multilevel algorithm for graph clustering and community structure detection. In: Aiello, W., Broder, A., Janssen, J., Milios, E.E. (eds.) WAW 2006. LNCS, vol. 4936, pp. 117–128. Springer, Heidelberg (2008)

28. Blondel, V.D., Guillaume, J.L., Lambiotte, R., Lefebvre, E.: Fast unfolding of communities in large networks. Journal of Statistical Mechanics: Theory and Experiment, 10008 (2008)

29. Zachary, W.W.: An information flow model for conflict and fission in small groups. Journal of Anthropological Research 33, 452–473 (1977)

30. Pons, P., Latapy, M.: Computing communities in large networks using random walks. Journal of Graph Algorithms and Applications 10(2), 191–218 (2006)

31. Pujol, J.M., Béjar, J., Delgado, J.: Clustering algorithm for determining community structure in large networks. Physical Review E 74, 016107 (2006)

32. Donetti, L., Muñoz, M.A.: Detecting network communities: a new systematic and efficient algorithm. Journal of Statistical Mechanics: Theory and Experiment, 10012 (2004)

33. Arenas, A., Fernández, A., Gómez, S.: Analysis of the structure of complex networks at different resolution levels. New Journal of Physics 10, 053039 (2008)

34. Lusseau, D., Schneider, K., Boisseau, O.J., Haase, P., Slooten, E., Dawson, S.M.: The bottlenose dolphin community of Doubtful Sound features a large proportion of long-lasting associations. Behavioral Ecology and Sociobiology 54, 396–405 (2003)

35. Krebs, V.: A network of books about recent US politics sold by the online bookseller amazon.com (2008), http://www.orgnet.com/

36. Girvan, M., Newman, M.E.J.: Community structure in social and biological networks. Proceedings of the National Academy of Sciences 99(12), 7821–7826 (2002)

37. White, S., Smyth, P.: A spectral clustering approach to finding communities in graphs. In: Proceedings of the 5th SIAM International Conference on Data Mining (SDM 2005), pp. 274–285. SIAM, Philadelphia (2005)

38. Gleiser, P., Danon, L.: Community structure in jazz. Advances in Complex Systems 6(4), 565–573 (2003)

39. Guimerà, R., Danon, L., Díaz-Guilera, A., Giralt, F., Arenas, A.: Self-similar community structure in a network of human interactions. Physical Review E 68, 065103 (2003)

40. Grossman, J.: The Erdös number project (2007), http://www.oakland.edu/enp/

41. Boguñá, M., Pastor-Satorras, R., Díaz-Guilera, A., Arenas, A.: Models of social networks based on social distance attachment. Physical Review E 70, 056122 (2004)

42. Newman, M.E.J.: The structure of scientific collaboration networks. Proceedings of the National Academy of Sciences 98(2), 404–409 (2001)

43. Albert, R., Jeong, H., Barabási, A.L.: Diameter of the World-Wide Web. Nature 401(6749), 130–131 (1999)

44. Csárdi, G., Nepusz, T.: The igraph software package for complex network research. Inter. Journal Complex Systems 1695 (2006)

Bulk-Insertion Sort: Towards Composite Measures of Presortedness*

Riku Saikkonen and Eljas Soisalon-Soininen

Helsinki University of Technology, Finland
{rjs,ess}@cs.hut.fi

Abstract. Well-known measures of presortedness, among others, are
the number of inversions needed to sort the input sequence, or the minimal number of blocks of consecutive elements that remain as such in the
sorted sequence. In this paper we study the problem of possible composition of measures. For example, after determining the blocks in an
input sequence, it is meaningful to measure how many inversions of the
blocks are needed to finally sort the sequence. With composite measures
in mind we introduce the idea of applying bulk insertions to improve
adaptive binary-tree (AVL) sorting; this is done by combining local insertion sort with bulk-insertion methods. We show that bulk-insertion
sort is optimally adaptive with respect to the number of bulks and with
respect to the number of inversions in the original input. As to composite measures, we define a new measure that tells how many inversions
are needed when the extracted bulks form the input. Bulk-insertion sort
is shown to be adaptive with respect to this measure. Our experiments
show that applying bulk insertion in AVL-tree sorting considerably reduces the number of comparisons and time needed to sort nearly sorted
sequences.

1 Introduction

Adaptive sorting, or the sorting of nearly sorted sequences, is the problem of
sorting a sequence of values that is already "almost" in sorted order according
to some intuitive notion of sortedness [1, 2]. One of the main measures of the
presortedness is the number of pairs of elements in the input that are in wrong
order (number of inversions, Inv). The aim is to produce an algorithm that is
more efficient than a non-adaptive sorting algorithm when the input is nearly
sorted. A sorting algorithm is said to be optimally adaptive with respect to a
presortedness measure if it is within a constant factor of the lower bound. For
example, optimality with respect to Inv (Inv-optimality) is implied by the time
bound $O(n \log(Inv/n))$, where n denotes the number of keys in the input.

Many optimally adaptive sorting methods are based on using a search tree
[3,4,5,6,7,8] where inputs are inserted one by one. This paper introduces the idea
of applying bulk insertions [9,10] to improve adaptive binary-tree (AVL) sorting;

---

* This research was partially supported by the Academy of Finland.

this is done by combining local insertion sort with bulk-insertion methods. Once the location for inserting the next element of the input is decided in the AVL tree using a finger structure [11,12,13,3,5] a bulk of consecutive elements that fit in this insertion point is extracted from the input and then inserted using bulk insertion. We show that bulk-insertion sort is optimally adaptive with respect to the number of bulks and with respect to the number of inversions in the original input.

As to a composite measure, let $Bulk(X)$ denote the sequence of the last elements of the bulks in input X. (For a single measure, $Bulk$ is defined as the number of bulks.) The composite mapping $Inv \circ Bulk$ now measures the number of inversions needed to sort the bulks extracted from the input sequence. Our main result is that bulk-insertion sorting has worst-case time complexity $O(n + k \log \min\{k, 1 + Inv(Bulk(X))\})$, where X is a sequence of length n and k denotes the number of bulks. This means that bulk-insertion sorting indeed is adaptive with respect to $Inv \circ Bulk$, but we cannot prove optimality.

In the experimental work our aim was not only to demonstrate that using bulk insertion in binary-tree adaptive sorting is beneficial, but to show up to which level of disorder bulk operations perform better than individual insertions. Our results show that up to about 10^5 inversions in integer-key inputs of length about 10^7 single-insertion sorting was about six times slower than bulk-insertion sorting, and for larger numbers of inversions, when the average bulk contained only a few keys, single insertion was slightly faster. For string keys the speed difference for small amounts of inversions was smaller.

We also compared our AVL-tree sorting algorithms with several other adaptive and non-adaptive algorithms (see Section 5). Except for Insertion sort, our bulk-insertion-based algorithm was the best for up to 10^5 inversions, and up to 10^4 inversions the next best was 60% slower. Insertion sort with worst case time bound $O(n^2)$ is admittedly fast for small amounts of inversions, but it is intolerably slow for larger amounts of inversions. Our bulk-insertion sort is tolerably fast for all inputs, and is thus a reasonable choice for sorting when most of the inputs are nearly sorted, but inputs completely in disorder may exist.

2   Bulk-Insertion Sort

A sequence $X = \langle x_1, \ldots, x_n \rangle$ is divided into *bulks* as follows. (For simplicity, we assume, as usual, that all keys in X are pairwise different.)

(i) The first bulk b_1 is the longest ascending or descending prefix of X.
(ii) Let $X = YZ$ such that Y contains the keys of the first i bulks in sorted order. The $(i+1)$th bulk of X is the longest ascending or descending prefix Z' of Z such that all keys in Z' are between two consecutive keys in Y or they are all smaller (resp. larger) than the smallest (resp. largest) key in Y.

Figure 1 gives an example of how bulks are found.

Clearly, any balanced search-tree structure can be used in finding bulks efficiently and as a basis of bulk-insertion sort. We use a variation of the AVL-tree

4, 8, 9, 20, 22, 23, 7, 6, 5, 3, 2, 1, 19, 14, 13, 10, 15, 16, 17, 18, 21, 12, 11

Fig. 1. Example of bulks created in bulk-insertion sort. Bulk-insertion sort uses 7 bulk insertions to sort this sequence of 23 keys.

BULK-INSERTION-SORT($X[1..n]$):

1: $k \leftarrow 1$
2: $P \leftarrow$ empty binary saved path
3: $T \leftarrow$ empty AVL tree
4: **while** $k \leq n$ **do**
5:    Search bottom-up in the binary saved path P for the lowest position $P[s]$ with $key(P[s]) < X[k] < key(ip(P[s]))$ or $key(P[s]) > X[k] > key(ip(P[s]))$.
6:    Search for $X[k]$ in T starting from $P[s]$. Save this path to $P[s+1..d]$.
7:    Find the longest ascending or descending sequence of keys $X[k..l]$, where $key(P[d]) < X[i] < key(ip(P[d]))$ or $key(P[d]) > X[i] > key(ip(P[d]))$ for all $i \in [k..l]$.
8:    Bulk insert $X[k..l]$ in T at the position pointed by $P[d]$.
9:    $k \leftarrow l + 1$
10: **end while**
11: **return** T

Fig. 2. The bulk-insertion sorting algorithm

bulk-insertion algorithm given in [9]. We only give an overview of the algorithm here; details can be found in [9]. We use an internal, height-valued AVL tree (i.e., each node stores the height of the subtree below it, as opposed to the balance factor present in a more typical AVL tree), since the bulk-insertion algorithm is much simpler to implement using height values.

An outline of our *bulk-insertion* AVL-*tree sorting* algorithm is given in Figure 2. The algorithm maintains the current search path, called the *binary saved path* P, with each node p storing its indirect parent. By the *indirect parent* $ip(p)$ of p we mean the lowest node above p, whose key is larger (resp. smaller) than the key $key(p)$ of p, if the key of the parent of p is smaller (resp. larger) than $key(p)$. Thus node p covers key $X[k]$, i.e., the place of $X[k]$ is found in the subtree with root p, if $key(p) < X[k] < key(ip(p))$ (p is the left child of its parent) or $key(p) > X[k] > key(ip(p))$ (p is the right child of its parent).

To reduce the number of comparisons, key $X[k]$ is compared with the indirect parent first, because if this comparison fails, we can skip the saved path entries between the direct and indirect parent. None of these entries can cover $X[k]$ because of the failed comparison with the indirect parent.

We have:

Theorem 1. *Using the binary saved path P, the lowest entry in P that covers key $X[k]$ can be found using at most s comparisons, where s is the number of entries in the path.*

In finding a bulk of m new keys (line 7 of the algorithm of Figure 2) we must check that the keys are ascending (resp. descending) and that they are smaller (resp. larger) than the current maximum (resp. minimum). To ensure that the next key is ascending (resp. descending) one comparison per key is required, but the check with the maximum (resp. minimum) can be performed by exponential and binary search [5]. Here, however, we must take care that ascending (resp. descending) checking will not be done more than once for any pair of consecutive keys.

We have:

Theorem 2. *Assume that the sequence* $X = \langle x_1, x_2, \ldots, x_n \rangle$, $n \geq 1$, *is divided into* k *bulks of sizes* $m_1, m_2, \ldots, m_k$. *The number of comparisons required for finding the bulks (without finger searches) is* $n + 2 \sum_{i=1}^{k} \log_2 m_i$.

To insert a bulk of m new keys into a given position in an AVL tree T, bulk insertion first forms a new AVL tree S (called an *update tree*) from the new keys, and places its root in the given position. A balanced update tree is created by placing the middle ($m/2$th) key in the root of S and proceeding recursively. Finally, the tree is rebalanced on the path P from the root of S to the root of the original tree T. This is done using rotations that first move S upwards in T with $O(\log m)$ steps, until the height of the root S is close to the height of its neighbors. Some final rotations are needed close to the root of S to bring the tree in balance.

The following theorem is implied by Theorem 4 of [9].

Theorem 3. *Assume that bulk insertions are performed into an initially empty* AVL *tree. The amortized rebalancing complexity of a bulk insertion with bulk size* m *is* $O(\log m)$.

Theorem 3 states the amortized time bound for rebalancing after bulk insertion; creating the bulk with m nodes still requires $O(m)$ time. However, in the case where bulk-insertion sort produces mainly large bulks, the sorting algorithm does not actually visit most of the nodes in each bulk, so creating all of the nodes is not necessary.

We use this observation to reduce the running time of bulk-insertion sort as follows, assuming that keys $X[1..n]$ are to be sorted. Instead of creating an update tree from keys $X[i]$ to $X[i+m-1]$, we create a single special placeholder node that contains the values i and m instead of the normal key and child pointers.[1] We place this node, called a *lazy node* and the bulk pointed by a *lazy bulk*, in place of the root of the update tree. Later, whenever a lazy node p is reached by rebalancing or by searching for the position of a new bulk going inside the lazy bulk, p is expanded, in constant time, to a normal node (with key $X[i + \lfloor m/2 \rfloor]$) with two new lazy nodes as its children (one with keys $X[i]$ to $X[i + \lfloor m/2 \rfloor - 1]$, and the other with keys $X[i + \lfloor m/2 \rfloor + 1]$ to $X[i + m - 1]$).

---

[1] Our implementation uses a special, otherwise unused, address for the left child pointer to mark such a placeholder node. The right child pointer and key are used to store m and a pointer to $X[i]$, as well as a flag that notes whether the sequence is ascending or descending.

3   Adaptivity and Complexity

By *single-insertion (AVL-tree) sort* we refer to local insertion sort using AVL trees, when single keys instead of bulks are inserted using the binary saved path strategy. The binary saved path is not a full-fledged finger: it does not use worst-case $O(\log d)$ time to move a distance d in the tree. It is thus not optimal with respect to *logDist* or *Loc* [14, 2], which are two equivalent measures of sortedness used to characterize local insertion sort using a finger tree. However, the binary saved path is optimal with respect to a simpler measure, the number of inversions:

Theorem 4. *Single-insertion* AVL-*tree sort implemented using the binary saved path takes time $O(n \log(1 + Inv/n))$, which is optimal with respect to Inv (the number of inversions).*

Proof. It is known that an approach that keeps the finger always pointing to the largest key in the tree is *Inv*-optimal (see [6]). Consider an insertion to a position x elements away from the largest key l, and assume that the previous insertion was one at a position p elements away from the largest key. Moving the binary saved path from p to x costs at most as much as moving a finger from p to l and from l to x (the time complexity is $O(\log p + \log x)$). Therefore, the binary-saved-path approach implies M-optimality for any measure M for which the "finger at the largest key" approach is M-optimal.   □

We note that the other AVL-tree-based sorting algorithms [4, 6, 7] are also *Inv*-optimal but not *logDist*- or *Loc*-optimal.

Next consider the bulk-insertion sorting algorithm described in the previous section. Theorem 4 immediately implies (because bulk-insertion sorting does at most the same amount of work as single-insertion sorting):

Theorem 5. *The bulk-insertion* AVL-*tree sorting algorithm implemented by the binary saved path has worst-case time complexity $O(n \log(1 + Inv/n))$, where n is the length of the input X and Inv the number of inversions in X. Thus bulk-insertion sorting is Inv-optimal.*

However, although $O(n \log(1 + Inv/n))$ is the worst-case complexity reached for $n = k$ where k denotes the number of bulks, it would be more interesting to analyze the complexity in terms of k.

Consider the adaptivity measure *Block* [2, 15], which is defined as the number of blocks of consecutive keys in the original input that are present as such in the sorted sequence. In other words, *Block* is the number of keys in an input that receive a new successor in sorted order. For instance, the sequence of Figure 1 consists of the blocks $\{8, 9\}, \{22, 23\}, \{15, 16, 17, 18\}$, and all other elements as one-element blocks, for a total of 18 blocks.

Let $k = Block(X)$ for input X of n keys. Carlsson et al. [15] show that a sorting algorithm is optimal with respect to *Block* if it has worst-case time complexity $O(n + k \log k)$. (See [1, 2, 3] for definitions of an adaptivity measure and its lower bound. In brief, a measure of presortedness $m(X)$ maps a sequence X to a non-negative integer such that $m(X) < m(Y)$ if X is "closer" to being sorted than Y [2].)

We define a new measure called *Bulk* as the number of bulks in input X of n keys. The measure *Bulk* is a natural generalization of *Block*, which in turn generalizes two well-known measures, namely *Rem*, the number of elements that have to be removed from a sequence in order to leave a sorted sequence, and *Exc*, the minimum number of arbitrary exchanges needed to bring a sequence into sorted order [3, 15]. Clearly, $Bulk(X) \leq Block(X)$ for any input X of length n, since each bulk consists of one or more blocks.

Lemma 1. *Let $k = Bulk(X)$ for sequence X of length n. The lower bound for measure Bulk is*

$$\Omega(n + k \log k).$$

Proof. Carlsson et al. [15] show that if, for some measures M_1 and M_2, $M_1(X) \leq M_2(X)$ for all X, then the lower bound for M_2 is also a lower bound for M_1. Thus, as the lower bound for *Block* is $\Omega(n + k \log k)$ [15], we conclude the lemma. □

Lemma 2. *Let $k = Bulk(X)$ for sequence X of length n. The time complexity of bulk-insertion* AVL-*tree sorting is $O(n + k \log(1 + Inv/n))$.*

Proof. In the bulk-insertion algorithm the next key to be inserted is searched for using the binary saved path only once for each bulk. Thus the time consumed by the finger searches is altogether $O(k \log(1 + Inv/n))$. By Theorems 2 and 3 all other tasks of the algorithm altogether take time $O(n)$. Thus we conclude the lemma. □

Theorem 6. *Bulk-insertion* AVL-*tree sorting is optimal with respect to the measure Bulk.*

Proof. By Lemma 1 it is enough to show that the worst-case time complexity of bulk-insertion sorting is $O(n + k \log k)$, where $k = Bulk(X)$. By Lemma 2 we have the bound $O(n + k \log(1 + Inv/n))$.

Since $1 \leq k \leq n$, it follows that $n = ck$, where $1 \leq c \leq n$. Then

$$k \log(1 + Inv/n) \leq \frac{n}{c} \log(1 + ck)$$
$$\leq \frac{n}{c} \log(2ck) = \frac{n}{c} \log 2c + \frac{n}{c} \log k$$
$$\leq 2n + k \log k.$$

Thus $O(n + k \log(1 + Inv/n)) = O(n + k \log k)$, and we conclude the theorem. □

4   Composite Measures

Above we have considered single measures of adaptivity such as *Inv*, *Block* and *Bulk*. However, it would be tempting to compose these measures suitably; that is, why couldn't we first apply the measure *Bulk* or *Block* and then the measure *Inv*? It would be natural to think that input X is first divided into bulks or blocks, and then measure how many inversions bulks or blocks require (in terms of their last elements, for example) before yielding a sorted sequence. Such a composite measure *Inv∘Block* or *Inv∘Bulk* would be natural, because applying only *Block*, for instance, means that any optimal algorithm needs to sort the blocks in $O(k \log k)$ time, where k is the number of blocks. The time bound $O(k \log k)$ means that there is no adaptivity present in terms of the order of the blocks.

When defining composite measures *Inv ∘ Block* and *Inv ∘ Bulk* we need to change the mappings *Block* and *Bulk* slightly. Instead of mapping an integer sequence to the number of elements in the sequence, *Block* (resp. *Bulk*) now maps a sequence X of n elements to a sequence

$$x_{i_1} x_{i_2} \ldots x_{i_k} \, ,$$

where each x_{i_j} is the last element of the jth block (resp. bulk) in X. In this way *Inv∘Block* and *Inv∘Bulk* are correctly defined as the number of inversions needed for blocks or bulks in yielding a sorted sequence.

The lower bound for sorting sequences Y containing the last elements of bulks when measuring by *Inv* cannot be less than

$$\Omega(n + k \log(1 + Inv(Y)/k)) \, ,$$

where k denotes the number of bulks (or elements in Y) – this is because of the lower bound for *Inv*-optimality. The measure *Inv ∘ Bulk* thus has the lower bound $\Omega(n + k \log(1 + Inv(Bulk(X))/k))$.

We can prove the following upper bound for bulk-insertion sorting.

Theorem 7. *For sequence X of length n, let k denote the length of $Bulk(X)$ when defined as a sequence. Then the time complexity of bulk-insertion sort is*

$$O(n + k \log(1 + Inv(Bulk(X)))) \, .$$

Proof. By Lemma 2 and its proof we conclude that the work with more cost than $O(n)$ comes from finger searches. Moreover, we see that finger searches are required only k times. For proving the theorem we estimate the work done by these searches.

By Theorem 4 and its proof we conclude that it is enough to estimate finger searches when the finger is kept pointing to the largest key in the underlying AVL tree.

Denote by d_j, $j = 1, \ldots, k$, the distance of the first key x of the jth bulk from the largest key in the current tree (containing all keys of the first $j - 1$ bulks). Moreover, let b_j be the number of keys in $Bulk(X)$ that appear between the largest key and the place of x in the current tree. Then

$$\sum_{j=1}^{k} d_j \leq \sum_{j=1}^{k} b_j \cdot n = Inv(Bulk(X)) \cdot n \, .$$

The execution time of the algorithm is $O(n + \sum_{j=1}^{k}(1 + \log(d_j + 1)))$. Further, using the reasoning from [3,4,5] and the above estimation for $\sum_{j=1}^{k} d_j$, we conclude:

$$\sum_{j=1}^{k} 1 + \log(d_j + 1) = k + \log \prod_{j=1}^{k}(d_j + 1) = n + k \log \prod_{j=1}^{k}(d_j + 1)^{1/k}$$

$$\leq n + k \log(\sum_{j=1}^{k}(d_j + 1)/k)$$

$$\leq n + k \log((1 + Inv(Bulk(X))) \cdot n/k).$$

By substituting here $ck = n$, where $1 \leq c \leq n$, we obtain:

$$n + k \log((1 + Inv(Bulk(X))) \cdot n/k) = n + k \log((1 + Inv(Bulk(X))) \cdot c)$$
$$= n + k \log(1 + Inv(Bulk(X))) + k \log c$$
$$\leq 2n + k \log(1 + Inv(Bulk(X))).$$

Thus $O(n + \sum_{j=1}^{k}(1 + \log(d_j + 1))) = O(n + k \log(1 + Inv(Bulk(X))))$ as desired.
□

Theorem 7 and Theorem 6 imply:

Theorem 8. *The time complexity of bulk-insertion sort is*

$$O(n + k \log \min\{k, 1 + Inv(Bulk(X))\}).$$

Theorem 8 can be interpreted such that bulk-insertion sort indeed is adaptive with respect to $Inv \circ Bulk$. But Theorem 8 does not imply optimality with respect to $Inv \circ Bulk$, since it does not match the above lower bound of $\Omega(n + k \log(1 + Inv(Bulk(X)))/k))$.

5   Experiments

We implemented the bulk-insertion AVL-tree sorting algorithm, as well as AVL-tree sorting without bulk insertion but using the binary saved path.[2] We compared the performance to the *Inv*-optimal algorithms Splaysort [16] and Splitsort [17]. Splitsort is known to be efficient especially in terms of running time [18,1] and Splaysort in the number of comparisons [16,18]. We also compared to standard Quicksort, Insertion sort (see [5], for example), and the qsort function in the C library, which we found to be Merge sort.[3] We used the deterministic Quicksort from [19, Figure 1], changed to always use the middle element as

---

[2] Our implementation was written in C, and ran under GNU/Linux on an AMD Athlon XP at 2167 MHz. Each experiment was repeated 10 times using newly generated input; we report averages.

[3] The source code of the qsort function in the C library we used (the GNU C library version 2.3.6) reveals that it uses Merge sort, unless there is a problem allocating memory for the needed additional $O(n)$ space, in which case it falls back to an in-place Quicksort implementation. We avoided the fallback in our experiments.

the pivot – in [19], this variation had the smallest number of comparisons when the number of inversions was small. We also tried the textbook optimizations of median-of-3 partitioning and fallback to Insertion sort for $n \leq 10$, but this improved the performance only very slightly.

We used the Splaysort implementation from [16]. For our implementation of Splitsort, we tried the space optimization in [17] which uses only n pointers of extra space. However, using $2n$ extra pointers to avoid copying data back and forth in the various phases of the Splitsort algorithm was much faster, so we only report running times of the latter.

To obtain comparable running times, we always sorted arrays. That is, when using the tree-based algorithms (Splaysort, AVL single ins. and AVL bulk ins.), we always wrote the sorted result back into the original array.

In order to explore the full extent of inversion adaptivity, from sorted sequences to completely random ones, we used three methods to generate input. First, small numbers of inversions (0 to about n) were produced using the algorithm of [20] applied to Inv. This algorithm exchanges k randomly chosen pairs of adjacent elements, starting from a sorted sequence, thus generating about k inversions on average (unless k is too large). Second, the algorithm described in [18] was used to produce larger numbers of inversions (about n to $n^2/8$). This algorithm generates about $mn/2$ inversions on average, by first dividing a sorted sequence into $\lceil n/m \rceil$ equal-sized blocks and permuting the elements in each block into random order, and then selecting a random element from each of m equal-sized blocks and permuting the selected elements into random order. Finally, for the non-adaptive case we created random sequences where every permutation is equally likely (which gives about $n^2/4$ inversions on average).

We examined both integer keys (word-sized integers in the range $[1..n]$ with $n = 2^{25}$) and string keys ($n = 1971968$)[4], most of which contain similar prefixes so that string comparisons will often need to look for more than the first few characters to differentiate between the strings. Figure 3(a,b) shows the number of comparisons performed by the algorithms (divided by n for clarity) using integer keys. Figure 3(c,d,e,f) gives running times for integer and string keys.

The results show that bulk insertion greatly improved AVL-tree based sorting when the number of inversions was small: with less than 10^5 inversions, single-insertion sorting was about 6 times slower in the integer case (up to 2 times slower in the string case). In this range, the number of comparisons used by bulk-insertion sort was very close to $1n$, while single insertion needed about $2n$. For larger numbers of inversions, single-insertion sorting was slightly faster than bulk-insertion sort, but the difference was small.

Comparing to the other sorting algorithms, for up to about 10^5 inversions (10^4 in the string case) bulk-insertion sort was the fastest – except for Insertion sort, which was hopelessly slow when the number of inversions was larger than about $10^2 n$. With more than 10^6 inversions, Splaysort, Splitsort and Quicksort

---

[4] As string keys we used the list of file names in the Debian GNU/Linux distribution release 4.0r2, http://ftp.debian.org/dists/Debian4.0r2/Contents-i386.gz with duplicates removed.

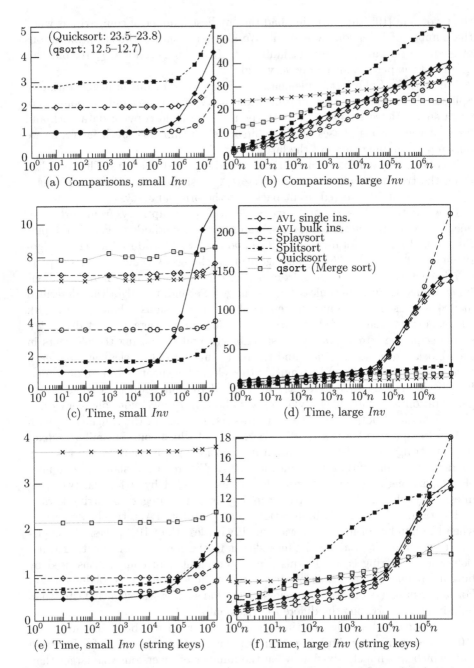

Fig. 3. Comparisons per element and total time (in seconds) used in sorting. **(a,b,c,d)** 4-byte integer keys ($n = 2^{25} \approx 34 \cdot 10^6$), **(e,f)** string keys ($n = 1971968$). In all figures, the x axis gives the number of inversions (*Inv*). The rightmost data point in **(b,d,f)** gives the non-adaptive case of a completely random sequence. The plot legend is given only in **(d)** for clarity.

were (variously) the fastest. With a very large number of inversions, about $10^6 n$ or more ($10^5 n$ in the string case), bulk-insertion sort was again faster than Splaysort – though here the array-based Splitsort, Quicksort and qsort are much faster.

The differences are smaller in the amount of comparisons. Splaysort and (for more than about $10^5 n$ inversions) qsort use the least comparisons. Up to about 10^5 inversions, bulk-insertion sort uses about as many comparisons as Splaysort (but is faster in running time, as noted above).

Reasons for the apparent slight adaptivity of Quicksort and qsort (Merge sort) are discussed in [19].

6 Conclusions

We introduced bulk-insertion AVL-tree sorting in order to improve local insertion sort. A bulk was defined as a sequence of elements of the input that fit between two consecutive elements of the sequence sorted thus far. We proved the optimality of bulk-insertion sort with respect to the number of inversions, and with respect to our new measure, the number of bulks. We also proved that bulk-insertion sort is adaptive with respect to the number of inversions when they are counted only for the last elements of the bulks. This result was obtained by introducing the concept of composite measures of adaptivity.

We compared bulk-insertion sort experimentally with various adaptive and non-adaptive sorting algorithms using inputs of length about 10^7 generated to contain different numbers of inversions. Of the algorithms compared, bulk-insertion sort is the best one when most inputs have few inversions (bulk-insertion sort was better up to 10^5 inversions than the others) but some inputs with very large *Inv* exist (thus ruling out Insertion sort).

Bulk-insertion sort is good also in many cases with a large number of inversions, one example being the sequence $\langle n+1, n+2, \ldots, 2n, 1, 2, \ldots, n \rangle$ that has a quadratic number of inversions but only two bulks. Our current experiments actually give an advantage to other methods than ours because only *Inv*-adaptivity is considered. We plan to experiment with input data adaptive to *Block* and the new measures *Bulk* and *Inv∘Bulk* in the future, thus providing an interesting contribution to instance generation and a way to see how competitive bulk-insertion sort is on its own turf.

References

1. Estivill-Castro, V., Wood, D.: A survey of adaptive sorting algorithms. ACM Computing Surveys 24(4), 441–476 (1992)
2. Petersson, O., Moffat, A.: A framework for adaptive sorting. Discrete Applied Mathematics 59(2), 153–179 (1995)
3. Mannila, H.: Measures of presortedness and optimal sorting algorithms. IEEE Transactions on Computers C-34, 318–325 (1985)
4. Mehlhorn, K.: Sorting presorted files. In: Weihrauch, K. (ed.) GI-TCS 1979. LNCS, vol. 67, pp. 199–212. Springer, Heidelberg (1979)

5. Mehlhorn, K.: Data Structures and Algorithms 1: Sorting and Searching. Springer, Heidelberg (1984)
6. Tsakalidis, A.K.: AVL-trees for localized search. Information and Control 67, 173–194 (1985)
7. Elmasry, A.: Adaptive sorting with AVL trees. In: IFIP 18th World Computer Congress, TC1 3rd International Conference on Theoretical Computer Science (IFIP TCS 2004), pp. 307–316. Kluwer, Dordrecht (2004)
8. Elmasry, A., Fredman, M.L.: Adaptive sorting: an information theoretic perspective. Acta Informatica 45, 33–42 (2008)
9. Soisalon-Soininen, E., Widmayer, P.: Amortized complexity of bulk updates in AVL-trees. In: Penttonen, M., Schmidt, E.M. (eds.) SWAT 2002. LNCS, vol. 2368, pp. 439–448. Springer, Heidelberg (2002)
10. Larsen, K.S.: Relaxed red-black trees with group updates. Acta Informatica 38, 565–586 (2002)
11. Blelloch, G.E., Maggs, B.M., Woo, S.L.M.: Space-efficient finger search on degree-balanced search trees. In: 14th Annual ACM-SIAM Symposium on Discrete Algorithms (SODA 2003), pp. 374–383. ACM Press, New York (2003)
12. Cole, R.: On the dynamic finger conjecture for splay trees, part II: The proof. SIAM Journal on Computing 30(1), 44–85 (2000)
13. Huddleston, S., Mehlhorn, K.: A new data structure for representing sorted lists. Acta Informatica 17, 157–184 (1982)
14. Katajainen, J., Levcopoulos, C., Petersson, O.: Local insertion sort revisited. In: Djidjev, H.N. (ed.) Optimal Algorithms. LNCS, vol. 401, pp. 239–253. Springer, Heidelberg (1989)
15. Carlsson, S., Levcopoulos, C., Petersson, O.: Sublinear merging and natural mergesort. Algorithmica 9, 629–648 (1993)
16. Moffat, A., Eddy, G., Petersson, O.: Splaysort: Fast, versatile, practical. Software, Practice and Experience 126(7), 781–797 (1996)
17. Levcopoulos, C., Petersson, O.: Splitsort – an adaptive sorting algorithm. Information Processing Letters 39, 205–211 (1991)
18. Elmasry, A., Hammad, A.: An empirical study for inversions-sensitive sorting algorithms. In: Nikoletseas, S.E. (ed.) WEA 2005. LNCS, vol. 3503, pp. 597–601. Springer, Heidelberg (2005)
19. Brodal, G.S., Fagerberg, R., Moruz, G.: On the adaptiveness of quicksort. In: 7th Workshop on Algorithm Engineering and Experiments (ALENEX 2005), Society for Industrial and Applied Mathematics, pp. 130–140 (2005)
20. Estivill-Castro, V.: Generating nearly sorted sequences – the use of measures of disorder. Electronic Notes in Theoretical Computer Science 91, 56–95 (2004)

Computing Elevation Maxima
by Searching the Gauss Sphere*

Bei Wang[1], Herbert Edelsbrunner[1,2], and Dmitriy Morozov[1]

[1] Department of Computer Science,
Duke University, Durham, North Carolina, USA
[2] Geomagic, Research Triangle Park, North Carolina, USA

Abstract. The elevation function on a smoothly embedded 2-manifold in $\mathbb{R}^3$ reflects the multiscale topography of cavities and protrusions as local maxima. The function has been useful in identifying coarse docking configurations for protein pairs. Transporting the concept from the smooth to the piecewise linear category, this paper describes an algorithm for finding all local maxima. While its worst-case running time is the same as of the algorithm used in prior work, its performance in practice is orders of magnitudes superior. We cast light on this improvement by relating the running time to the total absolute Gaussian curvature of the 2-manifold.

1   Introduction

This paper introduces a new algorithm for computing all local maxima of the elevation function defined on a 2-manifold embedded in $\mathbb{R}^3$. This function has been introduced by Agarwal et al. [4] for the purpose of improving the prediction of protein interaction through docking. The approach identifies protrusions (knobs) and cavities (wells) on the two surfaces and matches them up. This idea goes back to Connolly [12] who used a function that maps each point of the protein surface to the fraction of a fixed-radius sphere centered at the point that lies outside the protein volume. As shown by Cazals et al. [6], this function resembles the mean curvature at the point in the limit, when the radius approaches zero. The fixed radius makes a choice of the scale the function reflects.

The elevation function introduced in [4] serves the same purpose, but in contrast to Connolly's function, the elevation is scale independent and marks small as well as large protrusions of varying shape and direction. Its construction is based on the persistence structure of the 2-parameter family of height functions, as explained in the next section. The task at hand is then the computation of all local maxima for two proteins and the use of the type, size, and location of the marked topographic features to identify promising positions for interaction. The experimental study in [23] shows that this approach is effective in finding initial positions that can then be refined by local optimization. The computationally most expensive step in this study is the determination of the elevation maxima. Using the algorithm in [4], the running time for a triangulated 2-manifold with

---

* This research is partially supported by the Defense Advanced Research Projects Agency (DARPA) under grants HR0011-05-1-0007 and HR0011-05-1-0057.

J. Vahrenhold (Ed.): SEA 2009, LNCS 5526, pp. 281–292, 2009.

m edges is proportional to $m^5 \log_2 m$. Since typical proteins give rise to surfaces with hundreds of thousands of edges, the quintic dependence on m is a serious drawback that limits the practical deployment of the method.

In this paper, we give a new algorithm that is faster for triangulated surfaces approximating smooth surfaces that we typically find in practice. They are characterized by having dihedral angles at edges that are close to half the full angle (molecular skin surface [14]). We relate the running time of our algorithm to the total absolute Gaussian curvature of the surface and this way determine that we can expect roughly a ten-thousand fold improvement over the running time of the old algorithm. We note, however, that we offer no improvement in the worst-case performance.

Since we incorporate the surface complexity in terms of total absolute Gaussian curvature into the analysis of the algorithm, it is worth mentioning that there is a large literature on the notion of curvatures for triangulated surfaces. We refer to [2] and [17,22] for details.

Outline. In Section 2, we introduce the geometric and topological background needed to understand the elevation function. We do this in two steps, discussing the mathematically cleaner smooth case in Section 2.1 and the computationally more useful piecewise linear (PL) case in Section 2.2. In Section 3, we present the algorithm for computing all elevation maxima, along with some implementation details and the analysis. In Section 4, we present our experimental results, employing our software to compute elevation maxima for a number of triangulated protein surfaces. We gather statistics on critical regions, pairwise intersections, and elevation maxima. We use these statistics as evidence that our assumption is a reasonable approximation of the reality for our data and that the new algorithm runs about four orders of magnitude faster than the old one.

2 Preliminaries

2.1 The Smooth Case

Morse functions. The class of smooth, real-valued functions is a challenging object that simplifies considerably if we add genericity as a requirement. Letting $f : M \to \mathbb{R}$ be a smooth function on a 2-manifold, a point $x \in M$ is *critical* if the derivative at x equals zero. The value of f at a critical point is a *critical value*. All other points are *regular points* and all other values are *regular values* of f. A critical point is *non-degenerate* if the Hessian, that is, the matrix of second partial derivatives at the point is invertible. In this case, the matrix has two non-zero eigenvalues, $\lambda_1 \neq \lambda_2$, and the *index* of the non-degenerate critical point is the number of negative eigenvalues. A non-degenerate critical point of index 0 is a *minimum*, of index 1 is a *saddle*, and of index 2 is a *maximum*. Finally, f is a *Morse function* if all its critical points are non-degenerate and its values at the critical points are distinct. Given a value $a \in \mathbb{R}$, the corresponding *sublevel set* consists of all points with value at most a, $M_a = f^{-1}(-\infty, a]$. Sweeping the manifold in the direction of increasing function value, we get a 1-parameter family of sublevel sets. The topology of the sublevel set changes precisely when the sweep passes through a critical point. Let $t_1 < t_2 < \ldots < t_n$ be the ordered sequence of critical values and $-\infty = s_0 < s_1 < \ldots < s_n = \infty$ a sequence of interleaved values, that is, $s_i < t_{i+1} < s_{i+1}$, for all i. By assumption of f being Morse, we get from the sublevel set at s_i to the one at s_{i+1} by passing exactly one non-degenerate critical point. The change can be characterized in terms of the dimension of the handle we attach to go from M_{s_i} to $M_{s_{i+1}}$. For index 0, we add a 0-handle, that is, an isolated point which we then thicken to a disk. For index 1, we add a 1-handle, that is an interval attached to the boundary of the sublevel set at its endpoints which we then thicken to a strip. Finally, for index 2, we add a 2-handle, that is, a disk attached to the boundary of the sublevel set along its boundary circle.

Persistent homology. Looking at the homology groups [18] of the sequence of sublevel sets, we use the concept of persistence to measure the lengths of the intervals along which homology classes exist [15]. Since sublevel sets between two contiguous critical values are indistinguishable, we may consider the finite sequence

$$\emptyset = M_0 \subseteq M_1 \subseteq \ldots \subseteq M_n = M,$$

where we simplify notation by setting $M_i = M_{s_i}$. Fixing a dimension p ($p \geq 0$), each sublevel set has a p-th homology group and the sequence is connected from left to right by homomorphisms induced by inclusion, which we denote as $f_p^{i,j} : H_p(M_i) \rightarrow H_p(M_j)$. We have a *birth* at M_i if the map $f_p^{i-1,i}$ is not surjective, and we have a *death* at M_j if the map $f_p^{j-1,j}$ is not injective. Furthermore, the death at M_j corresponds to the birth at M_i if there is homology class γ in $H_p(M_i)$ that is not in the image of $f_p^{i-1,i}$, its image in $H_p(M_{j-1})$ is still not in the image of $f_p^{i-1,j-1}$, but its image in $H_p(M_j)$ is in the image of $f_p^{i-1,j}$. We call $f(t_j) - f(t_i)$ the *persistence* of this birth-death pair. As explained in [8], this method gives a pairing between births and deaths that has many interesting properties. Each death corresponds to a unique birth but not every birth corresponds to a death. To remedy this shortcoming, we extend the sequence of homology groups for extended persistence as described in [9]. Writing $M^a = f^{-1}[a, \infty)$ for the *superlevel set* of a, we go up with absolute homology groups of sublevel sets, as before, and we come back down with relative homology groups,

$$0 = H_p(M_0) \rightarrow H_p(M_1) \rightarrow \ldots \rightarrow H_p(M_n)$$
$$\rightarrow H_p(M, M^n) \rightarrow \ldots \rightarrow H_p(M, M^0) = 0,$$

where we simplify notation by setting $M^i = M^{s_i}$, $M^0 = M$ and $M^n = \emptyset$. Now every birth corresponds to a death. In fact, we have two events at every critical point, one going up and one coming down, but duality implies that we just get each pair twice, see [9]. As a consequence of duality, the birth-death pairs we get for the negative function, $-f$, are the same. This turns out to be important in the definition of the elevation function.

For 2-manifolds, there is a more elementary way to introduce extended persistence using the Reeb graph of the function. Instead of giving details, we refer to [4] and we mention that this approach leads to a fast algorithm. It consists of constructing the Reeb graph in a sweep [10] followed by deconstructing it in another sweep using cutting and linking trees [4,16]. We run this algorithm for a piecewise linear function on a triangulated 2-manifold. Letting m be the number of edges in the triangulation, as before, the running time computing the extended persistence for a given height function is bounded by some constant times $m \log_2 m$.

Elevation. To define elevation, we assume the 2-manifold $\mathbb{M}$ is smoothly embedded in $\mathbb{R}^3$. For a direction $u \in \mathbb{S}^2$, we consider the height function $h_u : \mathbb{M} \to \mathbb{R}$ defined by $h_u(x) = \langle x, u \rangle$. Generically, h_u is a Morse function, but for some directions u it is not, either because a critical point is degenerate or because two or more critical points map to the same height value. Considering the entire sphere of directions, we get a 2-parameter family of height functions.

For each $u \in \mathbb{S}^2$, we pair up births with deaths using the extended sequence of homology groups defined by the sublevel and the superlevel sets of h_u. In the Morse function case, each birth-death pair identifies two critical points, x and y, one giving birth and the other giving death, and we define the elevation at these two points as their persistence or, equivalently, the absolute height difference in the direction u, $E(x) = E(y) = |h_u(x) - h_u(y)|$. Each point of $\mathbb{M}$ is critical in two directions, u and $-u$, and is thus assigned two values, the absolute height difference to the paired critical point in the two directions. Since $h_{-u} = -h_u$, the paired point is the same so we get a unique value at every point. This is the *elevation function* of the 2-manifold, $E : \mathbb{M} \to \mathbb{R}$.

To get a feeling for this function, we consider a protrusion (a mountain) of the 2-manifold. To measure the height of the mountain, we measure from the top down, to the first saddle that separates it from an even higher mountain. We can do this in various directions, so we do it to maximize the height. This might be in a direction along which the first saddle is ambiguous. Perhaps there are three such saddles at the same height value in this direction, similar to the third type in Figure 1 in which we have a saddle with the same height difference to three minima. In this direction, we have two violations of genericity required for Morse functions, because there are three critical points with the same height value. Indeed, local maxima of E tend to arise along non-generic directions. An exception is the 1-legged maximum defined by only two critical points (with one leg between them). Besides this case, we have 2-legged maxima defined by three critical points, and 3- and 4-legged maxima defined by four critical points each; see Figure 1.

Curvature. We will later discover that the running time of our algorithm for finding all local maxima relates to the total absolute curvature of the surface. We introduce this concept using the *Gauss map*, $N : \mathbb{M} \to \mathbb{S}^2$, defined by mapping a point x of $\mathbb{M}$ to the outer unit normal, $N(x)$, at x. Assuming $\mathbb{M}$ is smoothly embedded in $\mathbb{R}^3$, the Gauss map is continuous and surjective but not necessarily injective. Indeed, the preimage of $u \in \mathbb{S}^2$ consists of all critical points of h_u with outer normal u, as opposed to $-u$. The multiplicity of N at u and $-u$ together is thus the number of critical points of h_u. We will see shortly that the total coverage of $\mathbb{S}^2$ is exactly the total absolute Gaussian curvature of $\mathbb{M}$.

Letting x be a point of $\mathbb{M}$ and $r > 0$ a radius, we define the *absolute Gaussian curvature* at x by taking the limit of a fraction of areas, $g(x) = \lim_{r \to 0} \frac{Area(N(A_r))}{Area(A_r)}$, where A_r is the neighborhood of points at distance at most r from x on $\mathbb{M}$. The *total absolute Gaussian curvature* is the integral of the local quantity, $G(\mathbb{M}) = \int_{x \in \mathbb{M}} g(x)dx$. It should be clear that $G(\mathbb{M})$ is the area of the total coverage of $\mathbb{S}^2$, taking multiplicity into account. For a given direction, the multiplicity is $|N^{-1}(u)|$. Hence, $G(\mathbb{M}) = \int_{u \in \mathbb{S}^2} |N^{-1}(u)|du$. Writing c_{avg} for the average number of critical points of the height functions, we thus have the total absolute Gaussian curvature equal to one half times

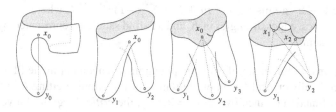

Fig. 1. The four generic types of local maxima of the elevation function. From left to right: the 1-, 2-, 3- and 4-legged maximum.

the area of the sphere times that average, $G(\mathbb{M}) = 2\pi c_{\mathrm{avg}}$. This integral geometry formula for the curvature will come handy in the analysis of our algorithm. For more information on the integral geometry formulation of curvature see Santaló [21].

2.2 The PL Case

Triangulated surfaces. We do all computations on a piecewise linear approximation of the smooth 2-manifold. To transport the smooth concepts to the PL category, we think of the PL surface as being approximated by a smooth surface. Tightening the approximation, we get a series and take the limit. This is the general intuition we have in the background guiding the formulation of definitions in the PL case.

A *triangulation* of a 2-manifold $\mathbb{M}$ is a simplicial complex, K, whose underlying space is homeomorphic, $|K| \approx \mathbb{M}$. It consists of vertices, edges, and triangles. To put K into $\mathbb{R}^3$, it suffices to map each vertex to a point; the edges and triangles are the convex hulls (of the images) of their vertices. This is a *geometric realization* if the triangles meet in shared edges and vertices but not in any other point sets. We call the result a *triangulated surface*, implicitly assuming that it is geometrically realized in $\mathbb{R}^3$. The *star* of a vertex is the set of simplices that contain it, and the *link* consists of all faces of simplices in the star that do not belong to the star, $\mathrm{St}\, v_i = \{\sigma \in K \mid v_i \in \sigma\}$; $\mathrm{Lk}\, v_i = \{\tau \subseteq \sigma \in \mathrm{St}\, v_i \mid \tau \notin \mathrm{St}\, v_i\}$. A PL function $f : |K| \to \mathbb{R}$ is determined by its values at the vertices. Assuming $f(v_i) \neq f(v_j)$ whenever $i \neq j$, we define the *lower link* as the subset of simplices in the link where f is smaller than at the vertex, $\mathrm{Lk}_-\, v_i = \{\sigma \in \mathrm{Lk}\, v_i \mid x \in \sigma \Rightarrow f(x) < f(v_i)\}$. Finally, v_i is *regular* if its lower link is contractible, and *critical*, otherwise. Since K triangulates a 2-manifold, every link is a circle and the only contractible closed subsets are points and closed paths. The lower link of a regular vertex is thus a single vertex or a path connecting two vertices. A *minimum* is characterized by $\mathrm{Lk}_-\, v_i = \emptyset$ and a *maximum* by $\mathrm{Lk}_-\, v_i = \mathrm{Lk}\, v_i$. In the remaining case, the lower link consists of $k+1 \geq 2$ paths and we call v_i a *k-fold saddle*, or a *simple saddle* if $k = 1$.

In contrast to the smooth case, it is not possible to turn a k-fold into a simple saddle by a small perturbation. We therefore treat them directly, without reduction to simple cases. As an example, consider the Euler-Poincaré Theorem which relates the topology of the 2-manifold with the critical point structure of its functions. Define the *index* of a simple critical point as index (v_i), index $(v_i) = 0$ if v_i is a minimum, 1 if v_i is a simple saddle, 2 if v_i is a maximum. Assuming K is connected, it is characterized by its *genus* and we have $2 - 2 \cdot \text{genus} = n - m + l = \sum_i (-1)^{\text{index}(v_i)}$, where n, m, l are the number of vertices, edges, triangles in K and a k-fold saddle is represented by k simple saddles in the sum.

Critical regions. Another significant complication we encounter in the PL case is that a vertex is generally critical for an entire region of directions. Letting $h_u : |K| \to \mathbb{R}$ be the height function defined by $h_u(x) = \langle x, u \rangle$, the *critical region* of a vertex is the closure of the set of directions along which v_i is critical,

$$R_i = \text{cl}\{u \in \mathbb{S}^2 \mid v_i \text{ is critical point of } h_u\}.$$

We construct it from the closed polygonal curve defined by the star of v_i. Specifically, we map each triangle in the star to its outer normal direction, a point on $\mathbb{S}^2$, and we connect the directions of two neighboring triangles by the shorter of the two connecting great-circle arcs. This gives a closed polygonal curve, π_i, which may or may not have self-intersections. To cope with the former, more complicated case, we orient π_i and define the *winding number* of a direction $u \in \mathbb{S}^2$ not on the curve as the number of times the curve goes around the directed line defined by u. Viewed along u, we count a counterclockwise turn as $+1$ and a clockwise turn as -1. Taking the sum we get the winding number, which are denoted as $w(u, \pi_i)$. For detailed study on polyhedron Gauss map, refer to [5]. Examples are shown in Figure 2. The winding number of u relates to the type of the vertex in the height function defined by u. Specifically, if v_i is regular then the winding number of u is 0, if v_i is a simple critical point then the winding number is $(-1)^{\text{index}(v_i)}$, and if v_i is a k-fold saddle then the winding number is $-k$.s

Curvature. Thinking of a vertex as a tiny region in an approximating smooth surface, we define its *Gaussian curvature* as the area of its critical region weighted by the winding number. More useful in this paper is its *absolute Gaussian curvature* defined as the area weighted by the absolute winding number, $g(v_i) = \int_{u \in \mathbb{S}^2} |w(u, \pi_i)| du$. The *total absolute Gaussian curvature* is then the sum over all vertices, $G(K) = \sum_i g(v_i)$. Equivalently, it is the area of the sphere times half the average number of critical vertices, taking multiplicities into account, as usual. The average is taken over all height functions, and we count half the critical vertices because v_i is critical for $u \in \mathbb{S}^2$ as well as $-u \in \mathbb{S}^2$.

3   Computation

In this section, we describe how we compute the elevation maxima for a given triangulated surface in $\mathbb{R}^3$. The algorithm is straightforward and the only new insight is in the analysis, relating the running time with the total absolute Gaussian curvature of the surface.

Types and filters. Recall that there are four types of elevation maxima for a generic smooth surface, as illustrated in Figure 1. We have the same four cases for a generic triangulated surface K in $\mathbb{R}^3$. Each maximum is given by a set of two, three, or four points. We consider the case in which all these points are vertices of K. The cases in

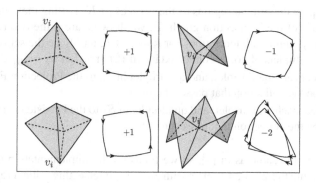

Fig. 2. Left: for a direction u with winding number $+1$ the corresponding vertex appears either as a maximum or a minimum. Right: for winding number -1 we have a simple saddle and for -2 we have a 2-fold or monkey saddle for the height function defined by the corresponding direction.

which some of the points in V lies on edges of K are similar. Let V be a set of vertices. A necessary requirement for V to define a maximum is that its vertices are critical for a common direction. More specifically, we need them critical in a particular direction that is determined by V. This direction, $u_V = (y - x)/\|y - x\|$, is slightly different for each type.

1-legged case, $V = \{x, y\}$. Here, u_V is the direction defined by the two points.

2-legged case, $V = \{x, y_1, y_2\}$. Letting y be the orthogonal projection of x onto the line passing through y_1 and y_2, u_V is defined if y lies between y_1 and y_2.

3-legged case, $V = \{x, y_1, y_2, y_3\}$. Letting y be the orthogonal projection of x onto the plane passing through y_1, y_2, y_3, u_V is defined if y lies in the triangle they span.

4-legged case, $V = \{x_1, x_2, y_1, y_2\}$. Letting x and y be the feet of the shortest line segment connecting the line passing through x_1 and x_2 with the line passing through y_1 and y_2, u_V is defined if x lies between x_1 and x_2 and y lies between y_1 and y_2.

PROJECTION FILTER. The direction u_V defined by the points in V is defined and belongs to the common intersection of critical regions, $u_V \in \bigcap_{v_i \in V} R_i$.

Note that the non-empty intersection of the critical regions is a necessary but not a sufficient condition for the set V to pass the Projection Filter. In turn, passing the Projection Filter is a necessary but not sufficient condition for the direction u_V to be an elevation maximum. For that, the set needs to satisfy another condition. To describe it, we write x_0 for x.

PERSISTENCE FILTER. For each pair x_i and y_j in V, there is an arbitrarily small perturbation u of u_V such that x_i, y_j is a birth-death pair for the height function h_u.

Algorithm. We compute the elevation maxima in three steps, starting with 2-, 3-, 4-tuplets V whose points have pairwise overlapping critical regions. The next two steps narrow down the selection using first the Projection and the Persistence Filter.

STEP 0. Compute the critical regions of the vertices of K. Letting the critical regions be the nodes of the intersection graph, R, we draw an arc if the two regions have a non-empty common intersection. For $k = 2, 3, 4$, let Q_k be the set of k-cliques, that is, the k-tuplets of nodes connected by all $\binom{k}{2}$ arcs. Let $S_0 = \bigcup_k Q_k$.

STEP 1. Subject each pair, triplet, and quadruplet in S_0 to the Projection Filter and let $S_1 \subseteq S_0$ be the collection that passes the filter.

STEP 2. Subject each pair, triplet, and quadruplet in S_1 to the Persistence Filter and let $S_2 \subseteq S_1$ be the collection that passes the filter.

Step 1 and 2 are the same as in [4], so we focus on the implementation of Step 0 in which we compute the 2-, 3-, 4-tuplets with pairwise intersecting critical regions.

Implementation. We break down Step 0 into three smaller steps, constructing the critical regions, finding the intersecting pairs, and computing the cliques of size 2, 3, 4 in the intersection graph. Implementation is done with Perl, C and CGAL [1]. All computations are exact except estimating the area and the bounding box of a critical region.

STEP 0.1. Recall that each critical region, R_i, is given by a closed polygon with m_i edges on the sphere. Those edges may intersect, and we take time $O(m_i^2)$ to construct the decomposition of the sphere [13], including winding numbers for all subregions. Reflecting R_i centrally through the origin in $\mathbb{R}^3$, we get the region $-R_i$ of inward normals along which v_i is critical. Constructing all critical regions takes time proportional to $\sum_i m_i^2$.

STEP 0.2. Most critical regions are small and simple. This suggests we use a bounding volume approach to find the intersecting pairs. Specifically, we find an axis-parallel box B_i in $\mathbb{R}^3$ that encloses the region R_i on $\mathbb{S}^2 \subseteq \mathbb{R}^3$. We do this in two steps, first computing the smallest enclosing sphere of R_i and second the smallest axis-aligned box that contains the sphere. Assuming that R_i fits inside a hemisphere of $\mathbb{S}^2$, the smallest enclosing sphere of its vertices also encloses R_i. To compensate for round-off errors, we increase the sphere slightly and compute the box B_i to enclose the enlarged sphere. Computing the smallest enclosing sphere of R_i takes randomized time $O(m_i)$, see [24]. Given the boxes B_i, we find the overlapping pairs using the segment-tree streaming algorithm as described in [25]. Writing b_i for the number of boxes that overlap B_i, we have a total of $b = \frac{1}{2} \sum_i b_i$ of overlapping pairs. The streaming algorithm takes time proportional to $n \log_2^3 n + b$ to find them. For each pair of overlapping boxes, we check whether or not the critical region they enclose have a non-empty intersection. Standard computational geometry methods allow us to determine whether or not R_i and R_j intersect in time $O(m_{ij} \log m_{ij})$, where $m_{ij} = m_i^2 + m_j^2$ [13].

STEP 0.3. The result of Steps 0.1 and 0.2 is a graph R. Its n nodes are the critical regions, and its q arcs are the pairs of critical regions with non-empty overlap. Writing $q = \frac{1}{2} \sum_i q_i$, where q_i is the degree of the i-th node, we compute the cliques of size 2, 3, 4 by checking all pairs and triplets of neighbors. Finding the cliques that include R_i thus takes time $O(\binom{q_i}{1} + \binom{q_i}{2} + \binom{q_i}{3})$.

Analysis. The time for Step 0 is dominated by the requirement for Step 0.2, which is some constant times $T_{\text{new}} = \sum_i \binom{q_i}{1} + \binom{q_i}{2} + \binom{q_i}{3}$. The time for Step 1 is some constant times $|S_0| \leq T_{\text{new}}$ and that for Step 2 is some constant times $T = |S_1| n \log_2 n$. This adds up to some constant times $T_{\text{new}} + T$, as compared to $T_{\text{old}} + T$ for the algorithm in [4], where $T_{\text{old}} = \binom{n}{2} + \binom{n}{3} + \binom{n}{4}$. Any improvement thus hinges on two properties, namely that T_{old} is significantly larger than T_{new} as well as T. We now show that the first property holds under grossly simplifying assumptions, and we provide evidence in the next section that both properties hold for data we encounter in practice.

CAP ASSUMPTION. The critical regions are spherical caps, all of the same size, and their centers are uniformly distributed on $\mathbb{S}^2$.

Recall that the areas of the critical regions add up to the total absolute Gaussian curvature, $\sum_i Area(R_i) = G(K)$. This sum is also half the area of the sphere times the average number of critical points of the height functions, $G(K) = 2\pi c_{\text{avg}}$. It follows the area of a single critical region is $Area(R_i) = 2\pi c_{\text{avg}}/n$, and because the cap is smaller than the flat disk of the same radius, its radius squared is $\rho^2 > 2c_{\text{avg}}/n$. Two caps overlap if and only if the center of one is contained in the cap of radius 2ρ around the center of the other. The area of the enlarged cap is less than four times $Area(R_i)$. Hence the probability for a region R_j to overlap R_i is $\text{Prob}[R_i \cap R_j \neq \emptyset] \leq 4 Area(R_i)/4\pi = 2c_{\text{avg}}/n$. Since expectations are additive even if the events are not independent, the expected number of k-tuplets of neighbors is $\text{Exp}[\binom{q_i}{k}] \leq \binom{n-1}{k} Area(R_i)^k/\pi^k \leq 2^k c_{\text{avg}}^k/k!$. Adding the expectations for $k = 1, 2, 3$ and all i gives

$$\text{Exp}[T_{\text{new}}] \leq n \cdot (2c_{\text{avg}} + 2c_{\text{avg}}^2 + \frac{4}{3}c_{\text{avg}}^3).$$

Recall that $c_{\text{avg}} = G(K)/2\pi$. It follows the average number of k-tuplets of critical regions overlapping a given one depends on the shape of the smooth surface and not on the size of the approximating triangulated surface. Similarly, the time for Step 0 depends on the shape and otherwise only linearly on the number of vertices in the triangulation.

4 Experiments

Input data. We use two types of triangulated surfaces approximating smooth models of biomolecular structures all listed in Table 1 Left. The first type is the molecular skin which uses hyperboloid and concave sphere patches to blend between the spheres that represent the atoms of a molecule [14]. An algorithm that constructs an approximating triangulated surface with guaranteed bounds on two- and three-dimensional angles is described in [7] and software written by Ho-lun Cheng is available at [3]. The second type is the molecular surfaces generated by Chimera [19]. The MSMS algorithm used in Chimera [20] constructs a triangulation of the solvent excluded surfaces initially computed by Connolly [11].

Critical point statistics. For each data set, we estimate the minimum, average, and maximum number of critical points of the height functions, which we sample at one thousand directions chosen from $\mathbb{S}^2$. The results are shown in Table 1 Middle. Comparing the estimated with the actual average, which we get using $c_{\text{avg}} = G(K)/2\pi =$

Table 1. Left: the triangulated surfaces used in our computational experiments together with their numbers of vertices, edges, and triangles. Middle: estimated minimum, average, and maximum of the number of critical points of the height functions. Right: minimum, average, and maximum of the number of triangles needed to triangulate the critical regions. Last column: percentage of non-simple critical regions. Top: molecular skin surfaces. Bottom: molecular Chimera surfaces.

id	name	n	m	ℓ	c_{min}	c_{avg}	c_{max}	$\frac{c_{avg}}{n}$	r_{min}	r_{avg}	r_{max}	%
0	1BRS-5to6	1,370	4,104	2,736	2	6.41	16	0.0047	2	3.99	8	12
1	1CLU-DBG	3,149	9,441	6,294	2	13.50	44	0.0043	2	4.01	12	15
2	1BRS-A-5to10	4,248	12,738	8,492	6	17.07	34	0.0040	2	4.01	10	17
3	1BRS-A-30to40	6,114	18,336	12,224	10	25.14	46	0.0041	2	4.01	10	16
4	1BRS-A-17to25	7,799	23,391	15,594	12	29.92	64	0.0038	2	4.01	10	20
5	1BRS-A-5to10	836	2,502	1,668	6	16.01	32	0.0192	2	4.08	11	29
6	1BRS-A-30to40	1,372	4,110	2,740	10	27.13	46	0.0198	2	4.13	15	30
7	1BRS-A-17to25	1,595	4,119	3,186	14	31.02	54	0.0194	2	4.09	10	33

$\sum_i Area(R_i)/2\pi$, we see that the error is small. For example, for data set 4, the estimated c_{avg} is 29.92 while the actual average is 29.94. Since all our skin triangulations approximate a smooth surface to about the same accuracy, for different surfaces, the average number of critical points scales linearly with n. Indeed, c_{avg}/n is between 0.003 and 0.005 for all our skin data sets.

As mentioned earlier, each vertex of K is critical for a region of directions, in fact two antipodal regions. Most of these regions are simple, that is, defined by a polygon without self-intersections. As shown in the last column in Table 1, the percentage of non-simple polygons is indeed rather small. Besides checking for self-intersections, we measure the complexity of a critical region by counting the triangles we need to triangulate it on the sphere. The minimum, average, and maximum of this number are given in the right half of Table 1.

Intersection statistics. The following statistics were collected for the finer molecular skin surfaces only. Recall that we compute the pairs of intersecting critical regions in two steps, first finding the intersections among the bounding boxes and second among the critical regions. Table 2 Left gives the statistics for both.

Given a pair of intersecting boxes, we test whether or not the corresponding critical regions intersect by checking the overlap among the triangles in their triangulations. The average number of triangle-triangle checks is consistently between 11 and 12, which justifies the use of this brute-force over a more sophisticated method.

Similar to the number of critical points, we expect that the average number of boxes intersecting a given box and the average number of critical regions intersecting a given critical region scale linearly with n. Indeed, b_{avg}/n is between 0.04 and 0.07 and q_{avg}/n is between 0.02 and 0.03 for all our skin data sets. The latter is about six times the average number of critical points; compare this with the factor two we got under the Cap Assumption. The observed relation between these two quantities is only about three times as loose, which is reasonable considering that real data necessarily violates the Cap Assumption to some extent (due to irregular shapes and different orientations of the critical regions). The new algorithm starts with T_{new} tuplets. A back-of-the-

Table 2. Left: the minimum, average, and maximum number of boxes intersecting a given box; the minimum, average, and maximum number of critical regions intersection a given critical region. Middle: the number of cliques before and after the Projection Filter and the Persistence Filter. Right: dominant terms in the running time of the old and the new algorithms.

id	b_{min}	b_{avg}	b_{max}	$\frac{b_{avg}}{n}$	q_{min}	q_{avg}	q_{max}	$\frac{q_{avg}}{n}$	$|S_0|/10^3$	$|S_1|$	$T_{old}/10^{10}$	$T_{new}/10^6$	$T/10^6$
0	12	94	207	0.069	9	40	97	0.029	1,608	2,373	15	24	33
1	27	204	626	0.065	11	82	250	0.026	32,119	20,521	410	508	749
2	52	236	556	0.056	20	92	201	0.022	43,572	17,175	1,356	720	882
3	95	243	859	0.040	29	134	330	0.022	198,023	56,797	5,820	3,327	4,368
4	99	423	1,276	0.054	35	160	543	0.021	433,116	94,300	15,411	7,354	9,508

envelope calculation suggests that T_{new} is roughly $n\binom{q_{avg}}{3}$, which is roughly a factor of ten thousand smaller than $\binom{n}{4}$, independent of the value of n. We thus might expect the new algorithm runs about four orders of magnitude faster than the old one.

Running time. Recall that S_0 is the set of cliques of size 2, 3, or 4 in the intersection graph of the critical regions. The subset $S_1 \subseteq S_0$ contains all cliques that pass the Projection Filter, and the subset $S_2 \subseteq S_1$ contains all cliques that also pass the Persistence Filter. The sizes of the first two sets are given in the middle of Table 2.

Most relevant to the running time of the algorithms for computing elevation maxima is S_1. Indeed, both the old and the new algorithm start with sets of 2-, 3-, and 4-tuplets that contain the cliques in S_0 and much more. As shown in Table 2 on the right, the overestimate by the old algorithm is about ten thousand times that of the new algorithm. Furthermore, in the new algorithm, the time for Step 0 and Steps 1 and 2 is fairly balanced. This implies a speed-up of about four orders of magnitude, which is consistent with back-of-the-envelope calculation mentioned above.

Conclusions. The main result of this paper is a new algorithm for computing all elevation maxima of a triangulated surface in $\mathbb{R}^3$. We provide experimental evidence that for practical data, the new algorithm runs about four orders of magnitude faster than the old one. The improvement is achieved by making the running time dependent on the total absolute Gaussian curvature of the surface and to a lesser extent on the number of vertices in the approximating triangulation.

References

1. Computational Geometry Algorithms Library, http://www.cgal.org
2. Banchoff, T.F.: Critical points and curvature for embedded polyhedral surfaces. Amer. Math. Monthly 77, 475–485 (1970)
3. The biogeometry web-pages, http://www.biogeometry.duke.edu
4. Agarwal, P.K., Edelsbrunner, H., Harer, J., Wang, Y.: Extreme elevation on a 2-manifold. Discrete Comput. Geom. 36, 553–572 (2006)
5. Alboul, L., Echeverria, G.: Polyhedral Gauss maps and curvature characterization of triangle meshes. LNCS, vol. 3605, pp. 14–33. Springer, Heidelberg (2005)

6. Cazals, F., Chazal, F., Lewiner, T.: Molecular shape analysis based upon the Morse-Smale complex and the Connolly function. In: Proc. 19th Ann. Sympos. Comput. Geom., pp. 351–360 (2003)
7. Cheng, H.L., Dey, T.K., Edelsbrunner, H., Sullivan, J.: Dynamic skin triangulation. Discrete Comput. Geom. 25, 525–568 (2001)
8. Cohen-Steiner, D., Edelsbrunner, H., Harer, J.: Stability of persistence diagrams. Discrete Comput. Geom. 37, 103–120 (2007)
9. Cohen-Steiner, D., Edelsbrunner, H., Harer, J.: Extending persistence using Poincaré and Lefschetz duality. Found. Comput. Math. (to appear)
10. Cole-McLaughlin, K., Edelsbrunner, H., Harer, J., Natarajan, V., Pascucci, V.: Loops in Reeb graphs of 2-manifolds. Discrete Comput. Geom. 32, 231–244 (2004)
11. Connolly, M.L.: Analytic molecular surface calculation. J. Appl. Crystallogr. 6, 548–558 (1983)
12. Connolly, M.L.: Shape complementarity at the hemoglobin albl subunit interface. Biopolymers 25, 1229–1247 (1986)
13. de Berg, M., van Kreveld, M., Overmars, M., Schwarzkopf, O.: Computational Geometry — Algorithms and Applications. Springer, Berlin (1997)
14. Edelsbrunner, H.: Deformable smooth surface design. Discrete Comput. Geom. 21, 87–115 (1999)
15. Edelsbrunner, H., Letscher, D., Zomorodian, A.: Topological persistence and simplification. Discrete Comput. Geom. 28, 511–533 (2002)
16. Georgiadis, L., Tarjan, R., Werneck, R.F.: Design of data structure for mergeable trees. In: Proc. 17th Ann. ACM-SIAM Sympos. Discrete Algorithm, pp. 394–403 (2006)
17. Morvan, J.: Generalized Curvatures. Springer, Heidelberg (2008)
18. Munkres, J.R.: Elements of Algebraic Topology. Addison-Wesley, Reading (1984)
19. Petterson, E.F., Goddard, T.D., Huang, C.C., Gouch, G.S., Greenblatt, D.M., Meng, E.C., Ferrin, T.E.: UCSF Chimera — a visualization system for exploratory research and analysis. J. Comput. Chem. 25, 1605–1612 (2004)
20. Sanner, M.F., Olson, A.J.: Reduced surface: an efficient way to compute molecular surfaces. Biopolymers 38, 305–320 (1996)
21. Santaló, L.: Integral geometry and geometric probability. Addison-Wesley, Reading (1976)
22. Cohen-Steiner, D., Morvan, J.: Second fundamental measure of geometric sets and local approximation of curvatures. J. Differential Geom. 74(3), 363–394 (2006)
23. Wang, Y., Agarwal, P.K., Brown, P., Edelsbrunner, H., Rudolph, J.: Course and reliable geometric alignment for protein docking. In: Proc. Pacific Sympos. Biocomputing 2005, pp. 64–75. World Scientific, Singapore (2005)
24. Welzl, E.: Smallest enclosing disks (balls and ellipsoids). In: Maurer, H.A. (ed.) New Results and New Trends in Computer Science. LNCS, vol. 555, pp. 359–370. Springer, Heidelberg (1991)
25. Zomorodian, A., Edelsbrunner, H.: Fast software for box intersections. Internat. J. Comput. Geom. Appl. 12, 143–172 (2002)

Author Index